Inference and Asymptotics

O.E. Barndorff-Nielsen
Institute of Mathematics,
University of Aarhus,
Denmark

D.R. Cox
Nuffield College,
Oxford, UK

CHAPMAN & HALL/CRC

A CRC Press Company
Boca Raton London New York Washington, D.C.

Library of Congress Cataloging-in-Publication Data

Catalog record is available from the Library of Congress

Direct all inquiries to CRC Press LLC, 2000 N.W. Corporate Blvd., Boca Raton, Florida 33431.

First edition 1994
Reprinted 1995
First CRC Press Reprint 2001
Originally published by Chapman & Hall

No claim to original U.S. Government works
International Standard Book Number 0-412-49440-X
Printed in the United States of America 3 4 5 6 7 8 9 0
Printed on acid-free paper

MONOGRAPHS ON STATISTICS AND APPLIED PROBABILITY

General Editors

1 Stochastic Population Models in Ecology and Epidemiology *M.S. Barlett* (1960)
2 Queues *D.R. Cox and W.L. Smith* (1961)
3 Monte Carlo Methods *J.M. Hammersley and D.C. Handscomb* (1964)
4 The Statistical Analysis of Series of Events *D.R. Cox and P.A.W. Lewis* (1966)
5 Population Genetics *W.J. Ewens* (1969)
6 Probability, Statistics and Time *M.S. Barlett* (1975)
7 Statistical Inference *S.D. Silvey* (1975)
8 The Analysis of Contingency Tables *B.S. Everitt* (1977)
9 Multivariate Analysis in Behavioural Research *A.E. Maxwell* (1977)
10 Stochastic Abundance Models *S. Engen* (1978)
11 Some Basic Theory for Statistical Inference *E.J.G. Pitman* (1979)
12 Point Processes *D.R. Cox and V. Isham* (1980)
13 Identification of Outliers *D.M. Hawkins* (1980)
14 Optimal Design *S.D. Silvey* (1980)
15 Finite Mixture Distributions *B.S. Everitt and D.J. Hand* (1981)
16 Classification *A.D. Gordon* (1981)
17 Distribution-Free Statistical Methods, 2nd edition *J.S. Maritz* (1995)
18 Residuals and Influence in Regression *R.D. Cook and S. Weisberg* (1982)
19 Applications of Queueing Theory, 2nd edition *G.F. Newell* (1982)
20 Risk Theory, 3rd edition *R.E. Beard, T. Pentikäinen and E. Pesonen* (1984)
21 Analysis of Survival Data *D.R. Cox and D. Oakes* (1984)
22 An Introduction to Latent Variable Models *B.S. Everitt* (1984)
23 Bandit Problems *D.A. Berry and B. Fristedt* (1985)
24 Stochastic Modelling and Control *M.H.A. Davis and R. Vinter* (1985)
25 The Statistical Analysis of Composition Data *J. Aitchison* (1986)
26 Density Estimation for Statistics and Data Analysis *B.W. Silverman* (1986)
27 Regression Analysis with Applications *G.B. Wetherill* (1986)
28 Sequential Methods in Statistics, 3rd edition
G.B. Wetherill and K.D. Glazebrook (1986)
29 Tensor Methods in Statistics *P. McCullagh* (1987)
30 Transformation and Weighting in Regression
R.J. Carroll and D. Ruppert (1988)
31 Asymptotic Techniques for Use in Statistics
O.E. Bandorff-Nielsen and D.R. Cox (1989)
32 Analysis of Binary Data, 2nd edition *D.R. Cox and E.J. Snell* (1989)
33 Analysis of Infectious Disease Data *N.G. Becker* (1989)
34 Design and Analysis of Cross-Over Trials *B. Jones and M.G. Kenward* (1989)
35 Empirical Bayes Methods, 2nd edition *J.S. Maritz and T. Lwin* (1989)

36 Symmetric Multivariate and Related Distributions
K.T. Fang, S. Kotz and K.W. Ng (1990)

37 Generalized Linear Models, 2nd edition *P. McCullagh and J.A. Nelder* (1989)

38 Cyclic and Computer Generated Designs, 2nd edition
J.A. John and E.R. Williams (1995)

39 Analog Estimation Methods in Econometrics *C.F. Manski* (1988)

40 Subset Selection in Regression *A.J. Miller* (1990)

41 Analysis of Repeated Measures *M.J. Crowder and D.J. Hand* (1990)

42 Statistical Reasoning with Imprecise Probabilities *P. Walley* (1991)

43 Generalized Additive Models *T.J. Hastie and R.J. Tibshirani* (1990)

44 Inspection Errors for Attributes in Quality Control
N.L. Johnson, S. Kotz and X. Wu (1991)

45 The Analysis of Contingency Tables, 2nd edition *B.S. Everitt* (1992)

46 The Analysis of Quantal Response Data *B.J.T. Morgan* (1992)

47 Longitudinal Data with Serial Correlation—A State-Space Approach
R.H. Jones (1993)

48 Differential Geometry and Statistics *M.K. Murray and J.W. Rice* (1993)

49 Markov Models and Optimization *M.H.A. Davis* (1993)

50 Networks and Chaos—Statistical and Probabilistic Aspects
O.E. Barndorff-Nielsen, J.L. Jensen and W.S. Kendall (1993)

51 Number-Theoretic Methods in Statistics *K.-T. Fang and Y. Wang* (1994)

52 Inference and Asymptotics *O.E. Barndorff-Nielsen and D.R. Cox* (1994)

53 Practical Risk Theory for Actuaries
C.D. Daykin, T. Pentikäinen and M. Pesonen (1994)

54 Biplots *J.C. Gower and D.J. Hand* (1996)

55 Predictive Inference—An Introduction *S. Geisser* (1993)

56 Model-Free Curve Estimation *M.E. Tarter and M.D. Lock* (1993)

57 An Introduction to the Bootstrap *B. Efron and R.J. Tibshirani* (1993)

58 Nonparametric Regression and Generalized Linear Models
P.J. Green and B.W. Silverman (1994)

59 Multidimensional Scaling *T.F. Cox and M.A.A. Cox* (1994)

60 Kernel Smoothing *M.P. Wand and M.C. Jones* (1995)

61 Statistics for Long Memory Processes *J. Beran* (1995)

62 Nonlinear Models for Repeated Measurement Data
M. Davidian and D.M. Giltinan (1995)

63 Measurement Error in Nonlinear Models
R.J. Carroll, D. Rupert and L.A. Stefanski (1995)

64 Analyzing and Modeling Rank Data *J.J. Marden* (1995)

65 Time Series Models—In Econometrics, Finance and Other Fields
D.R. Cox, D.V. Hinkley and O.E. Barndorff-Nielsen (1996)

66 Local Polynomial Modeling and its Applications *J. Fan and I. Gijbels* (1996)

67 Multivariate Dependencies—Models, Analysis and Interpretation
D.R. Cox and N. Wermuth (1996)

68 Statistical Inference—Based on the Likelihood *A. Azzalini* (1996)

69 Bayes and Empirical Bayes Methods for Data Analysis
B.P. Carlin and T.A Louis (1996)

70 Hidden Markov and Other Models for Discrete-Valued Time Series *I.L. Macdonald and W. Zucchini* (1997)

71 Statistical Evidence—A Likelihood Paradigm *R. Royall* (1997)

72 Analysis of Incomplete Multivariate Data *J.L. Schafer* (1997)

73 Multivariate Models and Dependence Concepts *H. Joe* (1997)

74 Theory of Sample Surveys *M.E. Thompson* (1997)

75 Retrial Queues *G. Falin and J.G.C. Templeton* (1997)

76 Theory of Dispersion Models *B. Jørgensen* (1997)

77 Mixed Poisson Processes *J. Grandell* (1997)

78 Variance Components Estimation—Mixed Models, Methodologies and Applications *P.S.R.S. Rao* (1997)

79 Bayesian Methods for Finite Population Sampling *G. Meeden and M. Ghosh* (1997)

80 Stochastic Geometry—Likelihood and computation *O.E. Barndorff-Nielsen, W.S. Kendall and M.N.M. van Lieshout* (1998)

81 Computer-Assisted Analysis of Mixtures and Applications—Meta-analysis, Disease Mapping and Others *D. Böhning* (1999)

82 Classification, 2nd edition *A.D. Gordon* (1999)

83 Semimartingales and their Statistical Inference *B.L.S. Prakasa Rao* (1999)

84 Statistical Aspects of BSE and vCJD—Models for Epidemics *C.A. Donnelly and N.M. Ferguson* (1999)

85 Set-Indexed Martingales *G. Ivanoff and E. Merzbach* (2000)

86 The Theory of the Design of Experiments *D.R. Cox and N. Reid* (2000)

87 Complex Stochastic Systems *O.E. Barndorff-Nielsen, D.R. Cox and C. Klüppelberg* (2001)

88 Multidimensional Scaling, 2nd edition *T.F. Cox and M.A.A. Cox* (2001)

89 Algebraic Statistics—Computational Commutative Algebra in Statistics *G. Pistone, E. Riccomagno and H.P. Wynn* (2001)

90 Analysis of Time Series Structure—SSA and Related Techniques *N. Golyandina, V. Nekrutkin and A.A. Zhigljavsky* (2001)

91 Subjective Probability Models for Lifetimes *Fabio Spizzichino* (2001)

92 Empirical Likelihood *Art B. Owen* (2001)

93 Statistics in the 21st Century *Adrian E. Raftery, Martin A. Tanner, and Martin T. Wells* (2001)

94 Accelerated Life Models: Modeling and Statistical Analysis *Vilijandas Bagdonavičius and Mikhail Nikulin* (2001)

95 Subset Selection in Regression, Second Edition *Alan Miller* (2002)

96 Topics in Modelling of Clustered Data *Marc Aerts, Helena Geys, Geert Molenberghs, and Louise M. Ryan* (2002)

97 Components of Variance *D.R. Cox and P.J. Solomon* (2002)

98 Design and Analysis of Cross-Over Trials, 2nd Edition *Byron Jones and Michael G. Kenward* (2003)

99 Extreme Values in Finance, Telecommunications, and the Environment *Bärbel Finkenstädt and Holger Rootzén* (2003)

100 Statistical Inference and Simulation for Spatial Point Processes *Jesper Møller and Rasmus Plenge Waagepetersen* (2004)

Contents

Preface

Our book *Asymptotic Techniques for Use in Statistics* was originally planned as an account of asymptotic statistical theory, but by the time we had completed the mathematical preliminaries it seemed best to publish these separately. The present book, although largely self-contained, takes up the original theme and gives a systematic account of some recent developments in asymptotic parametric inference from a likelihood-based perspective.

Chapters 1–4 are relatively elementary and provide first a review of key concepts such as likelihood, sufficiency, conditionality, ancillarity, exponential families and transformation models. Then first-order asymptotic theory is set out, followed by a discussion of the need for higher-order theory. This is then developed in some generality in Chapters 5–8. A final chapter deals briefly with some more specialized issues.

The discussion emphasizes concepts and techniques rather than precise mathematical verifications with full attention to regularity conditions and, especially in the less technical chapters, draws quite heavily on illustrative examples. Each chapter ends with outline further results and exercises and with bibliographic notes.

Many parts of the field discussed in this book are undergoing rapid further development, and in those parts the book therefore in some respects has more the flavour of a progress report than an exposition of a largely completed theory.

We have attempted to follow a connected line of developments but inevitably some parts are more specialized than other. The following are some of the sections that might be largely omitted at a first reading: the later part of section 2.8, sections 3.8, 5.3–5.7, 6.3, 6.4, 7.3–7.6 and Chapter 9.

We are grateful to Dawn Peters and Neil Shephard for comments on some of the chapters.

Our collaboration has been supported throughout by a grant from the European Communities for which we wish to express our strong appreciation.

Susanne Slotsdal and Oddbjørg Wethelund, Aarhus University, did outstanding work typing a difficult manuscript.

O.E. Barndorff-Nielsen
D.R. Cox

CHAPTER 1

Preliminaries

1.1 Introduction

To give a systematic framework for the theoretical discussion of statistical methods and, in particular, to provide a basis for the development of new methods, we proceed as follows. Suppose that we want to analyse observations on variables which we call **response variables**; we collect these as an $n \times 1$ vector $y = (y_1, \ldots, y_n)^T$. Note that, for a reason that will appear later, connected with the use of tensor methods of calculation, we sometimes use superscripts to identify components rather than suffices. In this notation $y = (y^1, \ldots, y^n)^T$ and the square of y^j has to be written $y^j y^j$ or sometimes as $(y^j)^2$. Often, although not necessarily, the components correspond to individuals under study (people, experimental animals, plots, etc.) and there is available for each individual data on other variables, which we call **explanatory variables**. For the moment we regard these as fixed and given and not explicitly recognized in the notation. We suppose that there is variation, at least partly haphazard, in the response variable. Our object is to study that variation and in particular to examine how the response variable depends on the explanatory variables.

We formalize this in three steps:

1. We regard y as the observed value of a random variable $Y = (Y_1, \ldots, Y_n)^T$ having an unknown probability distribution conveniently specified by a probability density function $p(y) = p_Y(y)$ with respect to an appropriate measure μ, usually Lebesgue measure on $\mathbb{R}^n$ or counting measure.
2. We restrict the unknown density to a suitable family $\mathscr{F}$. We concentrate in the present book largely on the case where the density is of known analytical form but involves a finite number of real unknown parameters $\theta = (\theta_1, \ldots, \theta_d)^T$. Further, we specify

the region $\Omega_\theta \subset \mathbb{R}^d$ of possible values of θ; Ω_θ is called the **parameter space**. To indicate the dependency of the density on θ we write $p(y;\theta)$ and we refer to this as the **model function**.

3. We assume that the objective of the analysis is one or more of:
   (a) assessing some aspects of θ, for example the value of a single component θ_b, say;
   (b) predicting the value of some as yet unobserved random variable whose distribution depends on θ;
   (c) examining the adequacy of the model specified by $\mathscr{F}$ and Ω_θ.
   Mostly we shall consider (a), although in fact much of the discussion is relevant to (c) and to a lesser extent (b).

Clearly, from a viewpoint more applied than the one we shall take, there is much about points 1–3 that needs critical discussion. In particular, the choice of the family $\mathscr{F}$ is of central importance and serves a dual role, namely, that of defining precisely the objective of analysis (the primary aspects) and that of defining the further features necessary for the choice of method of analysis to have a satisfactory basis (the secondary aspects).

Parameterization of the primary aspects is very often essential to establish a link with subject-matter concepts. In some applications it may be fruitful to adopt weaker specifications of secondary aspects, leading to so called semiparametric models.

The objective of the present discussion is to provide a framework for the relatively systematic analysis of a very wide range of possible $\mathscr{F}$. We do not do this by aiming to satisfy various formal optimality criteria. Rather, we focus on the likelihood function and quantities derived from it. We regard the primary mode for summarizing evidence about a one-dimensional parameter of interest to be a nested series of upper and lower confidence limits based as directly as possible on the likelihood. Searching comments on the relation between a formal theory of inference and the careful analysis of scientific data would take us too far afield!

While we aim to deal with general $\mathscr{F}$, it is important to have some special families in mind, both to illustrate general concepts and because of their intrinsic importance. We now introduce three such families, returning later to more thorough discussion. These are the normal theory linear models, the exponential models and the transformation models. In a sense the latter two are generalizations of the first.

1.2 Normal theory linear model

In this section we use the superscript notation for the first time. Suppose that for each j there are q known constants $x_1^j, \ldots, x_q^j$, typically values of explanatory variables, and an unknown parameter vector $\beta = (\beta^1, \ldots, \beta^q)^T$ such that

$$E(Y^j) = x_r^j \beta^r, \quad E(Y) = x\beta. \tag{1.1}$$

In the first version, the summation convention is used; that is, we automatically sum with respect to a letter represented as a subscript and as a superscript. In the second version we give the equivalent and probably more familiar version in which Y is an $n \times 1$ vector, x an $n \times q$ matrix with elements (x_r^j) and β is a $q \times 1$ vector. We suppose, without essential loss of generality, that the rank of x is $q \leqslant n$.

A family of densities is specified by supposing that $Y^1, \ldots, Y^n$ are independently normally distributed with variance $\tau = \sigma^2$ and expectations (1.1) and by giving the parameter space. If τ is known, equal say to τ_0, the dimension, d, of the parameter space is q and the standard parameter space is $\{\beta \in \mathbb{R}^q\}$, whereas in the more common case in which τ is unknown the parameter space is $\{\beta \in \mathbb{R}^q, \tau > 0\}$ and $d = q + 1$. The standard parameter spaces may, of course, be modified for particular applications.

Note that when dealing with particular problems it is often, as here, convenient to depart from the general notation of section 1.1. We assume that the reader is familiar with the motivation for and of least the simpler properties of (1.1).

It is convenient to write

$$g_{rs} = x_r^j x_s^k \delta_{jk}, \quad g = x^T x \tag{1.2}$$

and to define least squares estimates $\hat{\beta}^r$ by

$$g_{rs}\hat{\beta}^s = x_r^k \delta_{jk} Y^j, \hat{\beta}^r = g^{rs} x_s^k \delta_{jk} Y^j, g\hat{\beta} = x^T Y, \hat{\beta} = g^{-1} x^T Y. \tag{1.3}$$

Here $\delta_{jk} = 1$ $(j = k)$, $\delta_{jk} = 0$ $(j \neq k)$ and (g^{rs}) is the matrix inverse of (g_{rs}). In some ways the most enlightening formulation of (1.3) is

$$x_r^k \delta_{jk}(Y^j - x_s^j \hat{\beta}^s) = 0, x^T(Y - x\hat{\beta}) = 0, \tag{1.4}$$

showing that $x\hat{\beta}$ is the orthogonal projection of Y on the space spanned by the columns of x. We have $\hat{\beta} = hY$ where $h = (x^T x)^{-1} x^T$ is referred to as the hat matrix.

It is convenient for later comparison with the exponential families to write the density of Y explicitly in two slightly different forms, one for known variance and one for unknown variance. In the former case we write for the density

$$\frac{1}{(2\pi)^{\frac{1}{2}n}\tau_0^{\frac{1}{2}n}}\exp\left\{-\frac{y^{\mathrm{T}}y}{2\tau_0}\right\}\exp\left\{\frac{y^{\mathrm{T}}x\beta}{\tau_0}-\frac{\beta^{\mathrm{T}}g\beta}{2\tau_0}\right\}$$
$$=\frac{1}{(2\pi)^{\frac{1}{2}n}\tau_0^{\frac{1}{2}n}}\exp\left\{-\frac{y^{\mathrm{T}}y}{2\tau_0}\right\}\exp\left\{\hat{\beta}^{\mathrm{T}}\left(\frac{g\beta}{\tau_0}\right)-\left(\frac{\beta^{\mathrm{T}}g\beta}{2\tau_0}\right)\right\}, \tag{1.5}$$

whereas when the variance is unknown we write

$$\frac{1}{(2\pi)^{\frac{1}{2}n}}\exp\left\{y^{\mathrm{T}}y\left(-\frac{1}{2\tau}\right)+\hat{\beta}^{\mathrm{T}}\left(\frac{g\beta}{\tau}\right)-\left(\frac{\beta^{\mathrm{T}}g\beta}{2\tau}\right)-\frac{1}{2}n\log\tau\right\}. \tag{1.6}$$

The significance of (1.5) and (1.6) lies in the occurrence of products of functions of observations and unknown parameters.

Example 1.1 Normal random sample

A very special case of the normal theory linear model is obtained when $Y^1,\ldots,Y^n$ are identically distributed with mean μ. That is $q=1$, $x_1^j=1$, $\beta^1=\mu$. The least squares estimate is $\bar{Y}=\sum Y^j/n$ and the factorizations (1.5) and (1.6) take the form

$$\frac{1}{(2\pi)^{\frac{1}{2}n}\tau_0^{\frac{1}{2}n}}\exp\left(-\frac{\Sigma(y^j)^2}{2\tau_0}\right)\exp\left(\bar{y}\frac{n\mu}{\tau_0}-\frac{n\mu^2}{2\tau_0}\right), \tag{1.7}$$

$$\frac{1}{(2\pi)^{\frac{1}{2}n}}\exp\left\{\Sigma(y^j)^2\left(-\frac{1}{2\tau}\right)+\bar{y}\left(\frac{n\mu}{\tau}\right)-\frac{n\mu^2}{2\tau}-\frac{1}{2}n\log\tau\right\}. \tag{1.8}$$

□

1.3 Exponential models

Suppose that there are m unknown parameters denoted by $\phi=(\phi^1,\ldots,\phi^m)^{\mathrm{T}}$ and to be called **canonical parameters**, and m statistics, i.e. functions of y, denoted by $s=(s_1,\ldots,s_m)^{\mathrm{T}}$ and to be called **canonical statistics** and such that the density of Y is

$$p_Y(y;\phi)=h(y)\exp\{s_i\phi^i-k(\phi)\}=h(y)\exp\{s^{\mathrm{T}}\phi-k(\phi)\}. \tag{1.9}$$

Here $h(y)$ and $k(\phi)$ are functions of the indicated arguments. We usually suppose that the components of s are not linearly dependent with probability one, and provided the canonical parameter space Ω_ϕ consists of all the ϕ such that

$$\int h(y)\exp(s^{\mathrm{T}}\phi)\mu(dy) < \infty$$

we refer to the family $\mathscr{F}$ as a **full exponential model**. Note that $\exp\{-k(\phi)\}$ is in effect a normalizing constant chosen to make (1.9) integrate to 1. Clearly the exponential form (1.9) is preserved if we apply any fixed nonsingular linear transformation to s provided we make the inverse transformation to ϕ, thus leaving $s^{\mathrm{T}}\phi$ invariant. If $0\in\Omega_\phi$, we can without loss of generality take $k(0)=0$ and then $h(y)=p_Y(y;0)$. In other cases we can in effect measure ϕ from some suitable origin $\phi_0\in\Omega_\phi$ by rewriting (1.9) in the form

$$p_Y(y;\phi) = p_Y(y;\phi_0)\exp[s^{\mathrm{T}}(\phi-\phi_0) - \{k(\phi)-k(\phi_0)\}]. \tag{1.10}$$

We call $p_Y(y;\phi)$ the (m,m) **exponential family** generated from the baseline $p_Y(y;\phi_0)$ by exponential tilting via s.

To interpret the function $k(\phi)$, we have from (1.9) that the moment generating function of the random variable S corresponding to s is

$$\begin{aligned} M(S;t,\phi) &= E\{\exp(S^{\mathrm{T}}t)\} \\ &= \int h(y)\exp\{s^{\mathrm{T}}(\phi+t)\}\mu(dy)\exp\{-k(\phi)\} \qquad (1.11) \\ &= \exp\{k(\phi+t)-k(\phi)\}. \qquad (1.12) \end{aligned}$$

The value of the integral in (1.11) follows provided $\phi+t\in\Omega_\phi$ because (1.9) has total integral one for that parameter value. Thus the cumulant generating function of S is

$$K(S;t,\phi) = k(\phi+t)-k(\phi) \tag{1.13}$$

and, provided $0\in\Omega_0$ and $k(0)=0$,

$$K(S;t,0) = k(t). \tag{1.14}$$

In any case we refer to $k(\phi)$ as the **cumulant function** of (1.9).

Example 1.2 Normal random sample

First let Y be a single observation normally distributed with mean μ and known variance τ_0, which we write as $Y\sim \mathrm{N}(\mu,\tau_0)$. Then, virtually as in (1.7), we can write the density as

$$\frac{1}{(2\pi\tau_0)^{1/2}}\exp\left(-\frac{y^2}{2\tau_0}\right)\exp\left(y\frac{\mu}{\tau_0}-\frac{\mu^2}{2\tau_0}\right). \tag{1.15}$$

Thus we can regard $N(\mu, \tau_0)$ as generated from $N(0, \tau_0)$ via the canonical statistic y and canonical parameter $\phi = \mu/\tau_0$. Note that the cumulant generating function of Y can be obtained by writing the final portion of (1.15), $-\mu^2/(2\tau_0)$ as $-\frac{1}{2}\phi^2\tau_0$ and applying (1.13). If the variance is unknown we use (1.8) with $n = 1$ to show the exponential family structure with canonical statistics $\{(y, y^2)\}$ and canonical parameter $(\mu/\tau, -\frac{1}{2}\tau^{-1})$.

When, instead of a single observation, we have a random sample of size n, the exponential family form is retained with canonical statistic $\bar{y}$ when τ_0 is known and $\bar{y}$ (or of course $\sum y_j$) and $\sum y_j^2$ when τ is unknown. □

Example 1.3 Normal theory linear model

It follows directly from (1.5) and (1.6) that when the variance is known we have the exponential family form with canonical statistic $\hat{\beta}$ and that when the variance is unknown the canonical statistic may be taken as $(\hat{\beta}, \sum y_j^2)$. The further special properties of the exponential family are easily verified. □

Example 1.4 Exponential distribution

Suppose first that Y is exponentially distributed with mean ρ^{-1} and density

$$\rho e^{-\rho y} = \exp\{y(-\rho) + \log \rho\}. \tag{1.16}$$

This is of exponential family form with canonical statistic y and canonical parameter $\phi = -\rho$. Note that the cumulant generating function of Y is

$$\begin{aligned} K\{Y; t, \phi\} &= -\log(-\phi - t) + \log(-\phi) \\ &= -\log(1 + t/\rho) \qquad (|t| < \rho), \end{aligned}$$

as is easily checked directly. The canonical parameter space is $(-\infty, 0)$, i.e. $\phi < 0$ and does not include $\phi = 0$. Thus if we wish to generate (1.16) by exponential tilting we must use (1.10) as starting point with $\phi_0 < 0$.

If $Y = (Y_1, \ldots, Y_n)$ is a random sample from (1.16) then

$$p_Y(y; \rho) = \rho^n e^{-n\bar{y}\rho} = \exp\{\bar{y}(-n\rho) + n \log \rho\},$$

again an exponential family with canonical statistic $\bar{y}$ and corresponding canonical parameter $-n\rho$. □

Both Examples 1.2 and 1.4 illustrate the important general point that the structure (1.9) of the exponential family is preserved under repeated sampling.

One very convenient and important choice of parameter in the exponential family is the canonical parameter, but other choices are possible either on mathematical or on subject-matter grounds. Indeed, in Example 1.4 expression in terms of ρ, ρ^{-1}, $\log \rho$ and possibly other choices may be fruitful. One important general reparameterization can be introduced as follows, provided Ω_ϕ has no boundary points or that the densities corresponding to such points are discarded from the full exponential family. Equation (1.13) for the cumulant generating function shows that, in a natural notation,

$$E(S_i; \phi) = \partial k(\phi)/\partial \phi^i \tag{1.18}$$

i.e.

$$E(S; \phi) = \nabla k(\phi),$$

where ∇ denotes the $m \times 1$ gradient operator $(\partial/\partial\phi^1, \ldots, \partial/\partial\phi^m)^{\mathrm{T}}$. We define the **mean parameter** η via

$$\eta = \nabla k(\phi). \tag{1.19}$$

The correspondence between η and ϕ is one-to-one.

There is clearly a connection between the sample space $\mathscr{S}$, say, of S and the parameter space Ω_η of η, associated with the canonical parameter space Ω_ϕ. In fact

$$\Omega_\eta \subset \operatorname{cl}\operatorname{conv}\mathscr{S}, \tag{1.20}$$

where cl conv denotes closed convex hull. Thus in the exponential distribution, Example 1.4, $\eta = \rho^{-1}$ and the parameter space of η is the same as the sample space of Y.

By the same argument as led to (1.13) and (1.18)

$$\operatorname{cov}(S_i, S_j; \phi) = \partial^2 k(\phi)/\partial\phi^i \partial\phi^j, \quad \operatorname{cov}(S; \phi) = \nabla\nabla^{\mathrm{T}} k(\phi), \tag{1.21}$$

with similar expressions for higher-order cumulants in terms of corresponding derivatives, cf. section 2.9. To use these results it is necessary only to extract the cumulant function $k(\phi)$ from the joint density.

In the standard situation discussed above both canonical statistic and canonical parameter lie in m-dimensional regions. In some problems, however, ϕ may be restricted to lie in a d-dimensional subspace, $d < m$. This is usually most conveniently expressed by writing $\phi = \phi(\theta)$, where θ is a d-dimensional parameter. We then have for the density of Y

$$p_Y(y;\theta) = h(y)\exp[s^T\phi(\theta) - k\{\phi(\theta)\}], \tag{1.22}$$

where $\theta \in \Omega_\theta \subset \mathbb{R}^d$. We call this system an **(m, d) exponential family** provided Ω_ϕ does not belong to a v-dimensional affine subspace of $\mathbb{R}^m$ with $v < m$, and we indicate this situation also by saying that the exponential family is **curved**. The restriction could be equally expressed by writing the mean parameter as $\eta(\theta)$.

Example 1.5 A simple (2,1) family, the normal parabola
Let (S_1, S_2) be independently normally distributed with unit variance and with mean $(\theta, c\theta^2)$ where θ varies in $\mathbb{R}$ and c is a known constant. This determines a $(2, 1)$ exponential family, unless $c = 0$ in which case we have a $(1, 1)$ family. The $(2, 2)$ family naturally associated with (S_1, S_2) has canonical and (in this case) also mean parameter space equal to $\mathbb{R}^2$ in which $(\theta, c\theta^2)$ defines a curve, in fact a parabola. □

We shall develop these issues further in sections 2.9 and 2.10.

1.4 Transformation models

There is a second important class of models of which again the normal theory linear models are a special case. These are the transformation models, best introduced via the special case of Example 1.1, the normal random sample.

Suppose first that $Y = (Y_1, \ldots, Y_n)$ are independently $N(\mu, \tau_0)$. Consider the family of transformations

$$Y_j \to Y_j + b, y_j \to y_j + b; \mu \to \mu + b, \tag{1.23}$$

acting respectively on the random variables, the observed values and the unknown parameter. Here b is an arbitrary real number. The model is preserved under these transformations, since if Y_j is $\mathrm{N}(\mu, \tau_0)$ then $Y_j + b$ is $\mathrm{N}(\mu + b, \tau_0)$.

We now have two methods of assessing evidence about μ. We may analyse $(y_1, \ldots, y_n)$, or we may analyse by the same method $(y_1 + b, \ldots, y_n + b)$, assess evidence about $\mu + b$, and then 'back-transform' to μ. An invariant procedure is one that yields the same result for all b. That one should use invariant procedures is reasonably compelling whenever there is no external evidence to indicate certain points in the parameter space as more likely in some sense than others. For example, to estimate μ by $c\bar{y}$, where $0 < c < 1$ is a constant, is not invariant but could be justified if the origin is 'special'.

Example 1.6 Location model

In the above discussion no appeal is, in fact, made to normality and the arguments would apply equally to the following. Let $(Y_1, \ldots, Y_n)$ be independent copies of a random variable having density $h(y - \mu)$, where $h(x)$ is a known density function and μ an unknown parameter. The double exponential or Laplace density $h(x) = \frac{1}{2}e^{-|x|}$ is an example. Then there is invariance under the transformations (1.23) on sample and parameter space. □

Example 1.7 Location–scale model

To cover in particular the normal random sample with both parameters unknown, suppose now that the density $h(y - \mu)$ of Example 1.6 is replaced by $\sigma^{-1}h\{(y - \mu)/\sigma\}$, where again $h(y)$ is a known density function, and both μ and σ are unknown, $\sigma > 0$. Now replace the transformations (1.23) by

$$Y_j \to b_1 Y_j + b_2, y_j \to b_1 y_j + b_2; \mu \to b_1 \mu + b_2, \sigma \to b_1 \sigma, \quad (1.24)$$

corresponding to change of origin and scale. Here $b_1 > 0, b_2$ are arbitrary real numbers. An invariant method of analysis for (μ, σ) is unaffected by the transformation (1.24) in the sense discussed above. □

Example 1.8 Non-normal linear model

To represent a linear model with possibly non-normal errors we retain the linear structure of expectations (1.1) and write, reverting to superscript notation,

$$Y^j = x^j_r \beta^r + \sigma \varepsilon^j, \quad (1.25)$$

where σ is an unknown parameter and $\varepsilon = (\varepsilon^1, \ldots, \varepsilon^n)^T$ are independent copies of a random variable with known density $h(\varepsilon)$ of zero mean.

Consider the transformation

$$Y^j \to b_1 Y^j + x_r^j b_2^r, y^j \to b_1 y^j + x_r^j b_2^r, \beta^r \to b_1 \beta^r + b_2^r, \sigma \to b_1 \sigma, \quad (1.26)$$

where $b_1 > 0$ and b_2^r are arbitrary. The requisite invariance structure is easily verified. □

Note that in all the examples there is a family of transformations, in fact a group, on the sample space and a closely associated family of transformations, also a group, on the parameter space.

The theory of transformation models will be developed in more generality in section 2.8. A further key idea can, however, conveniently be introduced at this stage via the location model of Example 1.6. The full family of transformations (1.23) taken for all real b, forming thereby a group, partitions the sample space into a collection of **orbits** such that:

1. any point on an orbit can be transformed into any other point on that orbit by suitable choice of b;
2. no two points on different orbits can be transformed into one another;
3. each point is on just one orbit.

In fact in this example if $y_{(n1)} \leqslant \cdots \leqslant y_{(nn)}$ denote the order statistics, each orbit is characterized by a constant value of

$$a = (y_{(n2)} - y_{(n1)}, \ldots, y_{(nn)} - y_{(n1)}),$$

sometimes called the **configuration** of the sample. Thus the observation y can be regarded as specifying two things, namely which particular orbit is occupied, and the precise position on the orbit. Now it can be shown that the distribution over possible orbits does not depend on μ and that in a certain sense the information about μ is contained in the conditional distribution of position given the specific orbit occupied.

1.5 Invariance

The previous section has sketched a few simple examples of the general type of situation in which there is a group of transformations of the observations and an associated group of transformations of the parameters. This both establishes the broad structure of the sample space and leads to the notion that, in the absence of some special considerations, the final conclusions should be invariant in the sense of not being dependent on which particular version of the data is employed. This is particularly compelling if the transformations contemplated, like scale changes, are ones that could be encountered physically rather than being solely a consequence of the mathematical structure of the model.

The notion of invariance can, however, be used much more widely, especially in connection with transformations of parameters. Thus in Example 1.4, concerning the exponential distributions $\rho e^{-\rho y}$, it would for many purposes be quite reasonable to take the mean ρ^{-1} or, say, $\log \rho$ as the parameter of interest. Invariance would require the conclusion of a statistical analysis to be unchanged by reformulation in terms of $\tilde{\rho}$, any reasonably smooth one-to-one function of ρ. That is, for example, the same answer would be reached by

1. formulations in terms of ρ, application of a method of analysis and the drawing of a conclusion about ρ;
2. formulations in terms of ρ^{-1}, application of the method of analysis, the drawing of a conclusion about ρ^{-1} followed by the taking of reciprocals.

In the approximations of asymptotic theory such invariance is nearly always achieved approximately, but it is much more satisfactory if the invariance can be achieved exactly, so that choice of parameterization is irrelevant. We shall see in Chapter 4 that achievement of exact invariance is a valuable basis for choosing between different procedures.

Of course the requirement is not always relevant. In a statistical decision problem with squared error loss, quite different answers would be obtained if the squared error is in terms of $\log \rho$ rather than ρ.

Invariance under reparameterization can usefully be formulated much more generally. Suppose that $\theta = (\psi, \chi)$, where ψ is a parameter of interest and χ a nuisance parameter. Then it is reasonable to consider one-to-one transformations from θ to $\tilde{\theta} = (\tilde{\psi}, \tilde{\chi})$, where $\tilde{\psi}$ is a one-to-one function of ψ and $\tilde{\chi}$ is a function of both ψ and χ. We call such transformations **interest-respecting reparameterizations.**

Further results and exercises

1.1 Prove that random samples from the following distributions form (m, m) exponential families with either $m = 1$ or $m = 2$: Poisson, binomial, geometric, gamma (index known), gamma (index unknown), inverse Gaussian. Identify the canonical statistics and the canonical and mean parameters.

The negative binomial distribution with both parameters unknown provides an example of a model that is not of exponential model form.

[Section 1.3]

1.2 Observations are available on random variables Y_{is} ($i = 1, \ldots, m$; $s = 1, \ldots, r$) having the structure $Y_{is} = \mu + \xi_i + \eta_{is}$, where ξ_i, η_{is}, are independent unobserved random variables normally distributed with zero mean and variances τ_ξ, τ_η. Thus there are unknown parameters $(\mu, \tau_\xi, \tau_\eta)$. By obtaining the covariance matrix of the Y_{is}, write down their joint density and show that it forms a (3, 3) exponential family. Identify the canonical statistics and canonical parameters. Obtain the parameter space for the canonical parameter and show that it is not equivalent to $\{(\mu, \tau_\xi, \tau_\eta)\colon \mu \in \mathbb{R}, \tau_\xi > 0, \tau_\eta > 0\}$.

[Section 1.3]

1.3 The $p \times 1$ random vector Y has a multivariate normal distribution (a) of zero mean and covariance matrix Σ, (b) of mean μ and covariance matrix Σ. Verify membership of the exponential family, obtain the canonical statistics, the canonical and mean parameters and verify the relation (1.21).

[Section 1.3]

1.4 The $p \times 1$ vector Y has a multivariate normal distribution of mean μ, covariance matrix Σ with elements σ_{ij} and inverse covariance matrix Σ^{-1} with elements σ^{ij}. Show that $\sigma^{ij} = 0$, $i \neq j$, if and only if Y_i and Y_j are conditionally independent given all the other components Y_k ($k \neq i, j = 1, \ldots, p$). Show next that the full exponential family form is retained if Σ is constrained by requiring that a specified set of the σ^{ij}, $i \neq j$, are zero, whereas the same is not true if a specified set of σ_{ij} are zero. The first set of models are called covariance selection models (Dempster, 1972) whereas the second form a subset of linear covariance structure models (Anderson, 1973).

[Section 1.3]

1.5 In the normal theory model of simple linear regression through the origin $Y_1, \ldots, Y_n$ are independently normally distributed with variance σ^2 and with $E(Y_j) = x_j\beta$, where $x_1, \ldots, x_n$ are known constants and β is an unknown parameter. Considering separately the cases where σ is known and unknown, describe carefully the orbits and the analogues of the configuration of the sample. What form do the results take when $x_1 = \cdots = x_n$? Prove that the conditional distribution of the least squares estimate of β given the orbit is the same as the unconditional distribution.

[Section 1.4]

1.6 Discuss some of the standard procedures of statistical inference, such as maximum likelihood estimation, likelihood ratio testing and the Wald test, with respect to whether they are invariant as regards (interest-respecting) reparameterizations.

[Section 1.5]

Bibliographic notes

General discussions of the relation between data and probability models, the formulation of such models and their broad role in the interpretation of data would take us outside the scope of the present book. For general discussions of the role of models, see Cox (1990) and Lehmann (1990). The normal-theory linear model is the subject of many books.

The idea of exponential models or families was used by Fisher (1934) and developed somewhat further in the late 1930s by Koopman, Pitman and Darmois after whom it is sometimes named. For a thorough account emphasizing connections with convex analysis, see Barndorff-Nielsen (1978) and also, from a different viewpoint, Brown (1988). Various further developments of the theory are discussed by Barndorff-Nielsen (1988a).

The analysis by likelihood-related methods of transformation models stems from Fisher's (1934) discussion of scale and location problems. For a general account, in particular of non-normal linear models, see Fraser (1968; 1979). Lehmann (1983; 1986) discusses certain aspects of invariance and transformation models in detail, largely in the context of Neyman–Pearson theory. Cox and Hinkley (1974) give a broad introduction to statistical theory from an eclectic viewpoint. Hinkley, Reid and Snell (1991) have edited a series of essays largely on recent work on these topics from a standpoint more applied than but broadly consonant with the present book.

CHAPTER 2

Some general concepts

2.1 Introduction

In this chapter we review and discuss a number of key elements of the basic theory of parametric statistical inference: likelihood and likelihood quantities, the closely associated ideas of sufficiency, conditioning and ancillarity, model adequacy and parameter orthogonality; and the main types of statistical models: transformation models, the nicest exponential models, which we term prime exponential models, and curved sub-models of these.

The elements are at least as important in the higher-order asymptotic theory, to be developed later in this book, as they are in the more classical theory of inference.

2.2 Likelihood

2.2.1 *Definitions*

A parametric model in the sense discussed in Chapter 1 involves a model function $p(y; \theta)$, sometimes for clarity denoted $p_Y(y; \theta)$, for a random variable Y and parameter $\theta \in \Omega_\theta$, the parameter space. Sometimes the emphasis is in effect on the function as a function of y for each fixed θ, that is on the probability density determined by θ. If we switch emphasis and study the function as a function of θ for fixed y, in particular for the y actually observed, we have the **likelihood function**

$$L_Y(\theta; y) = L(\theta; y) = L(\theta) = p_Y(y; \theta). \tag{2.1}$$

Usually we work with the log-likelihood

$$\ell_Y(\theta; y) = \ell(\theta; y) = \ell(\theta) = \log p(y; \theta). \tag{2.2}$$

Sometimes we study this as a random variable

$$\ell_Y(\theta; Y) = \ell(\theta; Y) = \log p(Y; \theta). \tag{2.3}$$

In (2.1), (2.2) and (2.3) we have given first a very explicit notation; where the context is clear it is helpful to use one of the more condensed notations.

For many purposes two likelihood functions that are proportional to one another, i.e. such that their ratio is independent of θ, $\theta \in \Omega_\theta$, are equivalent. Note, in particular, that if we choose a different dominating measure to determine the density functions $p(y; \theta)$ the expression (2.1) for $L(\theta)$ changes by a factor depending on the y only, while the problem of inference on θ remains the same. Thus, in principle, we have many equivalent versions of the likelihood function and, accordingly, in likelihood calculations, we shall often drop factors depending on the data y only. Similarly, additive terms depending only on y may be dropped from log-likelihoods.

Another way of expressing the same idea is to concentrate on changes of the log-likelihood function. This can be done in several useful ways, by considering a particular version of the likelihood, such as $L_0(\theta) = L(\theta)/L(\theta_0)$ for some fixed θ_0, or the **normed likelihood** $\bar{L}(\theta) = L(\theta)/L(\hat{\theta})$ where $\hat{\theta}$ is the value of θ maximizing $L(\theta)$, and also by defining the **score function** by

$$u_r(\theta; y) = \partial\ell(\theta; y)/\partial\theta_r, \tag{2.4}$$

$$u_Y(\theta; y) = u(\theta; y) = \nabla_\theta \ell(\theta; y), \tag{2.5}$$

where $\nabla_\theta = (\partial/\partial\theta_1, \ldots, \partial/\partial\theta_d)^T$ is the gradient operator. Later we shall introduce the more compact notation ℓ_r for (2.4) and ℓ_θ for (2.5). To study the score function as a random variable we write

$$u_Y(\theta; Y) = u(\theta; Y) = U(\theta) = U = \nabla_\theta \ell(\theta; Y). \tag{2.6}$$

2.2.2 *Independent components*

The discussion in section 2.2.1 is for arbitrary vector random variables Y. Often, however, the components Y_j are mutually independent. Then both log-likelihood and score function are sums of contributions. It is sometimes convenient to write, for example,

$$\ell(\theta; y) = \ell(\theta) = \sum_{j=1}^{n} \ell(\theta; y_j), \tag{2.7}$$

$$u(\theta; y) = u^+(\theta) = \sum_{j=1}^{n} u_j(\theta; y_j), \tag{2.8}$$

where + denotes summation and $\ell(\theta; y_j)$ is found from the density of Y_j.

Quite generally, i.e. even for dependent random variables, if $Y_{(j)} = (Y_1, \ldots, Y_j)$, we may write

$$\ell(\theta; y) = \sum_{j=1}^{n} \ell_{Y_j|Y_{(j-1)}}(\theta; y_j | y_{(j-1)}), \tag{2.9}$$

each term being computed from the conditional density given all previous values in the sequence. Here $y_{(0)}$ specifies whatever is needed to determine the distribution of the first observation y_1. This version is relevant, for instance, for time-series models.

Example 2.1 Exponential distribution and Cauchy distribution
If $Y = (Y_1, \ldots, Y_n)$, where the Y_j are independent copies of a random variable with density $\rho e^{-\rho y}$, then the log-likelihood and score functions for the whole vector y are

$$\ell(\rho) = n \log\rho - \rho \sum y_j, \tag{2.10}$$

$$u(\rho) = u^+(\rho) = n/\rho - \sum y_j. \tag{2.11}$$

If, on the other hand, the component density is the Cauchy distribution $\pi^{-1}\{1 + (y - \theta)^2\}^{-1}$, then

$$\ell(\theta) = -\sum_{j=1}^{n} \log\{1 + (y_j - \theta)^2\}, \tag{2.12}$$

$$u(\theta) = u^+(\theta) = 2\sum_{j=1}^{n} (y_j - \theta)\{1 + (y_j - \theta)^2\}^{-1}. \tag{2.13}$$

Note that in accordance with the convention for likelihood versions a constant term $-n\log\pi$ has been dropped from $\ell(\theta)$. □

Example 2.2 Exponential models
The contrast between (2.10) and (2.12) illustrates the simplifications gained within the class of exponential models. In the former, but not the latter special case, the data enter only through the sum (or mean), whereas the only analogous simplification for the Cauchy distribution is that a sample point y and all $n!$ sample points obtained by permutation of coordinates have equal log-likelihood.

In general if $Y = (Y_1, \ldots, Y_n)$ and the components are independent with Y_j having density

$$h_j(y_j)\exp\{s_j^{\mathrm{T}}\phi - k_j(\phi)\} \tag{2.14}$$

with common ϕ, where s_j is a function of y_j, then

$$\ell(\phi) = \sum\{s_j^{\mathrm{T}}\phi - k_j(\phi)\}, \tag{2.15}$$

$$\begin{aligned} u^+(\phi) &= \sum\{s_j - \nabla_\phi k_j(\phi)\} \\ &= \sum(s_j - \eta_j), \end{aligned} \tag{2.16}$$

where η_j is the mean parameter associated with Y_j, i.e.

$$\eta_j = E(S_j; \phi). \tag{2.17}$$

□

Example 2.3 Gaussian autoregressive process

Suppose that $(y_1, \ldots, y_n)$ are generated by a stationary first-order Gaussian autoregression with correlation parameter ρ, mean μ and innovation variance τ. That is Y_1 is $\mathrm{N}\{\mu, \tau/(1-\rho^2)\}$ and for $j = 2, \ldots, n$

$$Y_j = \mu + \rho(Y_{j-1} - \mu) + \varepsilon_j, \tag{2.18}$$

where $(\varepsilon_2, \ldots, \varepsilon_n)$ are independently distributed as $\mathrm{N}(0, \tau)$. The general form (2.9) simplifies because the distribution of Y_j given $Y_{(j-1)} = (Y_1, \ldots, Y_{j-1})$ depends only on Y_{j-1} and contributes to the log likelihood the term

$$-\tfrac{1}{2}\log(2\pi\tau) - \frac{\{y_j - \mu - \rho(y_{j-1} - \mu)\}^2}{2\tau}$$

so that the full log-likelihood is

$$\begin{aligned} & -\frac{n}{2}\log(2\pi\tau) + \tfrac{1}{2}\log(1-\rho^2) - \frac{(y_1-\mu)^2(1-\rho^2)}{2\tau} \\ & - \sum_{j=2}^{n} \frac{\{y_j - \mu - \rho(y_{j-1} - \mu)\}^2}{2\tau} \\ & = -\frac{n}{2}\log(2\pi\tau) + \tfrac{1}{2}\log(1-\rho^2) - \frac{(y_1-\mu)^2 + (y_n-\mu)^2}{2\tau} \\ & - \sum_{j=2}^{n-1} \frac{(y_j-\mu)^2(1+\rho^2)}{2\tau} + \frac{\rho}{\tau}\sum_{j=2}^{n}(y_j-\mu)(y_{j-1}-\mu). \end{aligned} \tag{2.19}$$

Note that the first and last observations enter symmetrically, in line with the time-reversibility of the stochastic process. If μ is known and equal to zero, say, (2.19) has (3,2) exponential family form with canonical statistic

$$y_1^2 + y_n^2, \quad \sum_{j=2}^{n-1} y_j^2, \quad \sum_{j=2}^{n} y_j y_{j-1}. \tag{2.20}$$

The first component of (2.20) has the nature of an end-effect. See also *Further results and exercises* 2.2. □

2.2.3 Calculation of likelihoods for more complex examples

The same ideas apply to more complex problems, for example to stochastic processes in continuous time, to mixtures of discrete and continuous responses, and to optimal stopping. We shall ignore technical mathematical problems connected with continuous time; the principle in all cases is to write down the probability under the model of what is actually observed.

Example 2.4 Time-dependent Poisson process

Suppose that we observe over a time period $(0, t_0)$ a Poisson process with rate function $\rho(t;\theta)$ depending on an unknown parameter vector θ. For example, one might have $\rho(t;\theta) = \exp(\theta_0 + \theta_1 t)$. The formal definition of the Poisson process requires that the conditional probability of a point in $(t, t+\Delta)$ given the previous history of the process is $\rho(t;\theta)\Delta + o(\Delta)$. Thus if points are observed at times $0 < t_1 < \cdots < t_n < t_0$, on dividing $(0, t_0)$ into a large number of short intervals of length Δ and disregarding the possibility of multiple occurrences, we have a likelihood

$$\prod_{j=1}^{n} \rho(t_j;\theta)\Delta \prod{}' \{1 - \rho(t_k;\theta)\Delta\}\{1 + o(1)\},$$

where the product Π' is over all subintervals not containing a point, t_k being the centre of the kth such subinterval. It follows that the log-likelihood is

$$\sum_{j=1}^{n} \log \rho(t_j;\theta) - \sum{}' \rho(t_k;\theta)\Delta + o(1),$$

where $\sum'$ is a sum over subintervals not containing a point. If now we let $\Delta \to 0$ we have the log-likelihood

$$\ell(\theta) = \sum_{j=1}^{n} \log \rho(t_j;\theta) - \int_0^{t_0} \rho(t;\theta)dt. \tag{2.21}$$

If in particular the rate is constant, $\rho(t;\theta)=\rho$, then

$$\ell(\rho)=n\log\rho-\rho t_0. \tag{2.22}$$

There is a close formal similarity with (2.10) referring to independent observations from an exponential distribution. In (2.10), however, the coefficient of $\log\rho$ is fixed and that of ρ random, whereas in (2.22) these roles are interchanged. Thus as random variables (2.10) and (2.22) are quite different, so that there is no immediate reason why the inferences from them should be the same.

A mathematically careful account of the log-likelihood for this process, and more generally of any stochastic process in continuous time, demands more attention to the limiting processes involved in passing from discrete to continuous time. □

Example 2.5 Non-regular problems

In the above examples the log-likelihood is a well-behaved, in particular smooth, function of θ in Ω_θ for all possible observed y. The simplest example where this is not the case arises when $Y_1,\ldots,Y_n$ are independent and identically distributed copies of a random variable Y uniformly distributed on $(0,\theta)$. Then

$$L_Y(\theta;y)=\begin{cases}\theta^{-n} & y_{(nn)}<\theta,\\ 0 & \text{otherwise,}\end{cases}$$

where $\max(y_1,\ldots,y_n)=y_{(nn)}$. That is,

$$\ell(\theta;y)=\begin{cases}-n\log\theta & y_{(nn)}<\theta,\\ 0 & \text{otherwise.}\end{cases} \tag{2.23}$$

There are many generalizations of this simple example, in which the 'non-regular' parameter θ is multi-dimensional, as when both terminals of the rectangular distribution are unknown, and in which the non-regularity is possibly confined to certain components, as in the three-parameter log-normal distribution with density

$$\frac{1}{\sqrt{(2\pi)}\sigma(y-\gamma)}\exp\left\{-\frac{\{\log(y-\gamma)-\mu\}^2}{2\sigma^2}\right\}$$

if $y>\gamma$ and 0 otherwise.

Another type of non-regularity arises if the parameter space Ω_θ is discrete or more generally has an unusual topology. Two examples are provided via a normal distribution of unknown integer mean and via the simplest form of change-point problem in which $Y_1, \ldots, Y_n$ are independently distributed with density

$$p_{Y_j}(y;\theta) = \begin{cases} p_0(y) & (j = 1, \ldots, \theta), \\ p_1(y) & (j = \theta + 1, \ldots, n). \end{cases} \tag{2.24}$$

See Section 3.8 for further discussion of such problems. □

In the examples above the number of observations, or in continuous time the period of observation, is regarded as fixed. This is by no means always the case, however. For instance, if the number of observations is obtained by a random mechanism then the likelihood should contain a contribution from that source, being the probability of the full stochastic system as observed, although we shall see later that the contribution from random sample size can often, although not always, be ignored, in effect by conditioning on the number of observations in fact observed.

A simple but atypical illustration is where the number of observations n is the value of a random variable N having a fixed and known distribution. Then the full likelihood is the probability that $N = n$, multiplied by the conditional likelihood given $N = n$, i.e. the change from the previous form is the insertion of a factor $P(N = n)$ not depending on θ. If N is random with a distribution depending on a further parameter ξ functionally independent of θ, then the likelihood and parameter space are respectively

$$p_N(n;\xi)p_{Y|N}(y;\theta \mid n), \qquad (\xi,\theta) \in \Omega_\xi \times \Omega_\theta, \tag{2.25}$$

where the second factor is that calculated previously.

Example 2.6 Sequential sampling

A much more general form is as follows. Given observations $y_{(m)} = (y_1, \ldots, y_m)$ on $Y_{(m)} = (Y_1, \ldots, Y_m)$ for some $m = 1, 2, \ldots$, let the probability that at least one further observation is taken be $q_{m+1}(y_{(m)};\xi)$. Then if the data $(y_1, \ldots, y_n)$ result the likelihood is in generality

$$\begin{aligned} &q_1(y_{(0)};\xi)p_{Y_1}(y_1;\theta)q_2(y_{(1)};\xi)p_{Y_2|Y_1}(y_2 \mid y_1;\theta)\cdots q_n(y_{(n-1)};\xi) \\ &\quad \times p_{Y_n|Y_{(n-1)}}(y_n \mid y_{(n-1)};\theta)\{1 - q_{n+1}(y_{(n)};\xi)\} \\ &= \left\{\prod_{r=1}^{n} q_r(y_{(r-1)};\xi)\right\}\{1 - q_{n+1}(y_{(n)};\xi)\} \end{aligned}$$

$$\times \left\{ \prod_{r=1}^{n} p_{Y_r | Y_{(r-1)}}(y_r | y_{(r-1)}; \theta) \right\}, \qquad (\xi, \theta) \in \Omega_\xi \times \Omega_\theta. \qquad (2.26)$$

Here $y_{(0)}$ is a formal specification of whatever is required to determine q_1. If, given that they are observed, the Y_r are independent, the last factor in (2.26) takes a much simpler form. The essential point, however, is that (2.26) is a product of a function of ξ times a function of θ. The key assumption underlying (2.26) is that the probability of stopping given the data available so far does not depend on θ and that the distribution of a component observation given that it is indeed observed, is the distribution of direct interest. Subject to these non-trivial requirements we shall see below in section 7.5 that the nature of the stopping mechanism can be virtually ignored. □

Example 2.7 Pure birth process

To illustrate the calculation of the log-likelihood for a discrete-state Markov process in continuous time, we consider the pure birth (or binary fission) process. At time $t = 0$ there are n_0 individuals. In each time interval $(t, t + \Delta)$ any individual present at time t has, independently of its own past and of that of the individuals present at t, probability $\rho\Delta + o(\Delta)$ of giving birth to a new individual (splitting into two). Let M_t be the number of individuals present at time t.

The method of calculation of the likelihood, but not the final answer, depends on whether the data are values of M_t over, say, $0 \leqslant t \leqslant t_0$; or the history of each individual over that period.

To take the former approach, let t_r be the length of time for which $M_t = r$. Throughout this period births are liable to occur in a Poisson process of rate ρr so that, by the argument of Example 2.4, each such period contributes to the likelihood a factor $e^{-\rho r t_r}$ with a further factor ρr if the period ends with a birth and a factor 1 if the period ends with the completion of the observation interval. Thus, on ignoring terms independent of ρ, the log-likelihood is

$$\ell(\rho) = n \log \rho - \rho \sum_{r=1}^{\infty} r t_r,$$

where $t_r = 0$ if the value $M_t = r$ is not observed and n is the total number of births in the period of observation. Now

$$\sum_{r=1}^{\infty} r t_r = s_{t_0},$$

where s_{t_0} is the total lifetime of all individuals during the period of observation. That is, the log-likelihood is of the same form as for a homogeneous Poisson process; the reason is that throughout the total period of 'exposure to risk' births occur at a constant probability rate. □

Example 2.8 General discrete-state Markov process in continuous time

Essentially the same argument applies to a well-behaved general discrete-state Markov process with transition rate q_{rs}, say, where

$$q_{rs} = \lim[P\{M_{t+\Delta} = s \mid M_t = r\}/\Delta], \quad q_r = \sum_{s \neq r} q_{rs} < \infty, \tag{2.27}$$

as $\Delta \to 0+$, M_t denoting the state of the system at time t. Let t_r be the total time spent in state r, i.e. with $M_t = r$. Then essentially the argument used in Example 2.7 gives the log-likelihood

$$\sum n_{rs} \log q_{rs} - \sum q_r t_r, \tag{2.28}$$

where n_{rs} is the transition count, i.e. the observed number of transitions from r to s.

There are two rather different situations to which this argument applies. In the first there are a finite (usually small) number of states and the q_{rs} are arbitrary parameters. In the second the q_{rs} are expressed in terms of a small number of further parameters. For example in the Markov birth–death process

$$\begin{aligned} q_{r,r+1} &= r\rho, \\ q_{r,r-1} &= r\mu, \\ q_{rs} &= 0 \text{ (otherwise)}, \end{aligned}$$

where ρ and μ are respectively birth and death rates. □

Example 2.9 Censored failure data

A common kind of data arising in medical and other applications has the following form. A number of independent individuals are observed and for each there is a potential failure time, say X_i for the ith individual. Individuals are, however, subject to uniformative right censoring. That is, for the ith individual there is a time c_i such that we observe $Y_i = \min(X_i, c_i)$ together with an indicator $d_i = 1$ $(X_i < c_i)$, $d_i = 0\,(X_i \geqslant c_i)$ corresponding to individuals failing or being censored respectively. If the density and survivor functions of X_i are $p(x;\theta)$, $\bar{P}(x;\theta)$, where,

$$\bar{P}(x;\theta)=\int_x^\infty p(u;\theta)du,$$

the log-likelihood is

$$\sum d_i \log p(y_i;\theta)+\sum(1-d_i)\log \bar{P}(y_i;\theta), \tag{2.29}$$

where the c_i are treated as fixed constants and where the uninformativeness of the censoring is expressed via the assumption that given $Y_i=c_i$, $d_i=0$, the only information about X_i is contained in the event $X_i>c_i$, which has probability $\bar{P}(c_i,\theta)$, or in general notation, $\bar{P}(y_i;\theta)$.

If the c_i are to be regarded as random variables, say, with density $q_i(c_i;\lambda)$, survivor function $\bar{Q}_i(c_i;\lambda)$ then to (2.29) is to be added a complementary term

$$\sum(1-d_i)\log q_i(y_i;\lambda)+\sum d_i \log \bar{Q}_i(y_i;\lambda). \tag{2.30}$$

Provided that there are no common parameters between θ and λ and the corresponding parameter spaces are independent, then so far as the likelihood for θ is concerned there are general arguments for suggesting that the factor (2.30) can be disregarded and then the distinction between fixed and random censoring is immaterial; the key assumption is the uninformativeness of the censoring. In particular, the censoring times of individuals who fail, i.e. who are not censored, are irrelevant. This is both convenient and conceptually appealing, but there is the difficulty that study of the sampling distribution of derived statistics may require specification of such potential censoring times.

In the special cases when failure times are exponentially distributed, $p(x;\rho)=\rho e^{-\rho x}$, (2.29) reduces to the form involving the number of failures and the total 'time at risk', i.e. to the form (2.22).

Often the dependence of the failure time on explanatory variables is the focus of interest and there is an extensive literature on how to represent such dependencies. One approach is to consider for the ith individual a $q \times 1$ vector z_i of explanatory variables and to assume that, conditionally on failure not having occurred by time x, the probability of failure in $(x, x+\Delta)$ is $\rho_0(x)\exp(\beta^T z_i)\Delta + o(\Delta)$, where β is a vector of unknown parameters and ρ_0 refers to a 'baseline' individual with $z_i=0$. The explanatory variables may be time-dependent. Then the survivor function of failure time and the corresponding density are

$$\bar{P}_i(x) = \exp\left\{-\int_0^x \rho_0(u)\exp(\beta^T z_i)du\right\}, \quad p_i(x) = \rho_0(x)\exp(\beta^T z_i)\bar{P}_i(x), \tag{2.31}$$

so that the contribution to the log-likelihood is

$$d_i\{\log \rho_0(x_i) + \beta^T z_i\} - \int_0^{x_i} \rho_0(u)\exp(\beta^T z_i)du \tag{2.32}$$

and for independent individuals the full log-likelihood is obtained by addition.

Later, in section 8.6, we shall develop a factorization of the likelihood based on (2.32) which effectively allows the separation of the infomation about β from that about $\rho_0(\cdot)$. □

2.2.4 *Score function and information*

In section 2.2.1 we introduced the score function (2.6), namely

$$u(\theta; Y) = u_Y(\theta; Y) = U(\theta) = U = \nabla_\theta \ell(\theta; Y),$$

designed to describe changes in the log-likelihood with θ. For regular problems for which the order of differentiation with respect to θ and integration over the sample space can be reversed, we have the important property that

$$E\{U(\theta); \theta\} = 0. \tag{2.33}$$

It is essential, as the notation in (2.33) makes explicit, that the same value of θ is used in evaluating $U(\theta)$ as in evaluating the expectation. To verify (2.33) note that a component of the left-hand side is

$$\int \left\{\frac{\partial \log p_Y(y;\theta)}{\partial \theta^r}\right\} p_Y(y;\theta)\mu(dy) = \int \frac{\partial p_Y(y;\theta)}{\partial \theta^r}\mu(dy)$$

$$= \frac{\partial}{\partial \theta^r}\int p_Y(y;\theta)\mu(dy) = \frac{\partial}{\partial \theta^r}1 = 0. \tag{2.34}$$

Example 2.5, involving the uniform distribution, illustrates the failure of the penultimate step in (2.34) in a non-regular problem. For there

$$\frac{\partial}{\partial \theta}\left\{\int_0^\theta \frac{1}{\theta}dy\right\}$$

has two contributions, one from differentiating the integral and one via the upper limit of the range of effective integration, and (2.33) does not hold.

It follows from Jensen's inequality that

$$E\{\ell(\theta'; Y) - \ell(\theta; Y); \theta\} \leqslant 0 \tag{2.35}$$

with equality if and only if the probability measures under θ and θ' are equivalent. Since, taking the scalar parameter case for simplicity, we have (2.35) holding for $\theta' > \theta$ and $\theta' < \theta$, it follows that $E\{\ell(\theta'; Y); \theta\}$ has a maximum at $\theta' = \theta$.

We have, when (2.33) holds, that

$$\begin{aligned} \operatorname{cov}\{U_r(\theta), U_s(\theta); \theta\} &= E\{U_r(\theta)U_s(\theta); \theta\} \\ &= \int \frac{\partial \ell(\theta; y)}{\partial \theta^r} \frac{\partial \ell(\theta; y)}{\partial \theta^s} p_Y(y; \theta)\mu(dy) \end{aligned} \tag{2.36}$$

and by a repetition of the argument leading to (2.34), provided a second differentiation under the expectation sign is legitimate,

$$\operatorname{cov}\{U_r(\theta), U_s(\theta); \theta\} = E\left\{-\frac{\partial^2 \ell(\theta; Y)}{\partial \theta^r \partial \theta^s}; \theta\right\}. \tag{2.37}$$

More compactly, the covariance matrix of U is

$$\begin{aligned} \operatorname{cov}\{U(\theta); \theta\} &= E\{U(\theta)U^{\mathrm{T}}(\theta); \theta\} \\ &= E(\nabla \ell \nabla^{\mathrm{T}} \ell; \theta) \qquad (2.38) \\ &= E(-\nabla \nabla^{\mathrm{T}} \ell; \theta) \\ &= E\{-\nabla U^{\mathrm{T}}(\theta); \theta\}. \qquad (2.39) \end{aligned}$$

The matrix (2.39) arises from various points of view and it is useful to call it the **expected information matrix** for θ, or sometimes the **Fisher information matrix**, and to denote it by $i(\theta)$. Correspondingly, the Hessian $-\nabla\nabla^{\mathrm{T}}\ell$ is called the **observed information matrix**, and it is denoted by $j(\theta)$. A qualitative justification for use of the word 'information' is that if the matrices are 'large', the function $\ell(\theta)$ tends to be sharply peaked around its maximum. We have $i(\theta) = E\{j(\theta)\}$, and calculation of $i(\theta)$ is on the whole usually slightly more convenient from (2.39) than from (2.38).

A very important general example is provided by the (m, m) exponential family (1.9). Here

$$U(\phi) = \nabla \ell = S - \nabla k(\phi)$$

and $\nabla\nabla^{\mathrm{T}}\ell = -\nabla\nabla^{\mathrm{T}}k(\phi)$, so that

$$\operatorname{cov}\{U(\phi);\phi\} = \nabla\nabla^{\mathrm{T}}k(\phi), \tag{2.40}$$

the Hessian matrix of $k(\phi)$. This follows also because the cumulant generating function of S is $k(\phi+t)-k(\phi)$, by (1.13).

Example 2.10 Exponential distribution

The exponential distribution provides a simple special case where for n independent and identically distributed observations

$$\ell(\rho) = n\log\rho - \rho\sum Y_j,$$
$$\partial\ell/\partial\rho = U(\rho) = n/\rho - \sum Y_j,$$
$$-\partial^2\ell/\partial\rho^2 = n/\rho^2,$$

and it is easily checked directly that

$$E\{U(\rho);\rho\} = 0,$$
$$\operatorname{var}\{U(\rho);\rho\} = -\partial^2\ell/\partial\rho^2. \qquad \square$$

Note that a very special feature of the (m, m) exponential family is that $\nabla\nabla^{\mathrm{T}}k(\phi)$ does not depend on the data, so that the expectation in (2.39) is unnecessary and $i(\phi) = j(\phi)$.

Example 2.11 Laplace distribution

An instance where the second derivative version of the information is not applicable is provided by the Laplace density

$$\tfrac{1}{2}\exp\{-|y-\theta|\} \qquad (-\infty < y < \infty), \tag{2.41}$$

where

$$\ell(\theta) = -|y-\theta|,$$

so that

$$u(\theta;y) = \operatorname{sgn}(y-\theta) \qquad (y \neq \theta).$$

The second derivative is thus 0 except at $y=\theta$ where there is a singularity and thus (2.39) is not available. From (2.38) it follows immediately that $\operatorname{var}\{U(\theta);\theta\} = 1$ for all θ.

A range of types of anomalous behaviour can be illustrated via the family of densities

$$\tfrac{1}{2}a\{\Gamma(1/a)\}^{-1}\exp\{-|y-\theta|^a\}.$$

In particular if $a<1$ moments of $u(\theta;Y)$ above the first do not exist. $\square$

When $Y = (Y_1, \ldots, Y_n)^T$ consists of n independent components, $\ell(\theta)$, and hence also $U(\theta)$ and $i(\theta)$, consists of a sum of n terms, one from each component random variable. In an extension of the notation of Section 2.2.1:

$$u(\theta; Y) = U^+(\theta) = \sum u^j(\theta; Y^j), \tag{2.42}$$

$$i(\theta) = i^+(\theta) = \sum i^j(\theta) = n\bar{\imath}(\theta), \tag{2.43}$$

where $\bar{\imath}(\theta)$ is the average information per component observation. If, in addition to being independent, the component observations are identically distributed, then $\bar{\imath}(\theta)$ is the information from a single observation and $U^+(\theta)$ is the sum of independent and identically distributed random variables.

The notation used so far in this subsection is incomplete in that the score $u(\theta; y)$ and the information $i(\theta)$ depend not only, as indicated, on the value of the parameter θ but also on the parameterization. That is, if we were to change by a smooth one-to-one transformation from θ to ψ and calculate the score and information in terms of ψ then different, although related, values will be obtained. Two notations can be useful. If just two parameterizations are of concern, we can use (u, i) for one and (u^*, i^*) for the other. More explicitly, however, we may write $(u^{(\theta)}, i^{(\theta)})$, $(u^{(\psi)}, i^{(\psi)})$ for quantities in the θ- and ψ-parameterization. Thus, on moving to a superscript notation for components and using the convention whereby summation occurs over symbols occurring both as suffices and superscripts, we have that

$$\begin{aligned} u_a^{(\psi)}(\psi; Y) &= \frac{\partial \ell\{\theta(\psi); Y\}}{\partial \psi^a} \\ &= \frac{\partial \ell(\theta; Y)}{\partial \theta^r} \frac{\partial \theta^r}{\partial \psi^a} \\ &= u_r^{(\theta)}(\theta; Y) \frac{\partial \theta^r}{\partial \psi^a}, \end{aligned} \tag{2.44}$$

i.e.

$$u^{(\psi)}(\psi; Y) = \left[\frac{\partial \theta}{\partial \psi}\right]^T u^{(\theta)}(\theta; Y), \tag{2.45}$$

where $\partial\theta/\partial\psi$ is the Jacobian of the transformation from θ to ψ, with (r, a)th element $\partial\theta^r/\partial\psi^a$. Similarly

$$i_{ab}^{(\psi)}(\psi) = \frac{\partial \theta^r}{\partial \psi^a} \frac{\partial \theta^s}{\partial \psi^b} i_{rs}^{(\theta)}(\theta), \tag{2.46}$$

$$i^{(\psi)}(\psi) = \left[\frac{\partial\theta}{\partial\psi}\right]^{\mathrm{T}} i^{(\theta)}(\theta) \left[\frac{\partial\theta}{\partial\psi}\right], \tag{2.47}$$

so that u transforms as a covariant vector and i as a covariant 2-tensor, properties important for some of the later developments. Later we shall often abbreviate $\partial\theta/\partial\psi$ to $\theta_{/\psi}$, and similarly in other cases, and (2.46) and (2.47) may then be expressed more concisely as

$$i(\psi) = \theta_{/\psi}^{\mathrm{T}} i(\theta) \theta_{/\psi}. \tag{2.48}$$

Further developments connected with score and information functions will occur later, notably the study of higher-order cumulants of the score function and of properties evaluated under different distributions, e.g. $E\{U(\theta); \theta'\}$ for $\theta' \neq \theta$.

2.3 Sufficiency

The emphasis throughout this book is on statistical methods that derive ultimately from the log-likelihood function and more specifically from its linked dependence on parameter and data. At least in simple cases, the nature of this dependence is such that the data are involved only via a limited number of functions. Thus in (2.11), referring to samples from an exponential distribution, the sum of the observations determines the log-likelihood function.

An enlightening way of expressing the same idea is through the notion of a sufficient statistic. Let the data y correspond to a random variable Y with density $p_Y(y;\theta)$, $\theta \in \Omega_\theta$. Let s be a statistic, i.e. a function of y and hence observed, such that if S denotes the corresponding random variable, then the conditional density of Y given $S = s$ does not depend on θ, for all s. That is, in the usual notation, for all $\theta \in \Omega_\theta$,

$$p_{Y|S}(y \mid s; \theta) = g(y, s). \tag{2.49}$$

For continuous random variables, there is a technical difficulty with (2.49) in that the conditional density is not uniquely defined. We agree, however, always to take maximally 'well-behaved' determinations of conditional densities, this being justified in particular because all real data are recorded in discrete form.

The motivation for regarding sufficiency as important is that it represents a separation of the information in the data into two types, that concerned with inference about θ given the model and that concerned with the adequacy of the model. To make this separation vivid, consider the following:

1. suppose first that the investigator observes that $S = s$. Then some inference can be drawn about θ, assuming the adequacy of the family and using in some way the distribution of S as a function of θ;
2. suppose in a second stage that the investigator learns that $Y = y$. The additional information in the second stage is derived in effect by observing one realization of the conditional distribution of Y given $S = s$. Since this distribution does not involve θ it can throw no additional light on the value of θ. If, however, the observation is extreme in some relevant sense it can throw doubt on the adequacy of the family. In the present book the main emphasis will be on the direct use of s for inference about θ, but the second role, also, is of considerable importance.

The definition (2.49) in general does not define S uniquely, because it is always possible to take aspects of Y not defined by a given S and to use them to augment S, say to S'. We therefore usually take the minimal S for which (2.49) holds, yielding the so-called minimal sufficient statistic. In future, by sufficient statistic we normally mean the minimal form.

It is a consequence of the above arguments that it is often convenient to determine the (minimal) sufficient statistic and to concentrate on its distribution as a function of θ; we call this process reduction by sufficiency. Examples are given later.

The determination of S directly from the definition is often difficult. Instead we use a factorization theorem: essentially, a necessary and sufficient condition that S is sufficient for θ is that for all y, θ

$$p_Y(y; \theta) = g(s, \theta)h(y) \tag{2.50}$$

for some functions $g(s, \theta)$ and $h(y)$. Without loss of generality, $g(s, \theta)$ may be taken as the unconditional density of S for given θ.

If (2.50) holds then direct calculation of the conditional density verifies (2.49), whereas if (2.49) holds then the multiplication law yields (2.50).

As with the definition (2.49), there are some technical difficulties in giving precise mathematical form to the concept of minimal sufficiency and to the factorization theorem. For a rigorous treatment see, for instance, Barndorff-Nielsen (1978, section 4.2).

When the densities $p(y;\theta)$, $\theta\in\Omega_\theta$, have common support, as is usually the case in applications, a constructive way to finding the minimal sufficient statistic is to form the ratios $L_0(\theta) = p(y;\theta)/p(y;\theta_0)$ for an arbitrary but fixed θ_0 and to seek the minimal statistic s for which all the ratios depend on y through s only. Formulated otherwise, the likelihood function $L_0(\cdot)$ is minimal sufficient.

Example 2.12 Exponential models
The density (1.9)

$$p_Y(y;\phi) = h(y)\exp\{s^T\phi - k(\phi)\}$$

has the form (2.50), so that s is the sufficient statistic. Note that sufficiency does not depend on the particular choice of parameterization adopted and, moreover, that any one-to-one transformation of s can be used instead of s itself. A particular example is the normal theory linear model with independent errors of variance τ. In matrix form the direct version of the log-likelihood, namely

$$-\tfrac{1}{2}n\log\tau - \frac{(y-x\beta)^T(y-x\beta)}{2\tau}$$

does not make this apparent. The identity

$$\begin{aligned}(y-x\beta)^T(y-x\beta) &= \{(y-x\hat\beta)+x(\hat\beta-\beta)\}^T\{(y-x\hat\beta)+x(\hat\beta-\beta)\}\\ &= (y-x\hat\beta)^T(y-x\hat\beta)+2(\hat\beta-\beta)^Tx^T(y-x\hat\beta)\\ &\quad+(\hat\beta-\beta)^T(x^Tx)(\hat\beta-\beta)\\ &= (y-x\hat\beta)^T(y-x\hat\beta)+(\hat\beta-\beta)^T(x^Tx)(\hat\beta-\beta),\end{aligned} \tag{2.51}$$

in virtue of the least squares equations (1.3), shows that the data enter the log-likelihood only via the residual sum of squares

$$SSD = (y-x\hat\beta)^T(y-x\hat\beta) \tag{2.52}$$

and the least squares estimates $\hat\beta$, or equivalently x^Ty. Thus the sufficient statistic is $(SSD, \hat\beta)$. If τ is known, the term in the likelihood involving SSD can be separated off into the term $h(y)$ in (2.50) and the sufficient statistic is $\hat\beta$ alone.

Note that from the exponential family form (1.6) of the likelihood the sufficient statistic emerges as $(y^Ty, \hat\beta)$, illustrating that the property of sufficiency is preserved under one-to-one transformation. □

In a curved (m, d) exponential model the dimension m of the sufficient statistic exceeds that of the parameter.

Example 2.13 Idealized component of variance problem
A model closely related to a one-way unbalanced variance component problem with known (zero) mean involves independent random variables $Y_1, \ldots, Y_n, Z$, such that $Y_1, \ldots, Y_n$ are normal with zero mean and with $\text{var}(Y_j) = \tau/m_j + \psi$ and rZ/τ has a chi-squared distribution with r degrees of freedom. Here Z corresponds to the mean square within groups, Y_j is the mean of the jth group, m_j the corresponding sample size and τ and ψ are respectively the components of variance within and between groups. The log likelihood is thus

$$-\tfrac{1}{2}\sum \log(\tau/m_j + \psi) - \tfrac{1}{2} r \log \tau - \tfrac{1}{2}\sum \frac{y_j^2}{\tau/m_j + \psi} - \frac{rz}{2\tau}. \tag{2.53}$$

The only reduction in dimensionality arising from sufficiency occurs when some of the m_j are equal. Thus if there were only two sample sizes m' and m'', the reduction by sufficiency would be to $\sum' y_j^2$, $\sum'' y_j^2$, z, where $\sum'$, $\sum''$ denote respectively sums over the samples of sizes m', m''. This forms, in the terminology of (1.22), a (3, 2) curved exponential family. If, furthermore, ψ is known, then we have a (2,1) family with sufficient statistic $(\sum' y_j^2, \sum'' y_j^2)$. □

Example 2.14 Order statistics
If $Y_1, \ldots, Y_n$ are independently and identically distributed with an unknown density $f(y)$, assumed continuous for simplicity and belonging to some family $\mathscr{F}$ of densities, the likelihood is $f(y_1) \cdots f(y_n)$. This is the same for all permutations of $y_1, \ldots, y_n$ and hence is defined by the order statistics $y_{(n1)} < \cdots < y_{(nn)}$, where we may assume all observations distinct. Given the order statistics, the probability of a given sequence $y_1, \ldots, y_n$ is $1/n!$, independently of the form of $f(y)$. Hence by direct application of the definition of sufficiency, the set of order statistics is sufficient. Equivalently, the sample cumulative distribution function is sufficient. For particular families $\mathscr{F}$ of densities $f(y)$, for instance an exponential family, the order statistics will not be minimal sufficient. But in general, in particular provided the family is rich enough, minimality will be achieved. That the family need not, however be rich is shown by the Cauchy location family given

by $f(y) = f(y;\theta) = \pi^{-1}\{1 + (y - \theta)^2\}^{-1}$ for which $(y_{(n1)}, \ldots, y_{(nn)})$ is in fact minimal sufficient.

Example 2.15 Non-regular problems
For non-regular problems such as (2.23), repeated samples from the uniform distribution on $(0, \theta)$, direct examination of the log-likelihood function is often the simplest approach to determining a sufficient statistic. In fact in (2.23) the likelihood is determined by the maximum order statistic. For sampling the uniform distributions on $(\theta - 1, \theta + 1)$, or on $(-\theta, \theta)$ both the smallest and the largest order statistic are needed to form the sufficient statistic. □

Example 2.16 Transformation models
The essence of transformation models has been sketched in Section 1.4. The sample space is divided into orbits. The distribution of the particular orbit occupied does not depend on the parameter. Conditionally on the orbit occupied, the precise position has a distribution depending on the unknown parameter. Except in special cases, such as the normal distribution, where the model is also an exponential family model, there is no reduction of dimensionality by sufficiency. We need a different concept to achieve useful simplification. A detailed discussion will be given in section 2.8. □

2.4 Conditioning and some general concepts

First we examine the role of conditioning in the definition of the probability of rather specific events, in particular what in the philosophical literature is called the problem of the unique case. For this, consider a population of individuals and an event A of interest, for instance that an individual dies from heart disease before age 70. Suppose, further, that we have a very large data set of relevant past cases which can be used as a basis for a frequency definition of probability. Now suppose that a series of new individuals is drawn randomly from the population under study and for each it is required to calculate the probability of event A. R.A. Fisher was the first to stress that it is not enough that the right long-run frequency interpretation shall be achieved. If each probability is to be relevant to the individual in question, it must be conditional on observed relevant features, such as age, sex, smoking habits and blood pressure. The conditioning may be achieved directly by extracting the relevant section of the data base or indirectly, for example by fitting a logistic regression equation. Note, however, that, especially if we condition directly, we must limit the conditioning: otherwise we would reach the position where each individual is not only unique, but also uninformative about other individuals, a barrier to the summarizing of relevant empirical regularities and to the development of general concepts and conclusions about the topic under study.

In connection with methods of statistical inference, probability is used in two quite distinct ways. The first is, as noted in Section 1.1, to represent the haphazard behaviour of the empirical system under investigation, i.e. to define the stochastic model assumed to have generated the data. The second is to assess uncertainty in conclusions, via significance levels, confidence regions, posterior distributions and so forth. In essence, we consider methods of analysis to be analogous to measuring instruments and hence to be calibrated via their performance under repeated use. That is, we enquire, whether by analytical arguments or by simulation, as to how such and such a method would perform if, hypothetically, it were used repeatedly on data derived from the model under study. The probabilities used as a basis for inference are thus long-run frequencies under hypothetical repetition. The issue now arises, as before, how these long-run frequencies are to be made relevant to the data under study. The answer lies again in conditioning the calculations so that the 'long run' matches the particular set of data in important respects.

Now in a Bayesian development this issue is dealt with automatically. It is supposed that the particular value of the parameter θ is itself generated by a random mechanism giving a known density $p_\Theta(\theta)$ for θ, or, put differently, that the uncertainty about θ arising from all sources other than the data is encapsulated in a known density $p_\Theta(\theta)$. Then Bayes's theorem yields

$$p_{\Theta|Y}(\theta \mid Y = y) \propto p_\Theta(\theta) p_{Y|\Theta}(y \mid \Theta = \theta), \tag{2.54}$$

where now we write the model distribution $p_Y(y; \theta)$ as a conditional density, the second factor in (2.54). Here the insertion of a random element in the 'generation' of θ allows us to condition on the whole of the data, $Y = y$, and relevance to the data is emphatically accomplished. If a meaningful form for $p_\Theta(\theta)$ can be agreed, then this approach is uncontroversial. In many applications, especially to the careful interpretation of scientific and technological data, we see-major obstacles to the specification of a generally meaningful $p_{\Theta}(\theta)$ and hence are forced to adopt a less direct route.

We suppose first that the whole parameter vector θ is of interest. We reduce the problem by sufficiency. If, with parameter dimension $d = 1$, there is a one-dimensional sufficient statistic, then we have reduced the problem to that of one observation from a distribution with one unknown parameter and there is, except in very special cases, no choice but to use probabilities (leading to confidence intervals, etc.) calculated from that distribution. Quite similarly, if there is a d-dimensional parameter of interest and a d-dimensional sufficient statistic, then there may be difficulties in presenting the conclusions vividly, but there is generally no choice as to which distribution to use. If, however, the dimension of the (minimal) sufficient statistic exceeds that of the parameter, then there is scope and need for ensuring relevance to the data under analysis by appropriate conditioning. We therefore aim to:

1. partition the minimal sufficient statistic s in the form $s = (t, a)$, so that $\dim(t) = \dim(\theta)$ and the random variable A is distribution constant, i.e. has a distribution not involving θ;
2. use for inference the conditional distribution of T given $A = a$.

Conditioning on $A = a$ makes the distribution used for inference involve repetitions like the data in some respects and hence achieves a position closer to the Bayesian one than would the use of the unconditional distribution. It also removes some arbitrariness in the aspect of the distribution to be used and, although this is not of principal importance, it frequently leads to analytical simplifications. To condition on a random variable whose distribution depends on θ might, under some circumstances, be convenient and meaningful but would in some sense sacrifice information about θ.

Example 2.17 A special uniform distribution
Suppose that $Y_1, \ldots, Y_n$ are independently and identically uniformly distributed on $(\theta - 1, \theta + 1)$. The results for general n are outlined in *Further results and exercises* 2.7; here we concentrate on $n = 2$. The sufficient statistic is the pair of order statistics $Y_{(21)} = \min(Y_1, Y_2)$, $Y_{(22)} = \max(Y_1, Y_2)$, i.e. the order in which the observations arise is suppressed as irrelevant so long as the model is accepted. It is convenient to transform these to the mean $\bar{Y} = \frac{1}{2}(Y_{(21)} + Y_{(22)})$ and the range $R = Y_{(22)} - Y_{(21)}$. A direct calculation shows that R has a

distribution not depending on θ, as is obvious from the transformation structure of the model. We thus have the situation contemplated above, the dimension of the sufficient statistic exceeding the dimension of the parameter and the statistic R, being distribution constant, plays the role of A.

Now the density of $\bar{Y}$ given $R = r$ is easily shown to be uniform on $(\theta - 1 + \frac{1}{2}r, \theta + 1 - \frac{1}{2}r)$, so that the conditional inference corresponds exactly to that from one observation on this distribution; one concludes that all values of θ between $\bar{y} - 1 + \frac{1}{2}r$ and $\bar{y} + 1 - \frac{1}{2}r$ are possible, that there are no grounds from the data for choosing between them and that upper and lower confidence limits, although in a sense artificial, could, if required, easily be found. Note particularly that if $r \doteq 2$, θ is very precisely determined, whereas if $r \doteq 0$, θ could reasonably be anywhere between $\bar{y} - 1$ and $\bar{y} + 1$.

It might seem possible instead to use an unconditional distribution of some one-dimensional statistic extracted from $(\bar{Y}, R)$, and $\bar{Y}$ itself is an almost inevitable choice. The marginal distribution of $\bar{Y}$ is obtained by convolution as the triangular distribution

$$p_{\bar{Y}}(\bar{y}; \theta) = \begin{cases} \bar{y} - \theta + 1 & (\theta - 1 < \bar{y} < \theta), \\ \theta + 1 - \bar{y} & (\theta \leqslant \bar{y} < \theta + 1), \end{cases} \tag{2.55}$$

from which unconditional confidence limits are readily calculated. This would, however, lead to absurd inferences if $r \doteq 2$, where we know the value of θ virtually exactly and the unconditional confidence region would include values logically inconsistent with the data.

Conditioning on $R = r$ assures the relevance of the long-run probability calculation to the data under analysis. Whether we observe a small or a large value of R is an accident not in itself informative about θ but very relevant to the precision achieved. If we do not condition on R either explicitly or implicitly, we sacrifice the relevance of the probability calculation to the data at hand. □

The above discussion raises the following issues, among others:

1. When can a suitable A be found?
2. Is such an A unique?
3. How can the discussion be extended when there are nuisance parameters, i.e. when θ can be partitioned so that $\theta = (\psi, \chi)$, the focus of concern being ψ, the parameter χ being of no direct interest?

2.5 Ancillarity

We now develop some of the ideas of section 2.4 in a little more detail. First, for the present purpose of inference assuming the correctness of the model specification, we suppose that a reduction by sufficiency has been made. A component a of this statistic such that the random variable A is distribution constant is often called **ancillary**, or sometimes **ancillary in the simple sense**. In later parts of this book we use the term 'ancillary' to mean a distribution-constant statistic which, together with the maximum likelihood estimator, constitutes a sufficient statistic. The discussion of Example 2.18 is a good illustration. Unless explicitly stated otherwise, we take the maximal form of ancillary available.

Ancillary statistics arise most frequently in transformation models; see sections 1.4 and 2.8. Indeed, Example 2.17 is a very simple special case. In general, the distribution over orbits does not depend on the parameter θ of the model and plays the role of the ancillary statistic, whereas the distribution of position on the orbit does depend on θ.

Example 2.18 Location model

Let $Y_1, \ldots, Y_n$ be independent and identically distributed with density $h(y-\mu)$, where $h(\cdot)$ is a known function and μ is unknown. An orbit is characterized by values of $a=(a_1,\ldots,a_{n-1})=(y_{(n2)}-y_{(n1)}, y_{(n3)}-y_{(n1)},\ldots,y_{(nn)}-y_{(n1)})$, where, in our usual notation, $y_{(nr)}$ is the rth order statistic. Of course any set of $n-1$ independent differences among the order statistics serves equally to specify an orbit. Position on the orbit can be specified by $y_{(n1)}$ or by $\bar{y}=\sum y_j/n$ or by any other location statistic; in an inference in which a is held constant, the different location statistics differ only by a constant. Now it follows directly from the transformation structure of the model that a, being invariant under the transformation $y \to y+b$, $\mu \to \mu + b$ has a distribution not depending on μ and hence is ancillary. We are led to consider the distribution of, say, $Y_{(n1)}$ given $A=a$.

Now the joint density of the order statistics is

$$n!\,h(y_{(n1)}-\mu)\cdots h(y_{(nn)}-\mu)$$

and the transformation $y_{(n1)} \to y_{(n1)}$, $y_{(n2)} \to y_{(n1)}+a_1,\ldots,y_{(nn)} \to y_{(n1)}+a_{n-1}$ has unit Jacobian, so that the density of $(Y_{(n1)}, A)$ is

$$n!\,h(y_{(n1)}-\mu)h(y_{(n1)}+a_1-\mu)\cdots h(y_{(n1)}+a_{n-1}-\mu)$$

and the conditional density is

$$p_{Y_{(n1)}|A}(y_{(n1)}|a;\mu)=\frac{h(y_{(n1)}-\mu)h(y_{(n1)}+a_1-\mu)\cdots h(y_{(n1)}+a_{n-1}-\mu)}{\int dz\, h(z-\mu)h(z+a_1-\mu)\cdots h(z+a_{n-1}-\mu)} \tag{2.56}$$

$$=\frac{L(\mu;y)}{\int L(v;y)dv}, \tag{2.57}$$

since integration with respect to z is equivalent to integration with respect to μ and the denominator of (2.56) does not depend on μ. Also the likelihood of the observed values is, by (2.1),

$$L(\mu;y)=\prod h(y_j-\mu). \tag{2.58}$$

This establishes the remarkable result that if the likelihood function for μ is normalized, as if it were a probability density for μ, the resulting function is the conditional density of, say, $Y_{(n1)}$ given the ancillary statistic $A=a$. Note that (2.57) should be viewed as a function of a location statistic, e.g. $y_{(n1)}$ or $\bar{y}$, it is immaterial which, and of μ, for fixed a; (2.58) is a function of the difference $y_{(n1)}-\mu$. Because integration with respect to $y_{(n1)}$ and to μ are mathematically, although of course not conceptually, equivalent it follows that an upper $1-\varepsilon$ confidence limit for μ is obtained as that value of (2.57) beyond which the upper tail area as a function of μ is the nominal value ε. Substantial generalization of this to transformation models is possible; see section 2.8.

If this argument is applied when $h(y)=(2\pi)^{-\frac{1}{2}}e^{-\frac{1}{2}y^2}$, we first, if we follow the general prescription, reduce by sufficiency to the sample mean $\bar{y}$ and consider the distribution of $\bar{Y}$, namely $N(\mu, n^{-1})$. If, nevertheless, we use (2.58) as it stands the same result is achieved. In the normal distribution, $\bar{Y}$ is distributed independently of the ancillary statistic A.

Example 2.19 Non-normal linear model

A similar but technically more complicated example is provided by Example 1.8, the linear model with non-normal errors. We define least squares estimates and residual sum of squares in the usual way, characterize orbits by any set of $n-p-1$ standardized residuals, now using a superscript notation,

$$a^1 = \frac{y^1 - x_s^1 \hat{\beta}^s}{\sqrt{SSD}}, \ldots, a^{n-p-1} = \frac{y^{n-p-1} - x_s^{n-p-1} \hat{\beta}^s}{\sqrt{SSD}} \tag{2.59}$$

and position on the orbit by $(\hat{\beta}, SSD)$. Note that the choice of $(\hat{\beta}, SSD)$ in this way is convenient but not essential. The random variable $A = (A^1, \ldots, A^{n-p-1})$ can be shown to have a distribution not depending on β and the error scale σ, and the conditional distribution of $(\hat{\beta}, SSD)$ given $A = a$ is available for inference. The calculation, analytical and numerical, of that distribution is, however, more complicated. □

In section 2.8 the theory of transformation models is given in more detail.

So far we have considered problems in which the whole unknown parameter is of interest. In fact, however, in the previous discussion, the argument for conditioning on $A = a$ rests not so much on the distribution of A being known as on its being totally uninformative about the parameter of interest. Suppose, much more generally, that we can write $\theta^{\mathrm{T}} = (\psi^{\mathrm{T}}, \chi^{\mathrm{T}})$, where ψ is of interest, and where

1. $\Omega_\theta = \Omega_\psi \times \Omega_\chi$, so that the parameters ψ and χ are variation independent;
2. the minimal sufficient statistic $s = (t, a)$;
3. the distribution of T given $A = a$ depends only on ψ;
4. one or more of the following conditions holds:
   (a) the distribution of A depends only on χ and not on ψ;
   (b) the distribution of A depends on (ψ, χ) in such a way that from observation of A alone no information is available about ψ;
   (c) at some given $\psi = \psi_0$, the statistic A, which may be a function of ψ_0, has a distribution depending only on χ.

The most straightforward of these extensions corresponds to (a). In this case A is said to be a **cut** and to be **S-ancillary** for ψ and **S-sufficient** for χ. Conceptually the arguments for conditioning on $A = a$ when ψ is the parameter of interest are as compelling as in the simpler case where A has a fixed distribution.

Conditions (b) and (c) are more problematical. The difficulty with (b) is in making precise the notion of no information being available about ψ, although the condition $\dim(A) \leqslant \dim(\chi)$ will normally ensure that extraction from A of information about ψ, χ being unknown, will be difficult or impossible. Finally in (c) the required factorization of the likelihood into the density of T given $A = a$ multiplied by the density of A is considered only at $\psi = \psi_0$, when ψ_0 is a fixed, possibly arbitrary, value of ψ. We now illustrate the three conditions (a)–(c).

Example 2.20 Two Poisson variables
Let (Y_1, Y_2) be independent Poisson random variables with means $(1 - \psi)\ell, \psi\ell$ where ℓ is a known constant. The likelihood is

$$\frac{e^{-(1-\psi)\ell} e^{-\psi\ell}(1-\psi)^{y_1}\psi^{y_2}\ell^{y_1+y_2}}{y_1!\,y_2!}. \tag{2.60}$$

There is no reduction by sufficiency and we have a (2, 1) exponential family. The random variable $A = Y_1 + Y_2$ is, however, distribution constant, having a Poisson distribution of mean ℓ, and hence is ancillary in the simple sense. The conditional distribution of, say, Y_2 given $A = a$ is binomial with index a and parameter ψ (in fact not depending on ℓ) and is uniquely appropriate for inference about ψ. The justification is that the value of A, the total observed 'count', tells us nothing on its own about the value of ψ. If we are 'lucky' and observe a large value of a, then inference about ψ is relatively precise; if we are 'unlucky' and observe a small value of a, inference for ψ is relatively imprecise. Conditioning assures relevance of the long-run hypothetical probabilities to the situation actually encountered.

Now suppose that ℓ, instead of being known, is an unknown parameter χ, say, and that the parameter space for χ is $\mathbb{R}^+ = (0, \infty)$ and that for ψ (0, 1), independently. Now A has a distribution depending only on χ, i.e. is S-ancillary, and the conditional distribution for Y_2 given $A = a$ is exactly as before, i.e. binomial with index a and parameter ψ; it does not involve χ essentially because, for each fixed ψ, A is sufficient for χ. The arguments for using this conditional distribution for inference on ψ are exactly as before, so long as there are no considerations linking the values of ψ and χ. □

Example 2.21 Simple linear regression
Suppose that $(Y_1, \ldots, Y_n)$ are independently normally distributed with variance τ and with

$$E(Y_j) = \alpha + \beta x_j$$

where $(x_1, \ldots, x_n)$ are known constants. Then the standard least squares analysis can be justified, as above, by appeal to sufficiency. Suppose now that $(x_1, \ldots, x_n)$ are observed values of random variables $(X_1, \ldots, X_n)$ having a joint distribution $p_X(x; \chi)$ depending on the parameter χ and that the parameter spaces for χ and for (α, β, τ) are independent. Then $(x_1, \ldots, x_n)$ is S-ancillary and the inference for (α, β, τ) proceeds conditionally, exactly as if the x_j were constants not random variables. The same argument holds for multiple regression.

It is essential for the success of the argument that the parameter of interest should be defined in terms of the conditional distribution. To take a specific instance, if $(x_1, Y_1), \ldots, (x_n, Y_n)$ correspond to pairs $(X_1, Y_1), \ldots, (X_n, Y_n)$ of independent bivariate normally distributed random variables, conditioning on the observed $(x_1, \ldots, x_n)$ is appropriate if the parameter of interest is the regression coefficient of Y on X, but not if the parameter of interest is the correlation coefficient. □

We now turn to examples where the requirements (b) and (c) are satisfied but not the condition (a) for S-ancillarity.

Example 2.22 Two binomial variables (2 × 2 contingency table)
We consider the comparison of two binomial variables, the basis of the conditioning being more complicated than the corresponding problem for Poisson variables, Example 2.20. Let Y_1, Y_2 be independent binomial random variables corresponding to the numbers of 'successes' in m_1, m_2 independent trials with probabilities of success (θ_1, θ_2). Let the parameter of interest be

$$\psi = \log\left(\frac{\theta_2}{1-\theta_2}\right) - \log\left(\frac{\theta_1}{1-\theta_1}\right), \tag{2.61}$$

the log-odds ratio.

The joint probability that $Y_1 = y_1$, $Y_2 = y_2$ can be written in the form

$$\binom{m_1}{y_1}\binom{m_2}{y_2}\left(\frac{\theta_1}{1-\theta_1}\right)^{y_1+y_2} e^{\psi y_2}(1-\theta_1)^{m_1}(1-\theta_2)^{m_2}$$

so that for any given ψ, $a = y_1 + y_2$ is sufficient for the remaining parameter. It follows that the conditional distribution of, say, Y_2

given $A = a$ is

$$P(Y_2 = y_2 \mid A = a) = \frac{\binom{m_1}{a - y_2}\binom{m_2}{y_2} e^{\psi y_2}}{\sum_{x_2} \binom{m_1}{a - x_2}\binom{m_2}{x_2} e^{\psi x_2}}. \tag{2.62}$$

The marginal distribution of $A = Y_1 + Y_2$ depends in a complicated way on both θ_1 and θ_2, i.e. on ψ and whatever parameter χ is taken to complete the specification. Therefore, to apply the notion of ancillarity to justify the conditioning with $T = Y_2$, $A = Y_1 + Y_2$, we need the generalized definition involving condition (b). □

Example 2.22 illustrates a general argument applicable when the parameter of interest is a component of the canonical parameter of an (m, m) prime exponential model. In the exponential model with density

$$\exp\{s^{\mathrm{T}}\phi - k(\phi)\}a(y),$$

suppose that $\phi^{\mathrm{T}} = (\phi_1^{\mathrm{T}}, \phi_2^{\mathrm{T}})$, $s^{\mathrm{T}} = (s_1^{\mathrm{T}}, s_2^{\mathrm{T}})$ and that ϕ_1 is the parameter of interest; it is convenient to take the complementary parameter to be η_2, the mean parameter corresponding to S_2. We write $\psi = \phi_1$, $\chi = \eta_2$ and set $T = S_1$, $A = S_2$. Then the condition 1–3 are satisfied. The distribution of T given $A = a$ depends only on ψ and, while the distribution of A in general depends in a complicated way on ψ and χ, the result that $E(A) = \chi$ shows that extraction of information about ψ will be at least difficult so that (b) is close to being satisfied.

In principle, therefore, this argument allows inference for a parameter ψ that is a linear transformation of the canonical parameters in an (m, m) exponential family. This is because the property of being a canonical parameter is invariant under linear transformation.

If, however, the parameter of interest is a ratio of canonical parameters, we have in general to use condition (c) to justify conditioning. This can be seen by noting that if $m = 2$ and $\psi = \phi_2/\phi_1$ then the null hypothesis $\psi = \psi_0$ is equivalent to $\phi_2 - \phi_1\psi_0 = 0$, so that we may take $\phi_2 - \phi_1\psi_0$ as a new parameter, although this will involve the test and conditioning statistic depending on ψ_0.

From the viewpoint of Neyman–Pearson theory, conditioning for inference in the (m, m) exponential family is justified by a requirement of exact similarity achieved in the light of the completeness of the minimal sufficient statistic only by conditioning.

We defer until later discussion of when ancillary statistics exist, although a number of the examples above generalize fairly directly to give ancillaries in the simple sense within transformation models and in the extended sense within exponential family models. There is the possibility of non-uniqueness in the maximal ancillary, although rather few specific instances of this are known. Example 2.23 is one; see *Further results and exercises* 2.12 for a more puzzling example.

Example 2.23 A special multinomial distribution
Consider a multinomial distribution with four cells, the probabilities for which are

$$\pi_1(\theta) = \tfrac{1}{6}(1-\theta), \pi_2(\theta) = \tfrac{1}{6}(1+\theta), \pi_3(\theta) = \tfrac{1}{6}(2-\theta), \pi_4(\theta) = \tfrac{1}{6}(2+\theta), \tag{2.63}$$

where θ is an unknown parameter, $|\theta| < 1$. Corresponding to n independent trials the log-likelihood is, except for a constant,

$$n_1 \log(1-\theta) + n_2 \log(1+\theta) + n_3 \log(2-\theta) + n_4 \log(2+\theta),$$

where the cell frequencies are $(n_1, \ldots, n_4)$. There is no reduction of dimensionality by sufficiency.

Now the random variable $A' = N_1 + N_2$ has a binomial distribution with parameter 1/3; the random variable $A'' = N_1 + N_4$ has a binomial distribution with parameter 1/2. Hence A' is ancillary and so, too, is A''. Note, however that the pair (A', A'') is not ancillary, so that it is not ideal to condition on both A' and A''.

If we condition on A' we consider the data as essentially two independent binomial trials; $a'_1 = n_1 + n_2$ trials with probability of 'success' $\pi_1(\theta)/\{\pi_1(\theta) + \pi_2(\theta)\} = \frac{1}{2}(1-\theta)$ and $a'_2 = n_3 + n_4$ trials with probability of success $\pi_3(\theta)/\{\pi_3(\theta) + \pi_4(\theta)\} = \frac{1}{4}(2-\theta)$. There is a similar but different partition for conditioning on A''. □

To discuss a possible basis for choice between A' and A'' it is best to argue more generally.

If A is ancillary in the simple sense we may write

$$p_Y(y;\theta) = p_{Y|A}(y|a;\theta)p_A(a). \tag{2.64}$$

Taking the scalar parameter case for simplicity, we have for the conditional expected information for θ given $A = a$

$$i_A(\theta; a) = E\left\{ -\frac{\partial^2 \log p_{Y|A}(Y; a, \theta)}{\partial\theta^2} \middle| A = a; \theta \right\}$$

$$= E\left\{ -\frac{\partial^2 \log p_Y(Y; \theta)}{\partial\theta^2} \middle| A = a; \theta \right\}. \tag{2.65}$$

by (2.64). If now we take expectations over the distribution of A we have, by a basic property of conditional expectations, that

$$E\{i_A(\theta; A)\} = E\left\{ -\frac{\partial^2 \log p_Y(Y; \theta)}{\partial\theta^2}; \theta \right\}$$

$$= i(\theta), \tag{2.66}$$

the unconditional expected information.

Thus in Example 2.23 with two ancillaries competing for attention

$$E\{i_{A'}(\theta; A')\} = E\{i_{A''}(\theta; A'')\}$$

and expected conditional information is thus no basis for choice. An objective of the conditioning is, however, to sort possible sets of data into the relatively informative and the relatively uninformative. If this is not done effectively, relevance of long-run probability calculations to the data under analysis will be impaired. Thus if

$$\operatorname{var}\{i_{A'}(\theta; A')\} > \operatorname{var}\{i_{A''}(\theta; A'')\} \tag{2.67}$$

then A' is, in a sense, preferable to A''.

In Example 2.23

$$\operatorname{var}\{i_{A'}(\theta; A'); \theta\} = 2n^{-1}\{(1-\theta^2)(4-\theta)(4-\theta^2)\}^{-2},$$

$$\operatorname{var}\{i_{A''}(\theta; A''); \theta\} = \theta^2 n^{-1}\{(1-\theta)(4-\theta^2)\}^{-2},$$

establishing the superiority of A' in this sense.

Note that there is some arbitrariness in measuring the selectivity of an ancillary via the variance of the conditional information and that there is the disturbing possibility, not encountered in the present special case, that the choice might depend on the true and unknown value of θ.

As remarked in the general discussion of section 2.4, there is a danger of loss of effective information from overconditioning in that the random variable under study may become virtually degenerate.

This arises notably in discrete problems in which the number of points of support in the conditional distribution becomes very small. It is hard to quantify the loss of effective information involved.

Lloyd (1992) pointed out that a conflict may arise between conditioning on a statistic A that is exactly ancillary but which has a small value of $\mathrm{var}(i|A)$, versus conditioning on A' that has a distribution depending slightly on the parameter θ of interest but which has a large value of $\mathrm{var}(i|A')$. Lloyd's example involves a single observation Y normally distributed with mean θ and with known variance taking one of the following four values with the probabilities specified:

Variance	99.9	100.1	10.1	9.9
Probability	$(\frac{1}{4}-k_1\theta)$	$(\frac{1}{4}-k_2\theta)$	$(\frac{1}{4}+k_1\theta)$	$(\frac{1}{4}+k_2\theta)$
A	0	1	0	1
A'	0	0	1	1

Here k_1, k_2 are very small known constants, e.g. 10^{-6}, θ being appropriately restricted to make the above probabilities positive.

The statistic A is exactly ancillary but virtually useless for discriminating informative from uninformative samples, whereas A' is only approximately ancillary and yet highly effective in achieving the desired discrimination. It seems clear that an analysis conditioning on A' is to be preferred to one conditioning on A.

The example is yet another warning of the dangers of overemphasizing exact rather than approximate fulfilment of properties whenever competing requirements are involved.

Ancillarity is important also in considering observed information.

For this note that the log-likelihood is, except possibly for a term not depending on the parameter, a function of a sufficient statistic s and parameter θ. If the dimensions of s and θ are equal, the maximum likelihood estimate $\hat{\theta}$ is usually a one-to-one function of s and then $\hat{\theta}$ is minimal sufficient if and only if s is minimal sufficient. We can, accordingly, take the log-likelihood in the form $\ell(\theta;\hat{\theta})$, it being the same as if the data consisted solely of $\hat{\theta}$ or s. If $s=(t,a)$, where t has the dimension of θ and a is ancillary, then we can generally write the log-likelihood as $\ell(\theta;\hat{\theta},a)$.

Thus the observed information can, in the scalar parameter case, be written as

$$j(\theta; \hat{\theta}, a) = -\partial^2\ell(\theta; \hat{\theta}, a)/\partial\theta^2 \tag{2.68}$$

usually evaluated at $\theta = \hat{\theta}$, i.e. as $j(\hat{\theta}; \hat{\theta}, a)$. For a vector parameter we use $-\nabla_\theta\nabla_\theta^{\mathrm{T}}\ell(\theta; \hat{\theta}, a)$. Now the maximum likelihood equation, which we study in depth later, is

$$[\partial\ell(\theta; \hat{\theta}, a)/\partial\theta]_{\theta=\hat{\theta}} = 0$$

and on differentiating this with respect to $\hat{\theta}$, we have

$$\left[\frac{\partial^2\ell(\theta; \hat{\theta}, a)}{\partial\theta^2} + \frac{\partial^2\ell(\theta; \hat{\theta}, a)}{\partial\theta\,\partial\hat{\theta}}\right]_{\theta=\hat{\theta}} = 0,$$

so that

$$j(\hat{\theta}; \hat{\theta}, a) = \left[\frac{\partial^2\ell(\theta; \hat{\theta}, a)}{\partial\theta\,\partial\hat{\theta}}\right]_{\theta=\hat{\theta}}$$

or for a vector parameter

$$j(\hat{\theta}; \hat{\theta}, a) = [\nabla_\theta\nabla_{\hat{\theta}}^{\mathrm{T}}\ell(\theta; \hat{\theta}, a)]_{\theta=\hat{\theta}}.$$

In some theoretical discussions in Chapters 5–8 it will be useful to consider the function $j(\theta; \theta, a)$ as an observed information function depending on the data only via the ancillary statistic a. Under reparameterization $j(\hat{\theta}; \hat{\theta}, a)$, and hence $j(\theta; \theta, a)$, obeys a tensor law of transformation and thus defines a metric, whereas in general $j(\theta; \hat{\theta}, a)$ does not do so. We shall write $\not{j}(\theta; a)$ or just $\not{j}(\theta)$ or $\not{j}$ as abbreviations for $j(\theta; \theta, a)$, the slash through j indicating the substitution of θ for $\hat{\theta}$ in $j(\theta; \hat{\theta}, a)$. The transformation law of $\not{j}$ may be expressed analogously to (2.46)–(2.48); thus, for instance,

$$\not{j}(\psi) = \theta_{/\psi}^{\mathrm{T}}\not{j}(\theta)\theta_{/\psi} \tag{2.69}$$

where ψ denotes a one-to-one transformation of θ.

2.6 Examination of model adequacy

Throughout the book we put primary emphasis on inference about the unknown parameter θ in a probability model under the supposition that the model is at least an adequate approximation. Checking the adequacy of models, and quite often from that constructing improved models, is, however, of critical importance and we now comment briefly on that.

Models can be checked by:

1. fitting an extended model with additional parameters γ, say, such that when $\gamma = 0$ the original model is recovered;
2. fitting a model separate from the original one and seeing whether in some sense a better fit is achieved;
3. examining whether some feature, predicted by the original model as holding regardless of θ, is in accord with the data.

Approach 1 is covered formally by our main discussion, and approach 2 will not be discussed here; see, however, Cox (1961). For the third approach we need a distribution connected with the model and holding independently of θ, and we have encountered above two somewhat different ways of finding such a distribution: one via the conditional distribution of the whole data given the sufficient statistic and the other via an ancillary statistic in the simple sense. Both definitions do not depend on θ. There still remains the issue of how to compare the data with the fixed distribution, whether informally, perhaps via some graphical procedure, or more formally via the value of a test statistic chosen either on semi-intuitive grounds as likely to be sensitive to departures of substantive interest or via considerations of sensitivity to potential alternative distributions. The last approach is closely related to the first two above.

We give two illustrations.

Example 2.24 Conformity with a Poisson distribution

Suppose that, according to the original model, $Y_1, \ldots, Y_n$ are independently distributed in a Poisson distribution of unknown mean μ. The sufficient statistic for μ can be taken as $\sum y_j = s$ and the distribution of Y given $S = s$ does not involve μ. In fact because S has a Poisson distribution of mean $n\mu$, it follows immediately that

$$P(Y_1 = y_1, \ldots, Y_n = y_n | S = s) = \frac{\prod e^{-\mu}\mu^{y_j}/y_j!}{e^{-n\mu}(n\mu)^s/s!}$$

$$= \frac{s!}{\prod y_j!}\frac{1}{n^s}, \tag{2.70}$$

the probability being zero if $\sum y_j \neq s$. Equation (2.70) defines s trials in the multinomial distribution with n cells and equal probabilities $(1/n, \ldots, 1/n)$. How are we to examine consistency with (2.70)? If $n = 2$

there is no difficulty, but if $n > 2$ and especially if n is large we must determine which aspect or aspects of the $\{y_j\}$ are of most concern.

In some applications one is concerned not so much with whether the Poisson distribution itself is adequate as with whether the variance of the data is reasonably close to that given by the Poisson distribution. In that case large (or small) values of $T_2 = \sum(Y_j - \bar{Y})^2$ may be of concern. In other applications the prediction of zero values may be of especial interest, in which case $T_0 = \text{card}\{Y_j = 0\}$ may be studied. That is, in the second case we judge the significance of the observed t_0 by reference to the known combinatorial distribution of the number of empty cells in the multinomial model (2.70) (Fisher and Yates, 1963, Table XXII). The distribution of T_2 under (2.70) can be determined exactly by enumeration. Note, however, that because $S = n\bar{Y}$ is fixed in the conditional distribution we may equivalently take as test statistic $T'_2 = \sum Y_j^2$. Then under the multinomial distribution

$$\begin{aligned} E(T'_2) &= nE(Y_j^2) \\ &= n\{(sn^{-1})^2 + sn^{-1}(1 - n^{-1})\} \\ &= s^2n^{-1} + s(1 - n^{-1}), \end{aligned} \tag{2.71}$$

this result following because Y_j has a binomial distribution with parameter n^{-1} and index s.

Calculation of an approximate distribution for T'_2 or T_2 takes us into the realm of asymptotic theory outside the scope of this chapter. Note, however, that

$$T_2 = T'_2 - s^2/n$$

so that the variance to mean ratio $\{\sum(Y_j - \bar{Y})^2/(n-1)\}\bar{Y}^{-1}$ taken conditionally is

$$R = \frac{(T'_2 - s^2/n)n}{(n-1)s} \tag{2.72}$$

and

$$E(R) = 1, \qquad \text{var}(R) = 2(n-1)^{-1}(1 - 1/s), \tag{2.73}$$

suggesting that $(n-1)s(s-1)^{-1}R$ is treated as chi-squared with degrees of freedom $(n-1)s/(s-1)$ to match (2.73). The first part follows from (2.71) and the second from the result sketched in *Further results and exercises* 2.13.

It is common in applications to use degrees of freedom $n-1$, giving the chi-squared index of dispersion test; (2.73) suggests that the particular distributional approximation will not be good if s is small.

Similar arguments apply to other exponential family problems, although the details may be harder to implement. □

Example 2.25 Conformity with normal theory linear model
No doubt partly because the normal theory linear model (1.1) is so widely used, there is a very large literature on the testing of the adequacy of such a model. From the present viewpoint such tests can in principle be based on the conditional distribution of the data given the minimal sufficient statistic $(\hat{\beta}, SSD)$ and this is equivalent to examination of the standardized residuals $(Y^j - \hat{Y}^j)/\sqrt{SSD}$ or $(Y^j - \hat{Y}^j/s$, where $\hat{Y}^j = x_t^j\hat{\beta}^t$ is the fitted value and s^2 is an estimate of the error variance, $s^2 = SSD/(n-q)$.

Specific statistics, for example tests of dependence on additional explanatory variables, of nonlinear dependence or interaction, of non-constant variance or of non-normality or non-independence of errors, are usually derived by other routes but can be expressed in the above form.

To simplify further discussion, consider the case where $Y_1, \ldots, Y_n$ are independently and identically distributed normally with mean μ and variance σ^2. Then the standardized residual can be taken in the form $R_j = (Y_j - \bar{Y})/s$, where $s^2 = (n-1)^{-1}\sum(Y_j - \bar{Y})^2$. The $\{R_j\}$ satisfy the constraints $\sum R_j = 0$, $\sum R_j^2 =$ constant and under the assumed model are uniformly distributed on the $(n-2)$-dimensional hypersphere. The statistics for testing normality based on standardized third and fourth cumulants are multiples of $\sum R_j^3$ and $\sum R_j^4$. The standard t-statistic for testing for dependence on an explanatory variable x_j can be written in the form

$$\frac{\sum(Y_j - \bar{Y})x_j/\sum(x_j - \bar{x})^2}{\left[(n-2)^{-1}\left\{\sum(Y_j - \bar{Y})^2 - \dfrac{\{\sum(Y_j - \bar{Y})x_j\}^2}{\sum(x_j - \bar{x})^2}\right\}\right]^{\frac{1}{2}}\{\sum(x_j - \bar{x})^2\}^{-\frac{1}{2}}},$$

i.e. the regression coefficient on x_j divided by its estimated standard error. It is clear that this involves the Y_j only via the R_j. □

2.7 Parameter orthogonality

When θ is multi-dimensional there are a number of advantages, conceptual and mathematical, if the matrix $i(\theta)$ is diagonal. More generally, suppose that θ is partitioned into the components $\theta = (\theta^1, \dots, \theta^{d_1}; \theta^{d_1+1}, \dots, \theta^d)^{\mathrm{T}} = (\theta_{(1)}^{\mathrm{T}}, \theta_{(2)}^{\mathrm{T}})$, say, such that $i_{rs}(\theta) = 0$ for all $r = 1, \dots, d_1$; $s = d_1 + 1, \dots, d$. The matrix $i(\theta)$ is thus block diagonal. We then say that $\theta_{(1)}$ is orthogonal to $\theta_{(2)}$, globally if $i_{rs}(\theta) = 0$ for all $\theta \in \Omega_\theta$ and locally at $\theta = \theta_0$ if $i_{rs}(\theta_0) = 0$. Orthogonality implies, in particular, that the corresponding components of the score statistic are uncorrelated. We shall see further statistical consequences in later chapters.

An important special case arises if $d_1 = 1$, when it is convenient to write $\theta = (\psi, \lambda^1, \dots, \lambda^q)^{\mathrm{T}}$, with $q = d - 1$. If we start with an arbitrary parameterization $(\psi, \chi^1, \dots, \chi^q)$ with ψ given, it is always possible to find $\lambda^1, \dots, \lambda^q$ as functions of $(\psi, \chi^1, \dots, \chi^q)$ such that ψ is orthogonal to $(\lambda^1, \dots, \lambda^q)$. To see this let ℓ^* and i^* be the log- likelihood and information matrix in terms of $(\psi, \chi^1, \dots, \chi^q)$ and write $\chi^r = \chi^r(\psi, \lambda^1, \dots, \lambda^q)$. Then

$$\ell(\psi, \lambda) = \ell^*\{\psi, \chi^1(\psi, \lambda), \dots, \chi^q(\psi, \lambda)\}$$

and

$$\frac{\partial^2 \ell}{\partial \psi\, \partial \lambda^r} = \frac{\partial^2 \ell^*}{\partial \psi\, \partial \chi^s} \frac{\partial \chi^s}{\partial \lambda^r} + \frac{\partial^2 \ell^*}{\partial \chi^t\, \partial \chi^s} \frac{\partial \chi^s}{\partial \lambda^r} \frac{\partial \chi^t}{\partial \psi} + \frac{\partial \ell^*}{\partial \chi^s} \frac{\partial^2 \chi^s}{\partial \psi\, \partial \lambda^r}.$$

The summation convention here applies over the range $1, \dots, q$. On taking expectations, the last term vanishes and if we require the orthogonality of ψ and λ, we have that

$$\frac{\partial \chi^s}{\partial \lambda^t} \left(i^*_{\psi s} + i^*_{rs} \frac{\partial \chi^r}{\partial \psi} \right) = 0.$$

Assuming that the transformation from (ψ, χ) to (ψ, λ) has non-zero Jacobian, this equation is equivalent to

$$i^*_{rs} \partial \chi^r / \partial \psi + i^*_{\psi s} = 0. \qquad (2.74)$$

These partial differential equations determine the dependence of λ on ψ and χ, and they are solvable in general. However, the dependence is not determined uniquely and there remains considerable arbitrariness in the choice of λ.

The above argument hinges on the scalar character of ψ. If there were two components ψ^1, ψ^2 for which it was required to satisfy (2.74) there would in general be no guarantee that the values of $\partial\chi^r/\partial\psi^1$ and $\partial\chi^r/\partial\psi^2$ so obtained would satisfy the compatability condition

$$\partial^2\chi^r/\partial\psi^1\partial\psi^2 = \partial^2\chi^r/\partial\psi^2\partial\psi^1. \tag{2.75}$$

Of course, irrespective of the dimension of ψ the orthogonality could be achieved locally at $\theta = \theta_0$ via a linear transformation of parameters with components depending on $i(\theta_0)$. More generally, for a fixed value ψ_0 of ψ it is possible to determine the complementary parameterization so that $i_{\psi\lambda}(\psi_0, \lambda) = 0$ identically in the argument λ.

It follows immediately from the definition that if ψ is orthogonal to λ, then any one-to-one smooth function of ψ is orthogonal to any one-to-one smooth function of λ. Note also that if the vector parameter ψ is orthogonal to the vector parameter λ, then any function of ψ, or any component of ψ, is orthogonal to any function of λ, or any component of λ.

Example 2.26 Exponential distribution

Let (Y_1, Y_2) be independently exponentially distributed with means $(\chi, \psi\chi)$. The differential equation (2.74) becomes

$$2\chi^{-2}\partial\chi/\partial\psi = -(\psi\chi)^{-1}, \tag{2.76}$$

the solution of which is $\chi\psi^{1/2} = g(\lambda)$, where $g(\lambda)$ is an arbitrary function of λ. A convenient choice is $g(\lambda) = \lambda$, so that in the new parameterization the means are $\lambda/\sqrt{\psi}$ and $\lambda\sqrt{\psi}$ respectively. □

Example 2.27 Normal theory linear model with complex variance structure

An important example in which orthogonality can be verified directly is provided by the normal theory linear model (1.1) in which, instead of the errors being independently normally distributed with zero mean and unknown variance, the errors have a multivariate normal distribution with covariance matrix $D(\tau)$, with elements $d^{jk}(\tau)$ known except for the unknown parameter vector τ taking values in a parameter space Ω_τ, the full parameter space being $\Omega_\beta \times \Omega_\tau$, where, as in section 1.2, Ω_β is the parameter space for the linear model for the means. Note that for the following argument it is crucial that there are no component parameters common to the mean structure and the variance structure. Then the log-likelihood is

$$-\tfrac{1}{2}\log\det D(\tau) - \tfrac{1}{2}(y^j - x^j_r\beta^r)d_{jk}(\tau)(y^k - x^k_s\beta^s), \qquad (2.77)$$

where $d_{jk}(\tau)$ are the elements of the concentration matrix (inverse covariance matrix). Note that our use of subscripts and superscripts is the opposite of the usual one. Differentiation with respect to, say, β^r yields an expression linear in the zero mean 'errors' $y^k - x^k_s\beta^s$, so that the mixed derivatives with respect to β^r and a component of τ retain that structure and hence have zero expectation, verifying orthogonality. The result follows also from the remark preceding Example 2.26. In an arbitrary multivariate normal distribution the mean and inverse covariance matrix are orthogonal, and the model here involves restriction to special functions of these parameters. □

Example 2.28 Generalized location model
Consider a model function of the form

$$f(y - \mu; \chi), \qquad (2.78)$$

i.e. for fixed value of the parameter χ we have a location model.

We introduce a new parameter

$$\psi = \mu + w(\chi), \qquad (2.79)$$

where w is a suitable function determined so that ψ is orthogonal to χ. To see that such a function exists note that, due to the location character, the $i_{\psi\chi}$ block of the information matrix is of the form

$$i_{\psi\chi} = u(\chi) + v(\chi)w'(\chi),$$

where both u and v do not depend on ψ. Thus, choosing to solve the equation $u(\chi) + v(\chi)w'(\chi) = 0$ for w, we have $i_{\psi\chi} = 0$.

It follows that if we have m independent samples, the ith sample being from (2.78) with location μ_i particular to the sample but common χ, then the differences $\mu_i - \mu_j$ are orthogonal to χ. In particular, for a regression specification $\mu_i = \alpha + \beta^{\mathrm{T}}z_i$ with $\sum z_i = 0$ we have that β is orthogonal to χ. □

There is a development parallel to that set out above, using observed information j, as defined at the end of section 2.5, rather than the expected information i. In this, instead of the equations (2.74) we obtain

$$\jmath^*_{rs}\partial\chi^r/\partial\psi + \jmath^*_{\psi s} = 0 \tag{2.80}$$

as the set of equations whose solution $\lambda = \lambda(\psi, \chi)$ has the property that

$$\jmath_{\lambda\psi} = \jmath_{\lambda\psi}(\psi, \lambda; \psi, \lambda, a) = 0 \tag{2.81}$$

identically in (ψ, λ). As is the case for expected information, the equations (2.80) are generally solvable.

Example 2.29 Location–scale model
Suppose $Y_1, \ldots, Y_n$ is a sample from the location–scale model and let the location parameter be the parameter of interest. Writing g for $-\log f$ one finds that $\lambda = \sigma + u\mu$ is observed orthogonal to μ, where

$$u = \{\sum a_v g''(a_v)\}/\{n + \sum a_v^2 g''(a_v)\}$$

and $(a_1, \ldots, a_n)$ is the configuration, i.e. $a_v = (y_v - \hat{\mu})/\hat{\sigma}$. □

2.8 Review of transformation models

A first discussion of transformation models, including a few simple but prototypical examples, was given in section 1.4. Now we sketch the main results of the general theory of transformation models.

Let a parametric statistical model with a model function $p(y; \theta)$ be given. For simplicity we restrict attention to continuously varying data y. We suppose that there exists a family G of smooth one-to-one transformations g of the sample space $\mathcal{Y}$ onto itself, with the property that if Y is distributed according to $p(y; \theta)$ then the random variable gY obtained by transforming Y by a g in G will be distributed according to $p(y; \theta')$ for some θ' in the parameter space Ω. This is to hold for arbitrary g in G and $\theta \in \Omega$. The point θ' thus determined will be denoted by $g\theta$. On the added assumption that the class of transformations G forms a group we say that the statistical model is a **transformation model under the action of** G. The definition implies that G in effect also acts on the parameter space Ω. For clarity it is often best to think of G as an abstract group that may act in different ways on various spaces of interest.

It is, in general, a non-trivial problem to decide whether a given statistical model is a transformation model, that is, whether there exists some group as above.

As already indicated in section 1.4, the actions of G on $\mathcal{Y}$ and on

Ω partition these spaces into orbits, the orbits of y in $\mathcal{Y}$ and θ in Ω being respectively $Gy = \{gy : g \in G\}$ and $G\theta = \{g\theta : g \in G\}$. The action of G on $\mathcal{Y}$ is said to be **transitive** if $\mathcal{Y}$ consists of a single orbit, and similarly for the action on Ω. When the action of G on Ω is transitive we speak of a **pure** transformation model, otherwise the model is said to be a **composite** transformation model.

We will assume that, as can usually be arranged, the actions are **free**, i.e. on any orbit and for any pair of points on that orbit there is only one transformation taking the first point into the second. We may then, in particular, think of a point y in the sample space as being representable as a pair (u, z) where u specifies the orbit on which y is located and z specifies the position on the orbit. The statistic u is a **maximal invariant**, that is, it is **invariant** in the sense that its value does not change if y is transformed to gy and it is **maximal** in that any other invariant statistic is a function of u. Note further that z is **equivariant**, that is, if y is transformed to gy then z is transformed to gz. The assumption that the actions are free implies, moreover, that for a pure transformation model the parameter space Ω is in one-to-one correspondence with the group G and, hence, that we may consider z as varying in Ω. In practice, under such models, z can almost always be taken as an estimator of θ, typically the maximum likelihood estimator.

A generally useful way of setting up a pair (u, z) of this kind consists in first finding an equivariant statistic whose values are elements of G and then defining u by

$$u = z^{-1}y. \tag{2.82}$$

In particular, for a pure transformation model, the maximum likelihood estimator is an equivariant statistic whose values can be interpreted as belonging to G. The equivariance of the maximum likelihood estimator for such models is a consequence of the general parameterization invariance of the maximum likelihood method.

Note also that if z and z' are any two equivariant statistics with values in G then $z^{-1}z'$ is invariant, which means that on each orbit z' is just a fixed translation of z.

Example 2.30 Location–scale group and model
Let G be the set of pairs $g = [\mu, \sigma]$ where $\mu \in \mathbb{R}$ and $\sigma > 0$. We may think of G as a group with group operation

$$[\mu', \sigma'][\mu, \sigma] = [\mu' + \sigma'\mu, \sigma'\sigma],$$

unit element $[0, 1]$ and inverse elements

$$[\mu, \sigma]^{-1} = [-\sigma^{-1}\mu, \sigma^{-1}].$$

Equivalently, G could be considered as the class of one-to-one transformations of $\mathbb{R}^n$ onto itself given, for $y = (y_1, \ldots, y_n) \in \mathbb{R}^n$, by

$$\begin{aligned} gy &= [\mu, \sigma](y_1, \ldots, y_n) \\ &= (\mu + \sigma y_1, \ldots, \mu + \sigma y_n). \end{aligned} \tag{2.83}$$

For $n = 1$ the action of G on $\mathbb{R}^n$ is transitive but not free. For $n > 1$ the action is no longer transitive but it is free provided we exclude from consideration the set of y in $\mathbb{R}^n$ for which $y_1 = \cdots = y_n$. This set has probability 0 for all the applications we have in mind.

As one of the possible representations of y of the form (u, z) we may choose

$$u = \left(\frac{y_1 - \bar{y}}{s}, \ldots, \frac{y_n - \bar{y}}{s}\right), \qquad z = (\bar{y}, s), \tag{2.84}$$

where $\bar{y}$ and s are the mean and standard deviation of $y_1, \ldots, y_n$.

Alternatively supposing that $y_1, \ldots, y_n$ is a sample from a location–scale model with parameters μ and σ, we may take

$$u = \left(\frac{y_1 - \hat{\mu}}{\hat{\sigma}}, \ldots, \frac{y_n - \hat{\mu}}{\hat{\sigma}}\right), \qquad z = (\hat{\mu}, \hat{\sigma}), \tag{2.85}$$

where $\hat{\mu}$ and $\hat{\sigma}$ are the maximum likelihood estimates of μ and σ, which we assume to be unique.

As noted in the general context above, due to the equivariance of $(\bar{y}, s)$ and of $(\hat{\mu}, \hat{\sigma})$, the two choices of u are in one-to-one correspondence and for each given u-value, that is on each orbit, there exists an elements $[a(u), b(u)]$ of G such that

$$[a(u), b(u)][\bar{x}, s] = [\hat{\mu}, \hat{\sigma}]. \qquad \square$$

To formulate the key properties of transformation models it is further useful to invoke the concept of invariant measures. A measure ζ on a space $\mathscr{S}$ with points s on which a group G acts is called **invariant with respect to** G if $g\zeta = \zeta$ for all $g \in G$, where $g\zeta$ is the measure into which ζ is transformed by g; in symbols $g\zeta$ is determined by $g\zeta(ds) = \zeta(g^{-1}(ds))$. Invariant measures exist in wide generality, and they are unique up to a multiplicative constant when the action of G is transitive. Later on in this section we shall indicate how invariant measures may be constructed. In general, invariance of a measure ζ on $\mathscr{S}$ may be proved by showing that

$$\int f(gs)d\zeta = \int f(s)d\zeta \tag{2.86}$$

for a sufficiently broad class of functions f on $\mathscr{S}$.

Any group G can act on itself in several ways, the two most important being **left action** and **right action**. The left action transforms an arbitrary element g' of G into gg', while the right action sends g' into $g'g^{-1}$. The corresponding left and right invariant measures, denoted α and β respectively, are in general different.

Example 2.31 Location–scale group
Writing $g = [\mu, \sigma]$, the transformations of the left and right actions of the location–scale group are

$$g[\mu', \sigma'] = [\mu + \sigma\mu', \sigma\sigma'] \tag{2.87}$$

and, because $g^{-1} = [-\mu\sigma^{-1}, \sigma^{-1}]$,

$$[\mu', \sigma']g^{-1} = [\mu' - \sigma'\sigma^{-1}\mu, \sigma'\sigma^{-1}], \tag{2.88}$$

respectively. The corresponding left and right invariant measures take the form

$$\begin{aligned} d\alpha &= \sigma^{-2}d\mu\, d\sigma, \\ d\beta &= \sigma^{-1}d\mu\, d\sigma. \end{aligned} \tag{2.89}$$

To illustrate these expressions and the relation (2.86), let f be defined on the location–scale group, in other words $\mathscr{S} = \{(\tilde{\mu}, \tilde{\sigma}): \tilde{\mu} \in \mathbb{R}, \tilde{\sigma} > 0\}$. Then, with $g = [\mu, \sigma]$ acting on the left we have that

$$\begin{aligned} \int f(g(\tilde{\mu}, \tilde{\sigma}))\alpha(d\tilde{\mu}, d\tilde{\sigma}) &= \iint f(\mu + \sigma\tilde{\mu}, \sigma\tilde{\sigma})\tilde{\sigma}^{-2}d\tilde{\mu}\, d\tilde{\sigma} \\ &= \iint f(\mu + \sigma\tilde{\mu}, \sigma\tilde{\sigma})(\sigma\tilde{\sigma})^{-2}d(\sigma\tilde{\mu})d(\sigma\tilde{\sigma}) \\ &= \iint f(u, v)v^{-2}du\, dv \\ &= \int f(\tilde{\mu}, \tilde{\sigma})\alpha(d\tilde{\mu}, d\tilde{\sigma}) \end{aligned}$$

verifying the left invariance of the measure α. In comparison, if the action of G on itself is the right action and we try to see whether α is right invariant we find

$$\int f(g(\tilde{\mu},\tilde{\sigma}))\alpha(d\tilde{\mu},d\tilde{\sigma}) = \iint f(\tilde{\mu}-\tilde{\sigma}\sigma^{-1}\mu,\tilde{\sigma}\sigma^{-1})\tilde{\sigma}^{-2}d\tilde{\mu}\,d\tilde{\sigma}$$

$$=\sigma^{-1}\iint f(\tilde{\mu}-\tilde{\sigma}\sigma^{-1}\mu,\tilde{\sigma}\sigma^{-1})(\tilde{\sigma}\sigma^{-1})^{-2}d\tilde{\mu}d(\tilde{\sigma}\sigma^{-1})$$

$$=\sigma^{-1}\iint f(u-v\mu,v)v^{-2}du\,dv$$

$$=\sigma^{-1}\iint f(w,v)v^{-2}dw\,dv$$

$$=\sigma^{-1}\int f(\tilde{\mu},\tilde{\sigma})\alpha(d\tilde{\mu},d\tilde{\sigma})$$

showing that α is not, in fact, right invariant.

A calculation similar to the above verifies that β is indeed the required measure in this case. A derivation of the expressions (2.89) is given in Example 2.32 below. □

Now suppose the model is a pure transformation model. Then typically, under certain regularity assumptions that are usually met in practice, the following main properties hold.

Let ζ and ν be invariant measures on the sample space $\mathscr{Y}$ and the parameter space Ω, respectively. Further, assuming that $p(y;\theta)$ is a probability density with respect to Lebesgue measure, let

$$\bar{p}(y;\theta)=p(y;\theta)(d\zeta/dy)^{-1}, \tag{2.90}$$

i.e. $\bar{p}$ is the model function relative to the invariant measure on $\mathscr{Y}$. Then we have the following results, all of which generalize the simple conclusions applying to the location problem, Examples 1.6, 2.18.

1. For any fixed θ_0 in Ω and for g determined so that $\theta=g\theta_0$ we have

$$\bar{p}(y;\theta)=\bar{p}(g^{-1}y;\theta_0). \tag{2.91}$$

2. Under the one-to-one transformation of y into (u,z), where u is maximal invariant and z is equivariant and varies over Ω, the invariant measure ζ is transformed into a product measure $\kappa\otimes\nu$, ν being invariant measure on Ω.

3. The model function factorizes as

$$\bar{p}(y;\theta) = q(u)\bar{r}(z, w; \theta) \tag{2.92}$$

for certain functions q and $\bar{r}$ and where w denotes an invariant statistic and is therefore a function of u. In particular, then, (z, w) is a sufficient statistic. Note that (2.92) comprises both the cases where no sufficient reduction is possible, corresponding to q constant and $w = u$, and the case where z is sufficient, corresponding to w being degenerate.
4. The statistics u and z are conditionally independent given w.
5. The maximal invariant statistic u has a probability density function $p(u)$ relative to the measure κ, which is representable as

$$p(u) = q(u)\int \bar{r}(z, w; \theta_0)dv(z). \tag{2.93}$$

This formula reflects, in particular, the distribution constancy of u.
6. The conditional probability density function of z given w, relative to the invariant measure v, is of the form

$$\bar{p}(z;\theta|w) = c(w)\bar{r}(z, w; \theta) \tag{2.94}$$

where $c(w)$ is a norming constant. By result 4, the conditional distribution of z given w is the same as the conditional distribution given u, i.e. $\bar{p}(z;\theta|w) = \bar{p}(z;\theta|u)$.
7. The conditional density (2.94) may, alternatively, be expressed in terms of an arbitrary version L of the likelihood function as

$$\bar{p}(z;\theta|w) = c(w)\frac{L(\theta)}{L(z)}, \tag{2.95}$$

$c(w)$ being a norming constant.

Proofs of the above propositions are available in Barndorff-Nielsen (1988a) and, with more details, in Barndorff-Nielsen, Blæsild and Eriksen (1989).

To re-express the above results in terms of probability densities relative to the relevant Lebesgue measures we need to know how to express the Radon–Nikodym derivative of an invariant measure ζ on a space $\mathscr{S}$ relative to Lebesgue measure ds on $\mathscr{S}$. If, for instance, $\mathscr{S} = \mathscr{Y}$ then this derivative can usually be given in terms of the representation (u, z) of y and the Jacobian determinant $J_g(y)$ of the mapping of $\mathscr{Y}$ onto itself given by g. Specifically,

$$d\zeta/ds = J_z(u)^{-1}. \tag{2.96}$$

A proof of this result will be sketched after the following examples.

Example 2.32 Location–scale group and model
First, we wish to derive the expression (2.89) for the left and right invariant measures on the location–scale group G. The left and right actions are both transitive and hence for a (u, z)-representation of $[\mu, \sigma]$ we may take $u = [0, 1]$ and choose, respectively, $[\mu, \sigma]$ and $[\mu, \sigma]^{-1}$ for z. The Jacobian matrices of the left and right transformations (2.87) and (2.88) are

$$J_g(\mu', \sigma') = \begin{bmatrix} \sigma & 0 \\ 0 & \sigma \end{bmatrix},$$

$$J_g(\mu', \sigma') = \begin{bmatrix} 1 & 0 \\ -\sigma^{-1}\mu & \sigma^{-1} \end{bmatrix}.$$

The corresponding Jacobian determinants $J_z(u)$ are

$$J_{[\mu,\sigma]}(0, 1) = \sigma^2$$

and

$$J_{[\mu,\sigma]^{-1}}(0, 1) = \sigma.$$

The relations (2.89) now follow from (2.96).

To determine an invariant measure on $\mathbb{R}^n$ under the action of the location–scale group and using (2.96) we find from (2.83) that the Jacobian determinant $J_g(y)$ is, in this case,

$$J_{[\mu,\sigma]}(y) = \sigma^n. \tag{2.97}$$

As a (u, z)-representation of y we may, for instance, choose (2.84) or (2.85) and then the invariant measures become, respectively,

$$s^{-n}dy \quad \text{and} \quad \hat{\sigma}^{-n}dy. \tag{2.98}$$

□

Example 2.33 Special location–scale model
To illustrate results 1–7 we consider the location–scale model determined by the probability density function

$$\frac{c(\varepsilon)}{\sqrt{(2\pi)}} \exp\{-\tfrac{1}{2}(x^2 + \varepsilon x^4)\}$$

where $\varepsilon \geqslant 0$ is considered as known and where $c(\varepsilon)$ is a norming constant. As the (u, z)-representation of a sample $y = (y_1, \ldots, y_n)(n > 1)$ from this model we take (2.84).

As shown above, an invariant measure on the sample space $\mathbb{R}^n$ is given by $d\zeta = s^{-n}dy$ and the model function $\bar{p}$ relative to this measure is therefore

$$\bar{p}(y; \mu, \sigma) = c(\varepsilon)^n(2\pi)^{-n/2}\sigma^{-n}s^n \cdot\exp[-\tfrac{1}{2}\{\sigma^{-2}\sum(y_i - \mu)^2 + \varepsilon\sigma^{-4}\sum(y_i - \mu)^4\}].$$

Furthermore, the invariant measure ν on $\hat{\Omega} = \mathbb{R} \times \mathbb{R}_+$ is $s^{-2}d\bar{x}\,ds$ and the lifting of ζ by (u, z) results in a product measure $\kappa \otimes \nu$, where κ is a uniform measure on the sphere in $\mathbb{R}^n$ given by $\sum u^i = 0$, $\sum u_i^2 = n - 1$.

For a factorization of $\bar{p}(y; \mu, \sigma)$ as in result 3 we introduce the standardized empirical moments

$$v_k = \sum(y_i - \bar{y})^k / \{\sum(y_i - \bar{y})^2\}^{k/2} \qquad (k = 3, \ldots, n),$$

and we let $w = (v_3, v_4)$. Note, incidentally, that if ε were unknown our model would be a (4, 3) curved exponential model with $(\bar{x}, s, w)$ constituting a minimal sufficient statistic. For simplicity, we just state the factorization in the case $(\mu, \sigma) = (0, 1)$:

$$\bar{p}(y; 0, 1) = q(u)\bar{r}(z, w; 0, 1)$$

with

$$q(u) = c(\varepsilon)^n(2\pi)^{-n/2}$$

$$\bar{r}(z, w; 0, 1) = s^n \exp\left[-\frac{n}{2}\{s'^2 + \bar{x}^2 + \varepsilon s'^4 v_4 + \varepsilon 4s^3 v_3\bar{x} + \varepsilon 6s'^2\bar{x}^2 + \varepsilon\bar{x}^4\}\right]$$

and where $s'^2 = (n - 1)s^2/n$.

By result 4 we find that $(v_5, \ldots, v_n)$ is conditionally independent of $(\bar{x}, s)$ given (v_3, v_4). □

Example 2.34 Arbitrary location–scale family

For an arbitrary location–scale model with (u, z)-representation (2.85), the conditional probability density function of $(\hat{\mu}, \hat{\sigma})$ given u, with respect to Lebesgue measure $d\hat{\mu}\,d\hat{\sigma}$, is, by (2.95), (2.96) and (2.89), of the form

$$p(\hat{\mu}, \hat{\sigma}; \mu, \sigma)|u) = c(u)\sigma^{-n}\hat{\sigma}^{n-2}\prod_{i=1}^{n} f\left(\frac{y_i - \mu}{\sigma}\right).$$

Here $c(u)$ is a norming constant and we have used the fact that $\prod f((y_i - \hat{\mu})/\hat{\sigma})$ is a function of u to include this term in $c(u)$. □

To explain why (2.96) is valid and to indicate a somewhat more general technique for the construction of invariant measures, we introduce the concepts of multipliers and modulators. Let $\mathscr{S}$ be a set on which G acts. In the present context we shall denote the Lebesgue measure on $\mathscr{S}$ by $d\lambda(s)$, rather than ds, and as above we write $g\zeta$ for the lifting of a measure ζ by g, i.e. $g\zeta(B) = \zeta(g^{-1}B)$ for every measurable subset B of $\mathscr{S}$. On the assumption, which is implicit in our discussion, that $g\lambda$ is absolutely continuous with respect to λ for all $g \in G$ we may define a **multiplier** function $\pi(g, s)$ by

$$d(g^{-1}\lambda)(s) = \pi(g, s)d\lambda(s). \tag{2.99}$$

It follows immediately from this definition that the multiplier π has the property

$$\pi(g'g, s) = \pi(g', gs)\pi(g, s). \tag{2.100}$$

A function m on $\mathscr{S}$ is called a **modulator** with associated multiplier π if m is positive and

$$m(gs) = \pi(g, s)m(s). \tag{2.101}$$

Suppose m is a modulator and define the measure ζ by

$$\zeta = m^{-1}\lambda. \tag{2.102}$$

Then ζ is invariant. The proof consists in verifying (2.86) and runs as follows

$$\begin{aligned}\int f(gs)d\zeta &= \int f(gs)m(s)^{-1}d\lambda \\ &= \int f(gs)m(gs)^{-1}\pi(g, s)d\lambda \\ &= \int f(gs)m(gs)^{-1}d(g^{-1}\lambda) \\ &= \int f(s)m(s)^{-1}d\lambda \\ &= \int f(s)d\zeta.\end{aligned}$$

The verification that (2.96) determines an invariant measure consists then in showing that $J_z(u)$ is a modulator with the appropriate multiplier. We omit the verification of this. (For the proof see Barndorff-Nielsen 1988a, section 2.1.)

The value in introducing modulators is that it is sometimes possible to guess the form of a modulator, thereby avoiding the calculation of the Jacobian $J_z(u)$. However, (2.96) not only gives a systematic way of determining invariant measures but is also important because of its relation to likelihood theory and, in particular, to the p^*-formula; see section 6.2.

For composite transformation models the parameter θ can be represented as (λ, γ), where λ is a maximal invariant function on the parameter space that specifies the orbit on which θ lies, while γ determines the position of θ on that orbit. Note that since we have assumed that the actions of G are free we may identify γ with an element of G, as will be done for instance in formula (2.103) below. The maximal invariant statistic u has a distribution which depends on λ only and inference on λ should be performed in the marginal distribution of u. The model function for this marginal distribution is available by (2.93) and may be written

$$p(u; \lambda) = q(u; \lambda) \int \bar{r}(z, w; \lambda, \gamma) d\nu(z). \tag{2.103}$$

In particular, (2.103) provides a version of the marginal likelihood $L(\lambda; u)$ for λ based on u. There is an alternative expression for $L(\lambda; u)$, derivable by means of formula (2.91) and the theory of left invariant and right invariant measures on groups. This is

$$L(\lambda; u) = \int L(\lambda, g) d\beta(g), \tag{2.104}$$

where β denotes right invariant measure on the group G and where $L(\lambda, g)$ is an arbitrary version of the original likelihood function for θ. Note that in (2.104) the integration is over part of the parameter space, whereas in (2.103) it is over part of the sample space. (A proof of (2.104) may be found in Barndorff-Nielsen, 1988a, section 2.2.)

2.9 Review of prime exponential models

In section 1.3 we gave a first discussion of exponential families. In the present section, and the next which concerns curved exponential models, we summarize a number of further properties of such families.

Consider a full exponential model $\mathcal{M}$ with model function

$$p(y;\phi) = a(\phi)b(y)e^{s^{\mathrm{T}}(y)\phi} \tag{2.105}$$

relative to some dominating measure μ. The common dimension of the canonical parameter ϕ and the canonical statistic s is denoted by m, and the canonical parameter ϕ varies over the set

$$\Phi = \left\{\phi: \int e^{s^{\mathrm{T}}(y)\phi} b(y)\mu(dy) < \infty\right\}.$$

This set is a convex subset of $\mathbb{R}^m$ and $k(\phi) = -\log a(\phi)$ is a convex function of ϕ. We denote the interior of Φ by int Φ or by Φ^0. As noted in section 1.3, the mean parameter η defined for $\phi \in \operatorname{int}\Phi$ by $\eta = E(S;\phi)$, is in one-to-one correspondence with the canonical parameter ϕ, and its domain of variation $H = \eta(\operatorname{int}\Phi)$ satisfies $H \subset \operatorname{cl}\operatorname{conv}\mathcal{S}$, where $\mathcal{S}$ is the support of the distribution of the canonical statistic S and cl conv stands for the closed convex hull of $\mathcal{S}$. Denoting the interior of cl conv $\mathcal{S}$ by $\mathcal{S}^0$, we have, in fact, that $H \subset \mathcal{S}^0$. In general H is a proper subset of $\mathcal{S}^0$ and when that is the case it causes difficulties in the statistical inference. However, as will be discussed below, the (m, m) exponential models met in practice often have the property that $H = \mathcal{S}^0$.

It is mathematically most natural to consider the sets Φ and $\Phi^0 = \operatorname{int}\Phi$ on the one hand and the sets $\mathcal{S}$, cl conv $\mathcal{S}$, $\mathcal{S}^0$, and H on the other as belonging to two different m-dimensional Euclidean spaces, and this is indicated in Figure 2.1 in the next section.

Now, let $\mathcal{M}^0$ denote the model obtained from $\mathcal{M}$ by deleting any boundary points of Φ, if such exist. That is, $\mathcal{M}^0$ consists of those densities $p(y;\phi)$ for which $\phi \in \operatorname{int}\Phi$. As noted already, the model $\mathcal{M}^0$ may be parameterized by the mean parameter η as well as by the canonical parameter ϕ. A third type of parameterization that is often of interest is a **mixed parameterization**. Let $\phi^{\mathrm{T}} = (\phi^{1\mathrm{T}}, \phi^{2\mathrm{T}})$ and $\eta^{\mathrm{T}} = (\eta_1^{\mathrm{T}}, \eta_2^{\mathrm{T}})$ be similar partitions of ϕ and η into components of dimensions q and $m-q$ and let $\zeta^{\mathrm{T}} = (\eta_1^{\mathrm{T}}, \phi^{2\mathrm{T}}) = (\eta_1, \ldots, \eta_q, \phi^{q+1}, \ldots, \phi^m)$ be the vector whose first q coordinates are equal to those of η while the last $m-q$ coordinates come from ϕ. The mapping from ϕ to ζ is one-to-one and we can therefore parameterize $\mathcal{M}^0$ by the mixed parameter ζ.

The models $\mathcal{M}$ and $\mathcal{M}^0$ and the cumulant function $k(\phi)$ are said to be **steep** if $k(\phi)$ has the property that as $\lambda \uparrow 1$

$$\frac{d}{d\lambda} k(\lambda\phi_0 + (1-\lambda)\phi) \uparrow \infty$$

for any $\phi \in \operatorname{int} \Phi$ and any boundary point of ϕ_0 of Φ. We call $\mathscr{M}^0$ a **prime exponential model** if $\mathscr{M}^0$ is steep. The importance of the concept of steepness is that this property is necessary and sufficient for equality of H and $\mathscr{S}^0$. (For the proof of this and other main results mentioned in the present section, see Barndorff-Nielsen, 1978). Note in particular that, as follows trivially from the definition of steepness, the model $\mathscr{M}^0$ is steep if Φ is an open set and thus equal to $\operatorname{int} \Phi$. This is often the case, and when it is the exponential model $\mathscr{M}^0 (= \mathscr{M})$ is said to be **regular**. The property of steepness is obviously preserved under repeated sampling from the model $\mathscr{M}^0$.

Thus, in quick review, we have first the class of full (m, m) exponential family models in which the parameter space consists of all ϕ for which the relevant norming constant exists. Then we have the subclass of prime (or core) models in which (i) the above steepness condition is satisfied, or equivalently for which the domain of variation of η is the interior of the convex hull of the support of the distribution of the canonical statistic; and (ii) elements of the full family corresponding to possible boundary points of the canonical parameter space are deleted. Finally, we have the further subclass of regular models where the space of all possible ϕ is open. Most of the specific examples we consider are regular.

Prime exponential models have a considerable number of important and useful properties, the most important being the following.

1. The maximum likelihood estimate $\hat{\phi}$ of the canonical parameter exists if and only if $s \in \mathscr{S}^0$. It is then the unique solution of the likelihood equation, which has the form
$$E(S; \hat{\phi}) = s. \tag{2.106}$$
Alternatively, (2.106) may be written as $\eta(\hat{\phi}) = s$.
2. The cumulant function $k(\phi)$ is infinitely often differentiable and the first- and higher-order derivatives of $k(\phi)$ give the cumulants of the canonical statistic S. That is, denoting generic coordinates of ϕ, η and S by $\phi^a, \phi^b, \ldots, \eta_a, \eta_b, \ldots$ and $S_a, S_b, \ldots$, respectively, and using $k_{/a_1 \cdots a_n}$ as a shorthand for $\partial^n k / \partial \phi^{a_1} \cdots \partial \phi^{a_n}$ and K to indicate

joint cumulants we have

$$k_{/a} = K\{S_a\} = E(S_a; \phi) = \eta_a,$$

$$k_{/ab} = K\{S_a, S_b\} = \text{cov}\{S_a, S_b\}$$

and so on, the general form being

$$k_{/a_1 \cdots a_n} = K\{S_{a_1}, \ldots, S_{a_n}\}. \tag{2.107}$$

3. The expected information matrix $i(\phi)$ for the canonical parameter ϕ satisfies

$$i_{ab}(\phi) = \partial \eta_a / \partial \phi^b = k_{/ab} \tag{2.108}$$

and the expected information matrix $i(\eta)$ for the mean parameter η satisfies

$$i_{ab}(\eta) = \partial \phi^a / \partial \eta_b = i^{ab}(\phi), \tag{2.109}$$

where $i^{ab}(\phi)$ is the (a, b)-element of the expected **formation** (inverse of the expected information matrix) for ϕ.

4. For any mixed parameterization $\zeta^{\mathrm{T}} = (\eta_1^{\mathrm{T}}, \phi^{2\mathrm{T}})$ the components η_1 and ϕ^2 are variation independent and orthogonal.

The proof of orthogonality is simple and instructive and runs as follows. Let

$$\Sigma = \text{var}(S) = i(\phi) = \begin{bmatrix} \Sigma_{11} & \Sigma_{12} \\ \Sigma_{21} & \Sigma_{22} \end{bmatrix}, \tag{2.110}$$

$$\Delta = \Sigma^{-1} = \begin{bmatrix} \Lambda_{11} & \Lambda_{12} \\ \Lambda_{21} & \Lambda_{22} \end{bmatrix}, \tag{2.111}$$

where the partitions of Σ and Δ correspond to the partitions of ϕ and η. Using (2.108) we find

$$\zeta_{/\phi} = \begin{bmatrix} \Sigma_{11} & \Sigma_{12} \\ 0 & I \end{bmatrix}$$

and hence we have

$$\phi_{/\zeta} = \begin{bmatrix} \Sigma_{11}^{-1} & -\Sigma_{11}^{-1}\Sigma_{12} \\ 0 & I \end{bmatrix}. \tag{2.112}$$

The log-likelihood function is $\phi^{\mathrm{T}} s - k(\phi)$ where now we are to consider ϕ as a function of ζ. It follows, using (2.48), (2.110), (2.112) and the standard formula $\Lambda_{22}^{-1} = \Sigma_{22} - \Sigma_{21}\Sigma_{11}^{-1}\Sigma_{12}$, that

$$\begin{aligned} i(\zeta) &= \phi_{/\zeta}^{\mathrm{T}} i(\phi)\phi_{/\zeta} \\ &= \begin{bmatrix} \Sigma_{11}^{-1} & 0 \\ 0 & \Lambda_{22}^{-1} \end{bmatrix}. \end{aligned} \tag{2.113}$$

Hence, in particular, the components η_1 and ϕ^2 of ζ are orthogonal.

2.10 Review of curved exponential models

A key feature of the exponential family models discussed above is that the dimensions of the parameter and of the canonical statistic coincide. We now consider briefly the important extension in which the dimension, d, of the parameter is less than that, m, of the canonical statistic. Such models are usually best considered as embedded in an ambient (m, m) exponential family, the canonical parameter (or equivalently the mean parameter) being constrained to a d-dimensional submanifold. This motivates the following definition.

Consider a model $\mathcal{M}$ with model function of the form

$$p(y;\theta) = A(\theta)b(y)e^{s^{\mathrm{T}}(y)\phi(\theta)} \tag{2.114}$$

relative to some dominating measure λ. As usual, the dimension and domain of variation of the parameter θ are denoted by d and Ω, respectively, and $\phi = \phi(\theta)$ and $s = s(y)$ are of dimension $m(\geqslant d)$. On the assumption that the exponential representation (2.114) is minimal and that $d < m$ we speak of $\mathcal{M}$ as a **curved exponential model** or, more specifically, as an (m, d) **exponential model**.

If we start from such a model $\mathcal{M}$ there is associated with it an ambient (m, m) exponential model $\mathcal{M}^0$ defined as an exponential model with model function

$$a(\phi)b(y)e^{s^{\mathrm{T}}(y)\phi}, \tag{2.115}$$

where the canonical parameter ϕ varies over a set Φ^0 with

$$\Phi^0 \subset \Phi = \left\{\phi\colon \int e^{s^{\mathrm{T}}(y)\phi}b(y)\mu(dy) < \infty\right\}.$$

Note that $A(\theta) = a(\phi(\theta))$. For simplicity, we assume from now on the $\mathcal{M}^0$ is a prime exponential model.

The set $\Phi_0 = \phi(\Omega)$ constitutes a d-dimensional submanifold of the convex set Φ^0. This is illustrated in Fig. 2.1 for the case $(m, d) = (2, 1)$.

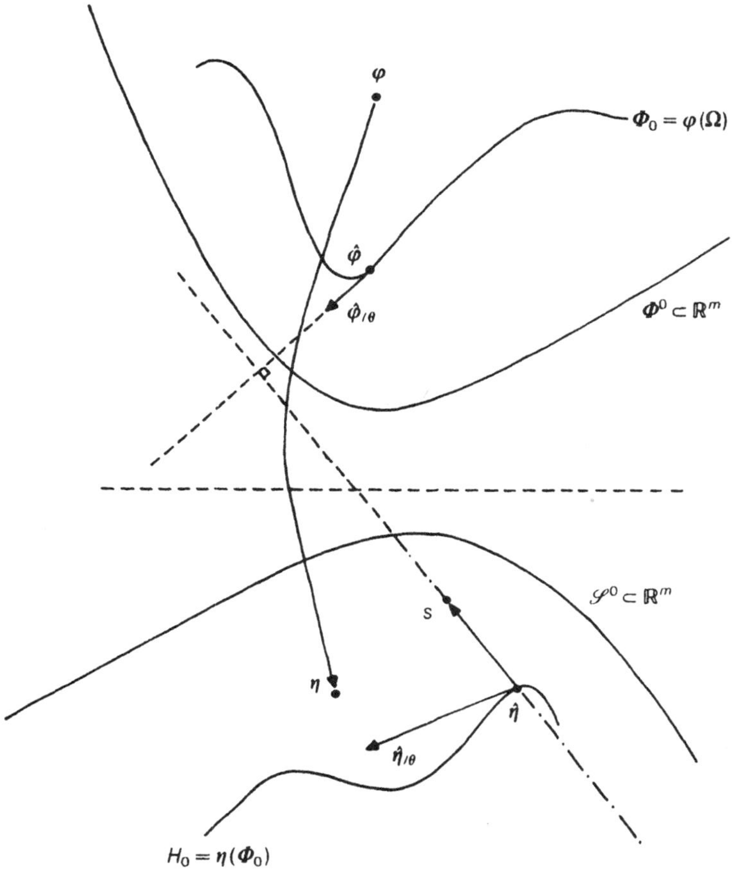

Figure 2.1 *The canonical parameter domain Φ^0 and the canonical open support domain $\mathscr{S}^0$ for the (m, m) model $\mathscr{M}^0$ associated with the curved (m, d) exponential model $\mathscr{M}$. The latter is determined by the submanifold $\Phi_0 = \phi(\Omega)$ of Φ^0 or, equivalently, by the submanifold $H_0 = \eta(\Phi_0)$ of $\mathscr{S}^0$. Also indicated are: (i) the mean value mapping η from Φ^0 to $\mathscr{S}^0$; (ii) the tangent vectors to the two submanifolds at the maximum likelihood points; (iii) the fact that the vector $s - \hat{\eta}$ is orthogonal to the tangent vector to Φ_0 with respect to the usual inner product of Euclidean geometry; and (iv) the affiine subset (signature –·–·–·–) of $\mathscr{S}^0$ determined by those values of the canonical statistic which give rise to the same solution to the likelihood equation for $\mathscr{M}$ as the actually observed value s.*

Also shown in Fig. 2.1 is the set $\mathscr{S}^0$ defined as the interior of the closed convex hull cl conv $\mathscr{S}$ of the support of the distribution of the canonical statistic s. It is mathematically most natural to consider the two sets Φ^0 and $\mathscr{S}^0$ as belonging to two different m-dimensional Euclidean spaces, and this is indicated in Fig. 2.1. Also indicated is the mean value mapping $\eta = \eta(\phi) = E(S; \phi)$ which gives a one-to-one correspondence between Φ^0 and $\mathscr{S}^0$. The figure illustrates several other points, discussed below.

A very simple example of a curved exponential family of order (2, 1) is provided by the normal parabola of Example 1.5.

Example 2.35 Normal parabola
Let $\mathscr{M}$ be the model specified by $Y = (Y_1, Y_2)$, where Y_1 and Y_2 are normally distributed with unit variance and with mean $(\theta, c\theta^2)$, where c is a known constant. The ambient exponential model $\mathscr{M}^0$ assigns (Y_1, Y_2) an arbitrary canonical parameter and $\mathscr{M}$ constrains the mean parameter to a curve (one-dimensional submanifold) in the two-dimensional space of the mean values of Y under $\mathscr{M}^0$. This space equals $\mathbb{R}^2$, and the curve is, in fact, a parabola, although very similar considerations would apply quite generally. The various general features illustrated in Fig. 2.1 can be identified in simple explicit form. □

Note that the exponential model remains of the (m, d) form under repeated sampling, and that in this situation the generic Fig. 2.1 remains, to a large extent, the same. There is however the important difference that the expected information matrices corresponding to Φ^0 and $\mathscr{S}^0$ change from, respectively, $i(\phi)$ and $i(\eta) = i^{-1}(\phi)$ to $ni(\phi)$ and $n^{-1}i(\eta)$. Correspondingly, typical statistical distances in Φ^0 space are of order $n^{-1/2}$ while in $\mathscr{S}^0$ space they are of order $n^{1/2}$.

We denote generic coordinates of θ by $\theta^r, \theta^s, \ldots$ and generic coordinates of ϕ, s, and η by $\phi^a, \phi^b, \ldots, s_a, s_b, \ldots$ and $\eta_a, \eta_b, \ldots$, respectively.

The log-likelihood function of (2.114) may then be written as

$$\ell = \ell(\theta) = s_a\phi^a(\theta) - k(\phi(\theta)) \tag{2.116}$$

where $k = k(\phi) = -\log a(\phi)$ is the cumulant function of the ambient (m, m) model (2.115). Differentiating with respect to θ via ϕ we find that the score vector is

$$\begin{aligned} \ell_r &= s_a \phi^a_{/r} - \eta_a \phi^a_{/r} \\ &= (s - \eta)_a \phi^a_{/r} \\ &= (s - \eta)^{\mathrm{T}} \phi_{/r} \end{aligned} \tag{2.117}$$

for $r = 1, \ldots, d$, where $\phi^a_{/r} = \partial \phi^a / \partial \theta^r$ (cf. section 2.2). At the maximum likelihood point, assuming it exists, we therefore have

$$(s - \hat{\eta})^{\mathrm{T}} \hat{\phi}_{/r} = 0. \tag{2.118}$$

The vectors $\phi_{/1}, \ldots, \phi_{/d}$ span (except in degenerate cases) the tangent space to the manifold Φ_0 at the point $\phi(\theta)$. Thus formula (2.118) tells us that the vector $s - \hat{\eta}$ is orthogonal to that tangent space at $\hat{\phi} = \phi(\hat{\theta})$ (cf. Fig. 2.1).

In the special case of Example 2.35 the maximum likelihood estimating equation has the simple interpretation that the vectors $(y_1 - \hat{\theta}, y_2 - c\hat{\theta}^2)$ and $(1, 2c\hat{\theta})$ are orthogonal and the geometrical interpretation follows directly.

The likelihood equation (2.118) shows, moreover, that the set $\mathscr{S}^0_{\hat{\theta}}$, say, of possible values in $\mathscr{S}^0$ that the canonical statistic may take and that would yield the same solution to the equation as the actually observed value s, constitutes an affine subset of $\mathscr{S}^0$, of dimension $m - d$.

Whereas, by our assumption, Φ_0 is a curved submanifold of Φ^0, it may happen that H_0 is an affine subset of $\mathscr{S}^0$. This means that the model $\mathscr{M}$ is determined by a linear specification of the mean parameter η. There are a considerable number of interesting examples of this.

Example 2.36 Behrens–Fisher problem

One Behrens–Fisher problem concerns inference on a common mean μ of two normal populations with separate variances σ_1^2 and σ_2^2 and sample sizes n_1 and n_2. The model is thus a (4, 3) curved exponential model with canonical parameter $\phi = (-\frac{1}{2}\sigma_1^{-2}, \mu\sigma_1^{-2}, -\frac{1}{2}\sigma_2^{-2}, \mu\sigma_2^{-2})$ and mean value parameter $\eta = (n_1(\sigma_1^2 + \mu^2), n_1\mu, n_2(\sigma_2^2 + \mu^2), n_2\mu)$, and it may be characterized by the linear mean value specification $n_2\eta_2 = n_1\eta_4$. □

Example 2.37 Proportionality of gamma means

If the mean values $\mu_1, \ldots, \mu_k$ of k gamma distributions are assumed to be identical or, more generally, to be mutually proportional with known coefficients of proportionality and if the shape parameters of the distributions are unknown then the distributions constitute a curved $(2k, k+1)$ exponential family. □

Under curved exponential models for continuously varying observations the likelihood function will in general with probability 1 have a unique global maximum. However, the likelihood function may have several local maxima, as can happen when the affine subsets $\mathscr{S}^0_{\hat{\theta}}$ of $\mathscr{S}^0$ intersect. Outside the class of curved exponential models that have the property of being also transformation models it is, in fact, rather typical that there is a positive probability of having several solutions to the likelihood equation but that this probability decreases exponentially fast with increasing sample size.

Example 2.38 Normal correlation coefficient
Let (Y_1, Y_2) have a two-dimensional normal distribution with mean value 0, unit variances and unknown correlation coefficient ρ. The model is thus a $(2, 1)$ exponential model with canonical statistic $s = (u, v) = (\frac{1}{2}y_1^2 + \frac{1}{2}y_2^2, y_1 y_2)$ and canonical parameter $\phi = (-(1-\rho^2)^{-1}, \rho(1-\rho^2)^{-1})$. Figure 2.2 shows the geometrical setting of this model,

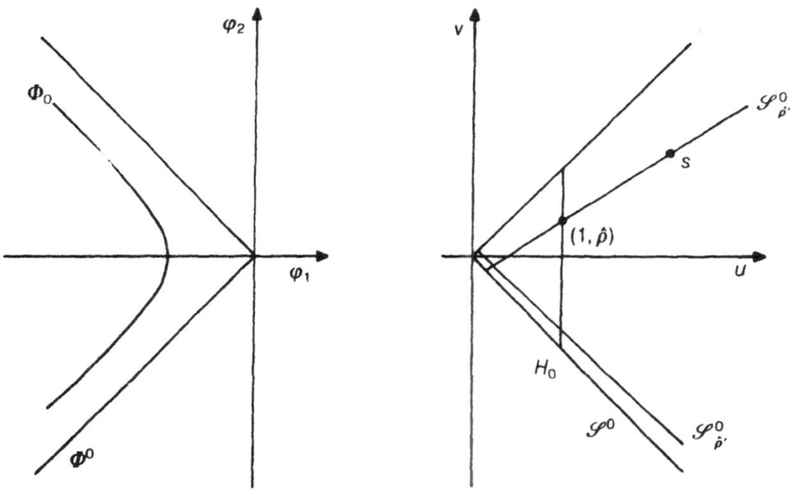

Figure 2.2 *The canonical parameter space and the canonical observation/mean space for the example of the normal correlation coefficient. The two half-lines $\mathscr{S}^0_{\hat{\rho}}$ and $\mathscr{S}^0_{\hat{\rho}'}$ are examples of the affine subsets $\mathscr{S}^0_{\hat{\theta}}$ with a constant likelihood equation solution.*

in line with the general picture of Fig. 2.1. Also shown in the figure are two instances $\mathscr{S}^0_{\hat{\rho}}$ and $\mathscr{S}^0_{\hat{\rho}'}$ of the affine subsets (here half-lines) of type $\mathscr{S}^0_{\hat{\theta}}$. Any line $\mathscr{S}^0_{\hat{\rho}}$ intersects the u-axis at the point $(\frac{1}{2}(1-\hat{\rho}^2), 0)$, and this point divides $\mathscr{S}^0_{\hat{\rho}}$ into a half-line and a line segment. For a value of s on the half-line part the likelihood equation has the unique solution $\hat{\rho}$ but if s is on the line segment the likelihood equation has a second solution. This is illustrated in the figure. □

Differentiating relation (2.117) with respect to θ^s, we have that the observed information is

$$j_{rs}(\theta) = \phi^a_{/r}\phi^b_{/s}k_{/ab} - (s-\eta)_{/a}\phi^a_{/rs} \tag{2.119}$$

which may be rewritten as

$$j_{rs}(\theta) = i_{rs}(\theta) - (s-\eta)_{/a}\phi^a_{/rs} \tag{2.120}$$

where i denotes the expected information. The second term on the right-hand side of (2.120) is an expression of the curvature of the model $\mathscr{M}$ and is essentially the Euler–Schouten or embedding curvature of the submanifold Φ_0 of Φ^0 with respect to the usual Euclidean metric on Φ^0 (Boothby, 1975).

There is an interesting connection between the geometry of curved exponential families and the **directed information distance** (or **Kullback–Leibler information measure**) I. If P and Q are two probability measures on the same sample space then, supposing that P and Q are mutually absolutely continuous, the directed information distance from P to Q is defined by

$$I(P,Q) = \int \log \frac{dP}{dQ} dP. \tag{2.121}$$

It should be noted that in general $I(Q,P)$ is different from $I(P,Q)$, i.e. the directed distance from Q to P is different from the directed distance from P to Q.

The directed information distance satisfies $I(P,Q) \geqslant 0$ with equality if and only if $P = Q$, as follows directly from Jensen's inequality. Furthermore, if Q_n is a sequence of probability measures then, under weak regularity conditions, we have that $I(P,Q_n) \to I(P,Q)$ if and only if $Q_n \overset{\to}{\to} Q$ (convergence in distribution).

Letting P_ϕ denote the probability measure given by (2.115) and writing $I(\phi',\phi)$ for $I(P_{\phi'}, P_\phi)$, we find that

$$I(\phi',\phi) = \eta^{\mathrm{T}}(\phi')(\phi'-\phi) + k(\phi) - k(\phi'). \tag{2.122}$$

In particular, letting $'\phi$ denote the maximum likelihood estimator of ϕ under the model $\mathscr{M}^0$, so that $'\phi$ satisfies $\eta('\phi) = s$, we have

$$I('\phi, \phi) = \ell('\phi) - \ell(\phi) \tag{2.123}$$

where $\ell(\phi) = s^T\phi - k(\phi)$ is the log-likelihood function under $\mathscr{M}^0$. It follows, in particular, that

$$I('\phi, \phi(\theta)) = \ell('\phi) - \ell(\theta), \tag{2.124}$$

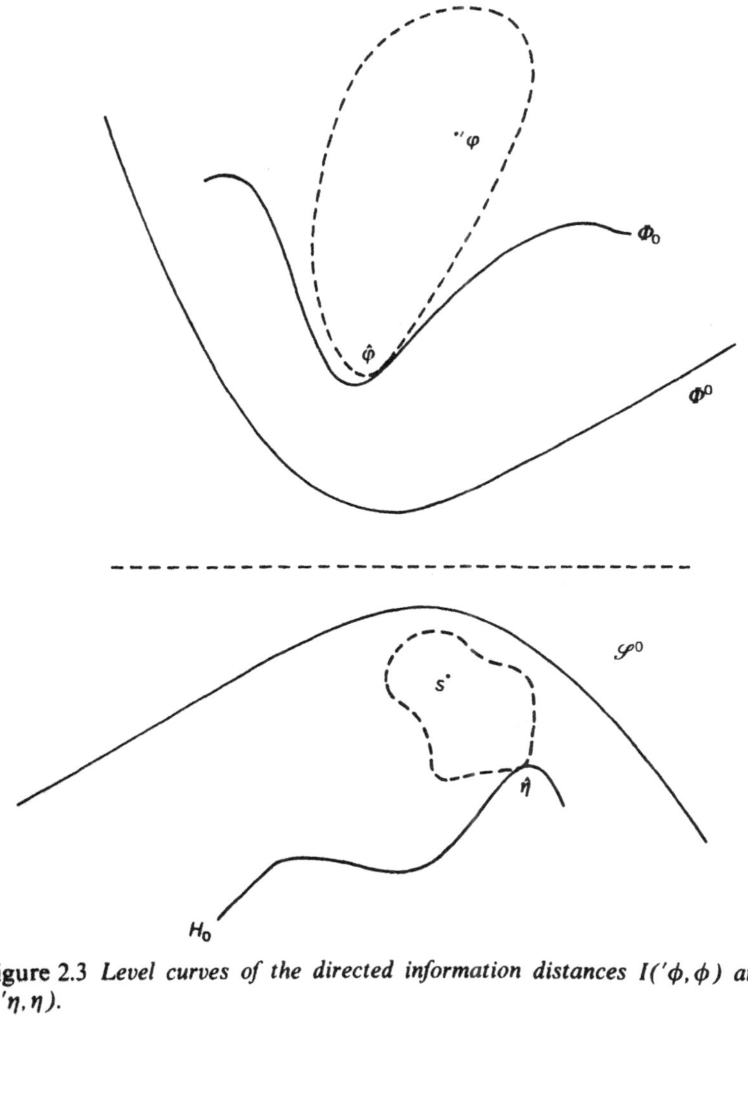

Figure 2.3 *Level curves of the directed information distances $I('\phi, \phi)$ and $I('\eta, \eta)$.*

where $\ell(\theta)$ is the log-likelihood function under the curved submodel $\mathscr{M}$. Hence, maximum likelihood estimation of θ is equivalent to minimization of $I('\phi, \phi(\theta))$ with respect to θ. In other words, the maximum likelihood estimate $\hat{\theta}$ has the property that the point $\hat{\phi} = \phi(\hat{\theta})$ is the point in Φ_0 that has the smallest I-distance from $'\phi$ (cf. Fig. 2.3).

As mentioned in section 2.9, the log-likelihood function for a full exponential model is concave. Formula (2.123) therefore shows that as a function of ϕ the directed information distance $I('\phi, \phi)$ is convex. In particular, the level curves of $I('\phi, \cdot)$ are convex. Note, however, that the mean value mapping η sends those level curves into curves that are generally not convex. These facts are indicated in Fig. 2.3. It follows further that

$$I('\phi, \phi)_{/\phi} = \eta - s \tag{2.125}$$

and

$$I('\phi, \phi)_{/\phi\phi} = i(\phi). \tag{2.126}$$

On the other hand, differentiation of $I('\eta, \eta)$, i.e. of $I('\phi, \phi)$ expressed as a function of $'\eta$ and η gives

$$I('\eta, \eta)_{/'\eta} = '\phi - \phi$$

and

$$I('\eta, \eta)_{/'\eta'\eta} = i('\eta) = i^{-1}('\phi).$$

Further results and exercises

2.1 For Example 2.10, observations from an exponential distribution, suppose that the distribution is parameterized in terms of the mean μ, i.e. in the form $\mu^{-1}e^{-y/\mu}$. Compute the score function and information in the new parameterization and verify the general transformation (2.46) and (2.47).

[Section 2.2]

2.2 Obtain the score function and information matrix for the Gaussian first-order autoregressive process from the log-likelihood (2.19), (a) with μ known and zero and (b) with μ unknown. Compute the log-likelihood when y_1, instead of having the stationary distribution of the process, takes an arbitrary deterministic value.

[Section 2.2]

2.3 Examine the log-likelihood function and its associated functions for the time-dependent Poisson process in which the rate function is $e^{\alpha+\beta t}$. Much more generally, a point process without coincident points is uniquely defined by its complete intensity function $\rho(t; \mathscr{H}_t)$. That is, if $\mathscr{H}_t$ denotes the history of the process up to but excluding t, then the probability of a point in $[t, t+\Delta)$ is $\rho(t; \mathscr{H}_t) + o(\Delta)$. Show that if points are observed at $t_1 < t_2 < \cdots < t_n$ in the period of observation $(0, t_0)$ then the likelihood is

$$\{\prod \rho(t_j, \mathscr{H}_{t_j})\} \exp\left\{-\int_0^{t_0} \rho(t; \mathscr{H}_t)dt\right\}.$$

Show that this yields a simple result, easily written down directly, for a renewal process in which the intervals between successive points are independently and identically distributed with density $f(y; \theta)$. Classify some of the main special types of point process according as to whether $\rho(t; \mathscr{H}_t)$ is or is not readily calculated.

[Cox and Isham, 1980; Section 2.2]

2.4 Verify the inequality (2.35) from first principles for Example 2.10, the exponential distribution.

[Section 2.2]

2.5 In a balanced one-way normal theory variance component problem there are observed random variables Y_{ij} $(i = 1, \ldots, m; j = 1, \ldots, r)$ such that $Y_{ij} = \mu + \xi_i + \eta_{ij}$, where $\{\xi_i\}$, $\{\eta_{ij}\}$ are mutually independent sets of independently normally distributed unobserved random variables of zero mean and variances τ_ξ, τ_η. Obtain the log-likelihood for $(\mu, \tau_\xi, \tau_\eta)$ and show that the sufficient statistic consists of the sample mean and sums of squares between and within 'rows'.

[Section 2.3]

2.6 Prove that for a random sample from the Cauchy location model the order statistic is minimal sufficient (the last statement in Example 2.14).

[Section 2.3]

2.7 For n random variables from the special uniform distribution of Example 2.17 show that the sufficient statistic is $(Y_{(n1)}, Y_{(nn)})$ and that the joint density of $Y_{(n1)}$ and $Y_{(nn)}$ is

$$(\tfrac{1}{2})^n n(n-1)(y_{(nn)} - y_{(n1)})^{n-2} \qquad (\theta - 1 < y_{(n1)} \leqslant y_{(nn)} < \theta + 1)$$

and zero otherwise. Show further that the conditional density of $M = \frac{1}{2}(Y_{(n1)} + Y_{(nn)})$ given $R = y_{(nn)} - y_{(n1)} = r$ is uniform over $(\theta - 1 + \frac{1}{2}r, \theta + 1 - \frac{1}{2}r)$.

[Section 2.4]

2.8 Prove Basu's theorem that if $S = s(Y)$ is a complete sufficient statistic for the parameter θ and if $A = a(Y)$ is distribution constant, i.e. has a distribution not depending on θ, then S and A are independent. Deduce that if $Y_1, \ldots, Y_n$ are independent and identically normally distributed with mean μ and variance σ^2, then the sample mean and any sample measure of dispersion (in a sense to be defined) are independent. Illustrate with other examples.

[Sections 2.3–2.5]

2.9 A statistic w is said to be L-sufficient (short for likelihood-sufficient) with respect to the interest parameter ψ provided

(i) the so-called normed profile likelihood $\bar{L}_P(\psi) = L(\psi, \hat{\chi}_\psi)/L(\hat{\psi}, \hat{\chi})$ depends on the data y only through w;
(ii) the marginal distribution of w depends only on ψ.

Show that this concept covers both the classical sufficiency concept and S-sufficiency and, moreover, that under a composite transformation model the maximal invariant statistic is L-sufficient with respect to the maximal invariant parameter. Illustrate by verifying that under the multivariate normal model the empirical correlation matrix is L-sufficient with respect to the theoretical correlation matrix and that under the von Mises–Fisher model the empirical precision is L-sufficient with respect to the theoretical precision.

[Rémon, 1984; Barndorff-Nielsen, 1985; Sections 2.5 and 2.8]

2.10 Show that the use of the REML estimates for unbalanced variance component problems can be 'justified' by L-sufficiency.

[Patterson and Thompson, 1971; Searle, Casella and McCulloch, 1992; Rémon, 1984]

2.11 *Some remarks on overconditioning.* Suppose that we have conditioned on certain features to obtain a probability θ_A. It is wise to condition on a further binary feature B in order to sharpen relevance? That is, should θ_A be separated into θ_{0A} and θ_{1A}?

Let $\phi = P(B = 0)$ and write $\theta_{0A} = \theta_A - \varepsilon(1 - \phi)$, $\theta_{1A} = \theta_0 + \varepsilon\phi$ so that $E(\theta_{BA}) = \theta_A$. We shall take ε to be of order $1/\sqrt{n}$, so that $\theta_{1A} - \theta_{0A}$ is of the same order as the sampling error. If we take a single estimate $\hat{\theta}$, i.e. ignore B, the resulting mean squared error is

$$\phi\varepsilon^2(1-\phi)^2 + (1-\phi)\varepsilon^2\phi^2 + \theta_A(1-\theta_A)/n = \varepsilon^2\phi(1-\phi) + \theta_A(1-\theta_A)/n.$$

If, on the other hand, we estimate θ_{0A}, θ_{1A} separately from, respectively, approximately $n\phi$ and $n(1-\phi)$ observations, the mean squared error, or in this case average variance, is

$$\begin{aligned} &n\phi\theta_{0A}(1-\theta_{0A})/(n\phi) + n(1-\phi)\theta_{1A}(1-\theta_{1A})/\{n(1-\phi)\} \\ &= 2\theta_A(1-\theta_A)/n + O(n^{-3/2}). \end{aligned}$$

Thus conditioning is advantageous from this viewpoint if and only if

$$\varepsilon^2\phi(1-\phi) > \theta_A(1-\theta_A)/n,$$

i.e. if and only if the mean squared difference in probabilities exceeds the mean squared sampling error. There is a close parallel with the condition for including a new variable in a multiple regression equation.

The quantitative implications of this for conditioning in discrete problems of statistical inference are unclear.

[Sections 2.4–2.5]

2.12 *Cauchy distribution.* Show that if Y has a Cauchy density of location μ and scale parameter σ, then $1/Y$ has a Cauchy density of location $\mu/(\mu^2+\sigma^2)$ and scale parameter $\sigma/(\mu^2+\sigma^2)$. From a set of independent and identically distributed random variables $Y_1, \ldots, Y_n$ from such a Cauchy density, show then that two (and, in fact, in terms of $1/(Y-k)$, an infinite number) of sets of maximal ancillary statistics can be obtained. Discuss the implications for inference for μ, for σ, and for μ/σ.

Show further that if Y has a density proportional to $\{1+(y-\mu)^2/\sigma^2\}^{-c}$ then, provided $b=(2c-1)^{-1}$ is an odd integer, Y^{-b} has a density of the same form with parameters transformed as in the Cauchy case. The values of c are particular rational numbers in $(0,1]$.

[McCullagh, 1992; Section 2.5]

2.13 Let $Y_1, \ldots, Y_n$ be independent Poisson random variables of mean μ, $S = Y_1 + \cdots + Y_n$ and let $g(Y_1, \ldots, Y_n)$ be any function of $Y_1, \ldots, Y_n$. By using the result that S has a Poisson distribution of mean $n\mu$, show that

$$E\{g(Y_1,\ldots,Y_n)|S=s\} = \frac{s!}{n^s}\mathscr{E}_s,$$

where $\mathscr{E}_s$ is the coefficient of μ^s in the unconditional expectation $e^{n\mu}E\{g(Y_1,\ldots,Y_n)\}$.

Hence or otherwise verify (2.73).

[Rao and Chakravarti, 1956; Section 2.6]

2.14 Develop the dispersion test analogous to R of Example 2.24 for conformity with a binomial distribution.

[Section 2.6]

2.15 The (2, 2) exponential family density

$$\exp\{y\phi + t(y)\tau - k(\phi,\tau)\}$$

may be said to be overdispersed relative to the family with $\tau = 0$ if, for given $E(Y)$, the variance of Y increases with τ. Show that a sufficient condition for this is that $t(y)$ is convex.

[Gelfand and Dalal, 1990; Section 2.6]

2.16 *Dispersion models.* The defining property of dispersion models is that their model function is of the form

$$a(\lambda, y)e^{\lambda t(y;\gamma)}$$

where $\lambda \in \mathbb{R}$ and $\gamma \in \mathbb{R}^k$ are parameters.

Show that λ and γ are orthogonal. Show further that if $\psi = \psi(\gamma)$ is a one-dimensional parameter of interest which depends on γ only, then it is possible to find a reparameterization (λ, ψ, χ) such that λ, ψ and χ are mutually orthogonal provided the matrix $E(\nabla_\gamma \nabla_\gamma^{\mathrm{T}} t)$ does not depend on λ.

[Section 2.7]

2.17 *Normal nonlinear regression.* Let $Y_1,\ldots,Y_n$ be independent normal with means η_i and a common variance σ^2. The means are assumed to be of the form $\eta_i = g(\gamma; x_i)$ where g is a known function, $\gamma \in \mathbb{R}^k$ $(k < n)$ is a parameter and $x_1,\ldots,x_n$ are known covariates. Show that the result of the previous exercise applies here, with $\lambda = \sigma^{-2}$.

The Michaelis–Menten model specifies η_i as

$$\eta_i = \alpha x_i/(\beta + x_i).$$

Show that the parameters $\lambda = \sigma^{-2}$, $\chi = \alpha^3 \sum x_i^2/(\beta + x_i)^3$, α are mutually orthogonal.

[Section 2.7]

2.18 The formula (2.95) for the conditional distribution of an equivariant estimator z given the maximal invariant statistic w (equivalently, given u) involves the class of possible likelihood functions and not just the one likelihood function, L_0 say, corresponding to the actually observed realization, y_0 say, of y. It is, however, possible to recast (2.95) so as to involve only L_0. Specifically, with $w_0 = w(y_0)$ and $z_0 = z(y_0)$ and considering z as an element of the group G, we have

$$\bar{p}(z; g \mid w_0) = c(w_0)\frac{L_0(z_0 z^{-1} g)}{L_0(z_0)},$$

this being a density relative to left invariant measure on G. This formula, which generalizes a similar expression for the location–scale model due to Fisher (1934), shows how the 'shape and position' of the conditional distribution of z may be simply determined by the observed likelihood function and the observed $z_0 \in G$, respectively.

[Section 2.8]

2.19 Examine the properties of the exponential model resulting from the balanced two-way normal theory components of variance model

$$Y_{is} = \mu + \xi_i + \eta_s + \zeta_{is},$$

$(i = 1, \ldots, m;\ s = 1, \ldots, r)$, where $\{\xi_i\}$, $\{\eta_s\}$, $\{\zeta_{is}\}$ are independent sets of normally distributed random variables of zero mean and variances respectively $\sigma_\xi^2, \sigma_\eta^2, \sigma_\zeta^2$. Specify the spaces H, $\mathscr{S}$, $\mathscr{S}^0$.

[Section 2.9]

2.20 A scalar random variable Y has an exponential family distribution. Show that the exponential family form is retained under

(a) left censoring at a and right censoring at b, where $a < b$ are known constants, i.e. values of Y less than a or greater than b are recorded as a or as b, respectively;

(b) truncation to the interval (a, b), i.e. only values of Y in this interval are observed.

Sketch a corresponding discussion for vector random variables.

[Section 2.9]

2.21 Let $\mathcal{M}$, $\mathcal{M}_0$, $\mathcal{M}_{00}$ be prime exponential models with $\mathcal{M}_{00} \subset \mathcal{M}_0 \subset \mathcal{M}$. (In other words, $\mathcal{M}_0$ is an affine hypothesis under $\mathcal{M}$ and $\mathcal{M}_{00}$ is an affine hypothesis under $\mathcal{M}_0$.) Furthermore, let $\hat{\phi}, \hat{\phi}_0$ and $\hat{\phi}_{00}$ be the maximum likelihood estimates of the canonical parameter under the three models. Then the directed distance satisfies the pythagorean relation

$$I(\hat{\phi}, \hat{\phi}_{00}) = I(\hat{\phi}, \hat{\phi}_0) + I(\hat{\phi}_0, \hat{\phi}_{00}).$$

[Section 2.9]

2.22 *Pure birth process.* This process was considered in Example 2.7. Show that there are two different sampling schemes under which it constitutes a $(1, 1)$ exponential model but that otherwise it is of the $(2, 1)$ type. Note that in any case the likelihood equation has a unique solution and discuss to what extent this will be typical of continuous-time Markov processes. (This is discussed further in Example 9.1.)

[Sections 2.2 and 2.10]

Bibliographic notes

The key concepts of this chapter, namely likelihood, sufficiency, information and conditioning, are due to R.A. Fisher; see especially Fisher (1922; 1925; 1934; 1935a; 1956). Bayesian approaches, in particular going back to Laplace, had in effect given likelihood a key role. Fisher (1956) suggested the direct use of likelihood for some issues, and this was set out in extended form by Barnard, Jenkins and Winsten (1962) and by Edwards (1972); see also Kalbfleisch and Sprott (1970). Formulations of likelihood principles and their implications due to Birnbaum (1962; 1969; 1970) have been extensively discussed; see in particular Berger and Wolpert (1984), who describe likelihood from a viewpoint very different from that employed here, and Durbin (1970), Kalbfleisch (1975) and Holm (1985). See also Basu (1975; 1977) and the collection edited by Ghosh (1988).

Rao (1947) and Bartlett (1953a; 1953b) particularly emphasized use of the score function. The tensor transformation properties of

Fisher information were given by Rao (1947); see also Jeffreys (1939).

The technique for calculation of likelihood functions under stochastic process models in continuous time indicated in section 2.2.3 can be made mathematically rigorous by use of the theory of product integration; see Gill and Johansen (1990) and, for some further informal discussion, Barndorff-Nielsen (1990d).

A measure-theoretic formulation of sufficiency was given by Halmos and Savage (1949); for later work see the references in the text. Definitions of sufficiency in the presence of nuisance parameters were given by Neyman and Pearson (1936), Fraser (1956), Barnard (1963) and Rémon (1984); see also Barndorff-Nielsen (1978; 1988a).

Conditioning was emphasized from the viewpoint of recovery of information lost in using the maximum likelihood estimate by Fisher (1925) and from a different angle by Fisher (1956). Welch (1939) pointed out the implicit clash with Neyman–Pearson notions of power, preferring the latter; see Barnard (1976) for a discussion of this. Cox (1958) gave a simple example designed to motivate the primary role of conditioning. Fraser's (1968) formulation of structural inference led to conditioning by a somewhat different route. Non-uniqueness of ancillaries was exemplified by Basu (1964), and a basis for choice in such cases discussed by Barnard and Sprott (1971) and Cox (1971). For more recent discussion, see Becker and Gordon (1983) and Lloyd (1992). Conditions for ancillarity in the presence of nuisance parameters have been given, *inter alia*, by Barnard (1963), Sverdrup (1966), Sandved (1972); see Barndorff-Nielsen (1978) for a comprehensive discussion. Applications of the concepts of cuts to the analysis of contingency tables and to mixed graphical models are given in Asmussen and Edwards (1983), Frydenberg (1990) and Frydenberg and Lauritzen (1989). For a different approach to ancillary statistics see Buehler (1982).

The use of conditional distributions for tests of model adequacy is examined by Fisher (1950) and Rao and Chakravarti (1956).

Parameter orthogonality was discussed in detail by Huzurbazar (1950; 1956) and Barndorff-Nielsen (1978), and Cox and Reid (1987a) gave a key property emphasized in the present discussion.

For general references on transformation models and exponential families, see the *Bibliographic Notes* to Chapter 1. For the curved exponential family, see Efron (1975; 1978).

CHAPTER 3

First-order theory

3.1 Introduction

In the present chapter we discuss first-order asymptotic theory arising in statistical inference in regular parametric models when the amount of information is large, so that techniques of local linearization and use of the central limit theorem are available. In particular, at least for models with independent component random variables, the score function, being a sum of independent components, is asymptotically normal. Local linearization relates the distribution of maximum likelihood estimates to that of the score function and the unknown parameters arising in the information matrix determining the limiting distribution can be consistently estimated.

This approach provides a collection of powerful and widely useful results associated with maximum likelihood estimates, score tests, likelihood ratio tests and confidence regions based thereon, and Chapter 3 develops these. There are two reasons for going further in the later chapters. One is to have more refined distributional approximations. The other is that first-order theory yields a variety of different procedures and test statistics equivalent to that order. While these often yield virtually identical results in applications, this is by no means invariably so and therefore it becomes important both to have a basis for choosing between the different statistics, mentioned above, and for developing modifications. Chapter 4 discusses these issues in broad terms as a preliminary to the detailed study in Chapters 5–8.

3.2 Main statistics: no nuisance parameters

The central definitions of score function, expected and observed information and of the relation between them are set out concisely in section 2.2. Notably if $\ell(\theta; y)$ is the log-likelihood function

$$\begin{aligned} u(\theta; Y) &= U(\theta) = \nabla_\theta \ell(\theta; Y), \\ E\{U(\theta); \theta\} &= 0, \\ \operatorname{cov}\{U(\theta); \theta\} &= E\{U(\theta)U^{\mathrm{T}}(\theta); \theta\} = E\{-\nabla_\theta \nabla_\theta^{\mathrm{T}} \ell(\theta; Y)\} = i(\theta). \end{aligned} \tag{3.1}$$

Later we shall use the more concise notation of ℓ_r for the rth component of $U(\theta)$, ℓ_{rs} for the (r, s) component of $\nabla_\theta \nabla_\theta^{\mathrm{T}} \ell$, etc. Further, the maximum likelihood estimate for given observations y is for regular problems defined as the solution, assumed unique, of

$$u(\hat{\theta}; y) = 0 \tag{3.2}$$

and the observed information

$$j(\theta) = -\nabla_\theta \nabla_\theta^{\mathrm{T}} \ell(\theta; y)$$

is usually evaluated at $\theta = \hat{\theta}$,

$$\hat{j} = j(\hat{\theta}; y) = -[\nabla_\theta \nabla_\theta^{\mathrm{T}} \ell(\theta; y)]_{\theta = \hat{\theta}}. \tag{3.3}$$

We regard the primary form for the summarization of evidence about θ to be the provision of a nested collection of confidence regions for θ at various confidence coefficients. Sometimes these are most conveniently presented via a point estimate and a measure of error (for example, standard error) together with a pivotal distribution such as the standard normal or Student t distribution. For our purposes, however, we shall mostly regard confidence regions as formed from test statistics, the confidence region consisting of all those parameter values that would not be 'rejected' at the significance level in question.

Thus when there are no nuisance parameters we consider testing the null hypothesis $\theta = \theta_0$, where θ_0 is an arbitrary known point in the parameter space. There are many ways of testing $\theta = \theta_0$ equivalent to the first order, i.e. based on test statistics that typically differ by $O_p(n^{-1/2})$. We shall concentrate on three such:

1. the Neyman-Pearson likelihood ratio statistic

$$w(\theta_0) = 2\{\ell(\hat{\theta}) - \ell(\theta_0)\}, \tag{3.4}$$

2. the Rao score statistic

$$w_u(\theta_0) = U^{\mathrm{T}}(\theta_0) i^{-1}(\theta_0) U(\theta_0); \tag{3.5}$$

3. the Wald, or maximum likelihood estimate statistic

$$w_p(\theta_0) = (\hat{\theta} - \theta_0)^T i(\theta_0)(\hat{\theta} - \theta_0), \tag{3.6}$$

where the suffix p warns that a particular parameterization is involved. We have given these in quadratic versions appropriate for multi-dimensional θ. For scalar θ all have a one-sided form, e.g. (3.4) may be replaced by

$$r(\theta_0) = \text{sgn}(\hat{\theta} - \theta_0)\sqrt{w(\theta_0)} \tag{3.7}$$

which we call the **directed likelihood**, and (3.5) and (3.6) may respectively be replaced by

$$r_u(\theta_0) = U(\theta_0)/\sqrt{i(\theta_0)}, \tag{3.8}$$

$$r_p(\theta_0) = (\hat{\theta} - \theta_0)\sqrt{i(\theta_0)}. \tag{3.9}$$

In the first-order asymptotic theory, discussed in section 3.3, the statistics (3.4)–(3.6) have the chi-squared distribution with d_θ degrees of freedom and the signed versions (3.7)–(3.9) have the standard normal distribution. Confidence regions at level $1 - \varepsilon$ are formed approximately, for example from (3.4), as

$$\{\theta : w(\theta) \leqslant \chi^2_{d_\theta, \varepsilon}\}, \tag{3.10}$$

where $\chi^2_{d_\theta,\varepsilon}$ is the tabulated upper ε point of the relevant chi-squared distribution. Upper and lower confidence limits for a scalar parameter can be formed similarly from the signed statistics, subject to some monotonicity requirements. If in (3.9) $i(\theta_0)$ is replaced by $i(\hat{\theta})$ or $j(\hat{\theta})$, we have confidence limits derived via a pivotal normal distribution centred on $\hat{\theta}$ and with standard error $1/\sqrt{i(\hat{\theta})}$ or $1/\sqrt{j(\hat{\theta})}$. In the multi-dimensional case the corresponding regions are sets of similar ellipsoids centred on $\hat{\theta}$.

In (3.8) $\sqrt{i(\theta_0)}$ is the exact standard deviation of $U(\theta_0)$ when $\theta = \theta_0$ and in (3.9) $1/\sqrt{i(\theta_0)}$ is the approximate standard deviation of $\hat{\theta}$ when $\theta = \theta_0$. Note that in the asymptotic calculations, because $U(\theta_0)$ and $i(\theta_0)$ refer to the total vector Y of dimension n, then as $n \to \infty$ and subject to general stability conditions,

$$\begin{aligned} U(\theta_0) &= \sqrt{n}\bar{U}(\theta_0) = O_p(n^{1/2}), \\ i(\theta_0) &= n\bar{\imath}(\theta_0), \\ \hat{\theta} - \theta_0 &= O_p(n^{-1/2}), \end{aligned} \tag{3.11}$$

where $\bar{\imath}(\theta_0)$ is the average information per observation and $\bar{U}(\theta_0)$ is a normalized score function. Also in (3.11), if the observations are independent and identically distributed random variables, then $\bar{\imath}$ is the information for a single observation. Rather generally, it is convenient to consider a hypothetical sequence of sets of data in which, as $n \to \infty$, (3.11) holds. Further, as $n \to \infty$ we have in probability that, provided $i(\theta)$ is continuous at $\theta = \theta_0$,

$$
\begin{aligned}
j(\hat{\theta})/n &\to \bar{\imath}(\theta_0), \\
j(\theta_0)/n &\to \bar{\imath}(\theta_0).
\end{aligned}
\tag{3.12}
$$

Therefore in (3.5), (3.6), (3.8) and (3.9), $i(\theta_0)$ can be replaced by any of $i(\hat{\theta})$, $j(\hat{\theta})$, $j(\theta_0)$ or indeed by $i(\theta^*)$ or $j(\theta^*)$, where θ^* is any consistent estimate of θ, in the following sense. If $\theta = \theta_0$ the various modified statistics differ typically by $O_p(n^{-1/2})$ and hence their limiting distribution is the same under the null hypothesis. Note, however, that if for some θ', relatively distant from θ_0, $\bar{\imath}(\theta') \ll \bar{\imath}(\theta_0)$, and if in fact $\theta = \theta'$, use of $i(\hat{\theta})$ or $j(\hat{\theta})$ would be equivalent to use of $i(\theta')$ and there would be a major loss of sensitivity.

The (m, m) exponential models are atypical, in particular because, directly for the canonical parameter, and in fact for any parameterization, $i(\theta) = j(\theta)$. Nevertheless, it is useful to illustrate the above possibilities on a simple special case which we do in Example 3.1.

The above discussion illustrates a notational point. In careful derivations of asymptotic distributional results it is very helpful to show explicitly the order of magnitude in n of various terms, in particular to ensure that terms of a particular order are consistently retained or rejected. For this an effective device is to express all quantities in terms of others that are $O(1)$, or for random variables $O_p(1)$, as $n \to \infty$. Thus we have written $i(\theta_0) = n\bar{\imath}(\theta_0)$ and $U(\theta_0) = \sqrt{n}\bar{U}(\theta_0)$, where $\bar{\imath}(\theta_0) = O(1)$ and $\bar{U}(\theta_0) = O_p(1)$. On the other hand, in stating and using results, rather than saying that $\sqrt{n}(\hat{\theta} - \theta_0)$ is asymptotically normal with zero mean and variance $\bar{\imath}^{-1}(\theta_0)$, it may be much simpler to write $\hat{\theta} - \theta_0 \stackrel{\sim}{\to} N\{0, i^{-1}(\theta_0)\}$.

Example 3.1 Exponential distribution
Let $Y_1, \ldots, Y_n$ be independently and identically distributed with density $\rho e^{-\rho y}$. The calculations in Example 2.1 and (2.7) show that

$$
\begin{aligned}
\ell(\rho; Y) &= n \log \rho - \rho \sum Y_j, \\
U(\rho) &= n/\rho - \sum Y_j, \\
i(\rho) &= j(\rho) = n/\rho^2
\end{aligned}
$$

and $\hat{\rho} = n/\sum Y_j = 1/\bar{Y}$. Thus to test $\rho = \rho_0$ the statistics (3.7)–(3.9) are respectively

$$
\begin{aligned}
r(\rho_0) &= \operatorname{sgn}(1 - \rho_0 \bar{Y})[2n\{\rho_0 \bar{Y} - \log(\rho_0 \bar{Y}) - 1\}]^{1/2}, \\
r_u(\rho_0) &= \sqrt{n}(1 - \rho_0 \bar{Y}), \\
r_p(\rho_0) &= \sqrt{n}\{(\rho_0 \bar{Y})^{-1} - 1\},
\end{aligned} \tag{3.13}
$$

with w, w_u and w_p formed by squaring. If, however, $i(\hat{\rho}) = j(\hat{\rho})$ is used in the last two statistics instead of $i(\rho_0)$, we obtain new versions $r'_u(\rho_0)$ and $r'_p(\rho_0)$ of the score and Wald statistics, where

$$
\begin{aligned}
r'_u(\rho_0) &= r_p(\rho_0), \\
r'_p(\rho_0) &= r_u(\rho_0),
\end{aligned}
$$

although this special relation is particular to the present example and parameterization; see *Further results and exercises* 3.3.

To study the relation between the various statistics (3.13) from first principles, note that $E(\bar{Y}) = 1/\rho_0$, $\operatorname{var}(\bar{Y}) = 1/(n\rho_0^2)$, and therefore we write $\rho_0 \bar{Y} = 1 + \tilde{Z}/\sqrt{n}$, where $\tilde{Z}$ is a random variable of zero mean and unit variance with an asymptotically normal distribution. Elementary expansions then show that under the null hypothesis

$$
\begin{aligned}
r(\rho_0) &= -\tilde{Z} + \tfrac{1}{3}\tilde{Z}^2/\sqrt{n} + O_p(n^{-1}), \\
r_u(\rho_0) &= -\tilde{Z}, \\
r_p(\rho_0) &= -\tilde{Z} + \tilde{Z}^2/\sqrt{n} + O_p(n^{-1}),
\end{aligned}
$$

showing the equivalence to first order of these test statistics and their asymptotically standard normal form. In this instance at least, $r(\rho_0)$ is likely to be intermediate between $r_u(\rho_0)$ and $r_p(\rho_0)$.

The calculations can be carried out in terms of the mean parameter $\mu = 1/\rho$ instead of the canonical parameter ρ. The statistics r and r_u are unchanged except for sign, whereas now $r_p = r_u$. If, as is natural in some circumstances, $\log \rho$ is taken as parameter, r_p becomes $-\sqrt{n}\log(\rho_0 \bar{Y})$.

Except for the disturbing lack of invariance of r_p under changes of parameterization, the relative advantages of the various statistics are unclear from the above discussion, although a more elaborate comparison with the 'exact' chi-squared distribution for $2n\rho_0 \bar{Y}$ would be possible. Application of the Cornish–Fisher inversion of the chi-squared distribution (Barndorff-Nielsen and Cox, 1989, p. 118) shows that

$$-\tilde{Z} + \frac{1}{3\sqrt{n}}(\tilde{Z}^2 - 1)$$

is close to having a standard normal distribution and this, except for a bias correction, matches most closely the signed likelihood ratio statistic $r(\rho_0)$, although this on its own is not an argument for preferring that statistic. Numerically the differences between $r(\rho_0)$, $r_u(\rho_0)$ and $r_p(\rho_0)$ will often be quite small. □

3.3 Distribution theory: no nuisance parameters

We now outline the asymptotic distribution theory that justifies the procedures sketched in the previous section.

A serious issue in principle concerns the asymptotic existence, uniqueness and consistency of the maximum likelihood estimate. Except in a prime exponential family setting, there are no very satisfactory general theorems on such questions and they are best approached on a case-by-case basis. A general result on the existence of a solution of the maximum likelihood estimating equation asymptotically close to the true parameter value is much easier to prove but is considerably less than is required. Here we assume throughout that $\hat{\theta}$ is well defined and consistent.

A key result is that in considerable generality $U = \nabla\ell(\theta; Y)$ is asymptotically normal with zero mean and variance $i(\theta)$. For independent component observations, or more generally for weakly dependent component observations, this follows from the additive form of U and a suitable central limit theorem. In particular for independent and identically distributed components it is a direct consequence of the simplest form of the central limit theorem and the assumed finiteness of var(U). Very generally, the asymptotic distribution is a consequence of the martingale property of the score vector.

More specifically, let $Y_{(1)}$, $Y_{(2)}, \ldots, Y_{(n)}$ be subsets of the full data set Y, the only restrictions on this specification being that $Y_{(i)}$ is a subset of $Y_{(j)}$ for $i < j$ and that $Y_{(n)} = Y$. We define $U_{(i)}$ by $U_{(i)} = \nabla\ell(\theta; Y_{(i)})$, where, as usual, $\ell(\theta; Y_{(i)})$ denotes the log-likelihood function for θ based on $Y_{(i)}$. Then $U_{(1)}, \ldots, U_{(n)}$ where $U_{(n)} = U$, form a martingale sequence under P_θ. This follows immediately from the general result that for any subset $Y_{(0)}$ of Y and with $U_{(0)} = \nabla\ell(\theta; Y_{(0)})$

we have

$$\begin{aligned}E_\theta\{U(\theta)\mid U_{(0)}(\theta)\} &= U_{(0)}(\theta) + E_\theta\{U(\theta) - U_{(0)}(\theta)\mid U_{(0)}(\theta)\}\\ &= U_{(0)}(\theta) + E_\theta\{\nabla \log p(Y;\theta\mid Y_{(0)})\mid U_{(0)}(\theta)\}\\ &= U_{(0)}(\theta) + E_\theta[E_\theta\{\nabla \log p(Y;\theta\mid Y_{(0)})\mid Y_{(0)}\}\mid U_{(0)}(\theta)]\\ &= U_{(0)}(\theta),\end{aligned} \tag{3.14}$$

where in the last step we have used formula (2.33).

The very general theory of (semi-)martingales and their central limit behaviour may thus be brought to bear on the question of the distribution of U and will often imply that the distribution is approximately normal, as above. However, by its very generality that theory yields little or no information on the accuracy of the normal approximation. For independent components a simpler form of the central limit theorem can be applied directly.

Suppose that $U(\theta) = U(\theta; Y) = [\ell_r(\theta)]$ has been shown to be asymptotically $N_d(0, i(\theta))$ distributed, i.e.

$$U(\theta) \mathrel{\overset{\sim}{\to}} N_d(0, i(\theta)). \tag{3.15}$$

To determine what this implies for the distribution of the maximum likelihood estimator $\hat{\theta}$, we move to superscript notation and expand the score $\ell_r(\theta)$ in a Taylor series around $\hat{\theta}$, writing in accord with our general convention

$$\begin{aligned}\ell_r(\theta) &= U_r(\theta) = \sqrt{n}\bar{\ell}_r(\theta) = \sqrt{n}\bar{U}_r(\theta),\\ \ell_{rs}(\theta) &= n\bar{\ell}_{rs}(\theta) = -j_{rs}(\theta),\\ \bar{\delta}^r &= \sqrt{n}(\hat{\theta}^r - \theta^r),\ \ell_{rst}(\theta) = n\bar{\ell}_{rst}(\theta),\\ i(\theta) &= n\bar{\imath}(\theta).\end{aligned}$$

Then $\ell_r(\hat{\theta}) = 0$ yields

$$\sqrt{n}\bar{\ell}_r(\theta) + n\ell_{rs}(\hat{\theta})\bar{\delta}^s/\sqrt{n} + n\bar{\ell}_{rst}(\hat{\theta})\bar{\delta}^s\bar{\delta}^t/n + \cdots = 0, \tag{3.16}$$

showing that we can to a first-order approximation ignore the third term and write

$$\begin{aligned}\bar{\delta}^r &= -\bar{\ell}^{rs}(\hat{\theta})\bar{\ell}_s(\hat{\theta}) + O_p(n^{-1/2})\\ &= \bar{\jmath}^{rs}(\hat{\theta})\bar{\ell}_s(\hat{\theta}) + O_p(n^{-1/2}).\end{aligned} \tag{3.17}$$

Because $\hat{j}^{rs}/i^{rs}$ converges in probability to one and $\bar{\ell}_s(\hat{\theta}) - \bar{\ell}_s(\theta) = O_p(n^{-1/2})$, it follows that

$$\bar{\delta}^r = \bar{\imath}^{rs}(\theta)\bar{\ell}_s(\theta) + O_p(n^{-1/2})$$

a linear function of asymptotically normal variables of zero mean. It follows from (3.15) that $\bar{\delta}^r$ is asymptotically normal with zero mean and covariance matrix $\bar{\imath}^{rv}\bar{\imath}_{vw}\bar{\imath}^{ws} = \bar{\imath}^{rs}$. In a notation without the explicit powers of n,

$$\hat{\theta} - \theta \simeq N_d(0, i^{-1}(\theta)). \tag{3.18}$$

There are two different steps involved in deriving (3.18): the local linearization leading to (3.17), which will fail if the log-likelihood is appreciably non-quadratic; and the replacement of observed information by expected information. This second step will be questionable if in repeated sampling the second derivative has large fluctuations around its expected value. It may then happen that the randomly normed deviation

$$\hat{j}^{\frac{1}{2}}(\hat{\theta} - \theta)$$

converges to the standardized normal distribution; see section 9.3.

A more rigorous discussion may proceed as follows, where for simplicity of notation we do not use the standardized versions $\bar{U}$, etc. Suppose it has been established, by some route, that as an underlying asymptotic index v tends to infinity we have $\hat{\theta} \xrightarrow{\text{p}} \theta$, where $\xrightarrow{\text{p}}$ denotes convergence in probability, and

$$U(\theta) \simeq \text{N}_d(0, i(\theta)),$$

$$j^{rs}(\hat{\theta}) i_{st}(\theta) \xrightarrow{\text{p}} \delta^r_t, \tag{3.19}$$

where δ^r_t denotes the Kronecker delta function. Using Taylor's theorem with a remainder term, we find

$$\ell_r(\theta) = \hat{j}_{rs}(\hat{\theta} - \theta)^s + H_{rs}(\hat{\theta} - \theta)^s, \tag{3.20}$$

where

$$H_{rs} = \int_0^1 \{ j_{rs}(\hat{\theta} + (\theta - \hat{\theta})\lambda) - j_{rs}(\hat{\theta}) \} \, d\lambda. \tag{3.21}$$

Equivalently,

$$(\sqrt{\hat{j}})^{rr'} \ell_{r'}(\theta) = (\sqrt{\hat{j}})_{r's}(\hat{\theta} - \theta)^s (\delta^{rr'} + K^{rr'})$$

where $\sqrt{\hat{j}}$ denotes a square root matrix of $\hat{j}$, $\delta^{rr'}$ is the Kronecker delta and

$$K^{rr'} = (\sqrt{\hat{j}})^{rr''} (\sqrt{\hat{j}})^{r's'} H_{r''s'}.$$

In view of the basic limit theorems set out, for instance, by Barndorff-Nielsen and Cox (1989, section 2.9), the asymptotic normality result (3.18) hinges on showing that.

$$K^{rr'} = o_p(1). \tag{3.22}$$

In fact, when that is the case the matrix $J^{rr'} = \delta^{rr'} + K^{rr'}$ can, except on a set whose probability tends to 0, be inverted and the inverse $J_{rr'}$ tends in probability to the identity matrix; thus

$$(\sqrt{\hat{j}})_{rs}(\hat{\theta} - \theta)^s = \delta_{rr'}(\sqrt{\hat{j}})^{r's}\ell_s + o_p(1) \tag{3.23}$$

and by (3.19) the vector whose rth coordinate is given by the right-hand side tends in distribution to $N_d(0, I)$.

To verify that $K^{rr'} = o_p(1)$ one may write H_{rs} as

$$H_{rs} = \int_0^1 \ell_{rst}(\tilde{\theta}(\lambda))\lambda d\lambda(\hat{\theta} - \theta)^t,$$

where $\tilde{\theta}(\lambda)$, which in fact depends on θ, $\hat{\theta}$, r and s, is a point on the line between θ and $\hat{\theta}$. Thus, since $(\hat{\theta} - \theta)^t \xrightarrow{p} 0$, for (3.22) and hence (3.18) to hold it suffices that there exists a neighbourhood $N(\theta)$ of θ such that

$$\sup_{\theta', \theta'' \in N(\theta)} |(\sqrt{j})^{ru}(\theta')(\sqrt{j})^{sv}(\theta')\ell_{uvt}(\theta'')| \leqslant M_t^{rs}(y), \tag{3.24}$$

where $M_t^{rs} = O_p(1)$. (The upper bound M_t^{rs} may depend on θ as well as on y.) In particular for scalar θ this requires local boundedness of $\{\partial^3\ell(\theta'')/\partial\theta^3\}/\{\partial^2\ell(\theta')/\partial\theta^2\}$. In the special case of Example 3.1 above, and in the notation of that example, the ratio is $-2\rho'^2/\rho''^3$ and the condition is satisfied.

The asymptotic normality relations (3.15) and (3.18) are asymptotically parameterization invariant. This means, in particular, that to show (3.15) for arbitrary parameterizations it is enough to do so for one parameterization, which may then be chosen so as to simplify the proof. For instance, for prime exponential models it is convenient to work with the mean value parameterization.

The asymptotic χ^2 behaviour of the log-likelihood ratio statistic $w = w(\theta) = 2\{\ell(\hat{\theta}) - \ell(\theta)\}$ is a simple consequence of the above considerations. In fact, by direct expansion in θ around $\hat{\theta}$ we find using (3.6) that

$$w(\theta) = \hat{j}_{rs}(\hat{\theta} - \theta)^r(\hat{\theta} - \theta)^s + o_p(1) \tag{3.25}$$

or equivalently

$$w(\theta) = i^{rs}\ell_r\ell_s + o_p(1), \tag{3.26}$$

whence $w \overset{\sim}{\to} \chi^2_d$. Alternative test statistics which, to first order, are asymptotically equal to w are the quadratic score statistic $w_u = i^{rs}\ell_r\ell_s$ and its variants $\hat{\imath}^{rs}\ell_r\ell_s$, $j^{rs}\ell_r\ell_s$ and $\hat{\jmath}^{rs}\ell_r\ell_s$, and the quadratic maximum likelihood statistic $w_p = i_{rs}(\hat{\theta}-\theta)^r(\hat{\theta}-\theta)^s$ and its variants $\hat{\imath}_{rs}(\hat{\theta}-\theta)^r(\hat{\theta}-\theta)^s$, $j_{rs}(\hat{\theta}-\theta)^r(\hat{\theta}-\theta)^s$ and $\hat{\jmath}_{rs}(\hat{\theta}-\theta)^r(\hat{\theta}-\theta)^s$. Note that of these only w and $i^{rs}\ell_r\ell_s$ are exactly parameterization invariant.

When the dimension d of θ is 1 we have by (3.26) that the directed likelihood

$$r = \operatorname{sgn}(\hat{\theta}-\theta)\sqrt{w(\theta)} \tag{3.27}$$

satisfies

$$r = \hat{\jmath}^{-\frac{1}{2}}U + o_p(1)$$

and hence

$$r \overset{\sim}{\to} N(0, 1). \tag{3.28}$$

To examine these results in more detail via higher-order theory the expansions, in particular (3.16), have to be taken to further terms, and this we study in Chapter 5.

3.4 Profile likelihood

Let $\psi = \psi(\theta)$ be a subparameter (or parameter function). The **profile likelihood** $L_P(\psi)$ for ψ is defined by

$$L_P(\psi) = \max_{\theta|\psi} L(\theta), \tag{3.29}$$

the maximum being over all θ that are consistent with the given value of ψ. We set $\ell_P = \log L_P$. Often ψ is a component of a given partition $\theta = (\psi, \chi)$ of θ into subvectors ψ and χ of dimension $d_\psi = d - d_\chi$ and d_χ, respectively, and one may then express the profile likelihood as

$$L_P(\psi) = L(\psi, \hat{\chi}_\psi),$$

where $\hat{\chi}_\psi$ denotes the maximum likelihood estimate of χ for a given value of ψ. Henceforth we assume this to be the case.

The profile likelihood may be used to illuminate various aspects of a full likelihood surface $L(\theta)$, for instance by plotting $L_P(\psi)$ for a variety of choices of the parameter function ψ. More importantly, $L_P(\psi)$ can to a considerable extent be thought of and used as if it was a genuine likelihood. In particular:

1. The maximum profile likelihood estimate of ψ equals $\hat{\psi}$.
2. The profile log-likelihood ratio statistic $2\{\ell_P(\hat{\psi}) - \ell_P(\psi_0)\}$ equals the log-likelihood ratio statistic for the hypothesis $\psi = \psi_0$, i.e.

$$2\{\ell_P(\hat{\psi}) - \ell_P(\psi_0)\} = 2\{\ell(\hat{\psi}, \hat{\chi}) - \ell(\psi_0, \hat{\chi}_0)\},$$

   where for simplicity of notation we write $\hat{\chi}_0$ for $\hat{\chi}_{\psi_0}$. Generally, we shall show that

$$w(\psi_0) = 2\{\ell_P(\hat{\psi}) - \ell_P(\psi_0)\} \overset{\sim}{\rightarrow} \chi^2_{d_\psi}, \tag{3.30}$$

   with in the scalar case, $d_\psi = 1$, a corresponding result for the directed likelihood.
3. A profile likelihood region

$$R_P = \{\psi : \ell_P(\hat{\psi}) - \ell_P(\psi) \leqslant c\}$$

   is, generally, an approximate confidence region for ψ, with approximate confidence level determined from (3.30). As for ordinary likelihood regions, the level is approximately 0.95, 0.99 and 0.999 according as $c = d_\psi + 1$, $d_\psi + 3$ and $d_\psi + 5$, provided $d_\psi \leqslant 5$.
4. The inverse of the observed profile information equals the ψ-aspect of the full observed inverse information evaluated at $(\psi, \hat{\chi}_\psi)$, i.e.

$$j_P^{-1}(\psi) = j^{\psi\psi}(\psi, \hat{\chi}_\psi) \tag{3.31}$$

   where j_P denotes observed profile information, that is minus the matrix of second-order derivatives of ℓ_P, and $j^{\psi\psi}$ is the $\psi\psi$-block of the inverse of the full observed information j.

For scalar ψ, formula (3.31) follows on differentiating $\ell_P(\psi) = \ell(\psi, \hat{\chi}_\psi)$ twice with respect to ψ and using

$$\partial\hat{\chi}_\psi / \partial\psi = -j_{\psi\chi}(\psi, \hat{\chi}_\psi) j_{\chi\chi}^{-1}(\psi, \hat{\chi}_\psi),$$

this latter formula being obtained by differentiating the partial likelihood equation $[\nabla_\chi \ell(\psi, \chi)]_{\chi = \hat{\chi}_\psi} = 0$ with respect to ψ. The vector parameter case follows similarly.

When ψ is scalar this implies that the curvature of the profile log-likelihood function is directly related to the precision of $\hat{\psi}$.

When there are no nuisance parameters a key property of the

log-likelihood function $\ell(\theta)$ is that the observed information $j(\hat{\theta}) = -[\nabla_\theta \nabla_\theta^{\mathrm{T}} \ell(\theta)]_{\theta=\hat{\theta}}$ can be used for the inverse asymptotic covariance matrix of $\hat{\theta}$. Now (3.31) shows that the corresponding function computed from the profile log-likelihood, namely

$$j_{\mathrm{P}}(\hat{\psi}) = -[\nabla_\psi \nabla_\psi^{\mathrm{T}} \ell_{\mathrm{P}}(\psi)]_{\psi=\hat{\psi}}$$

determines the inverse asymptotic covariance matrix for the component parameter ψ.

3.5 Further statistics: nuisance parameter case

For testing $\psi = \psi_0$ the forms of score statistic and maximum likelihood estimate statistic corresponding to the profile log-likelihood ratio statistic (3.30) are derived by first partitioning not only the maximum likelihood estimate but also the full score vector and the information matrix and its inverse

$$U(\theta) = \begin{pmatrix} U_\psi(\psi, \chi) \\ U_\chi(\psi, \chi) \end{pmatrix},$$

$$i(\theta) = \begin{pmatrix} i_{\psi\psi}(\psi, \chi) & i_{\psi\chi}(\psi, \chi) \\ i_{\chi\psi}(\psi, \chi) & i_{\chi\chi}(\psi, \chi) \end{pmatrix},$$

$$i^{-1}(\theta) = \begin{pmatrix} i^{\psi\psi}(\psi, \chi) & i^{\psi\chi}(\psi, \chi) \\ i^{\chi\psi}(\psi, \chi) & i^{\chi\chi}(\psi, \chi) \end{pmatrix}. \tag{3.32}$$

Because of the asymptotic normality of $\hat{\theta}$ with covariance matrix $i^{-1}(\theta)$, it follows under $\theta = \theta_0$ that $\hat{\psi}$ is asymptotically normally distributed with mean ψ_0 and covariance matrix $i^{\psi\psi}(\psi_0, \chi_0)$ which can be replaced by, for example, $i^{\psi\psi}(\psi_0, \hat{\chi}_0)$, $i^{\psi\psi}(\hat{\psi}, \hat{\chi})$, $j^{\psi\psi}(\psi_0, \hat{\chi}_0)$, $j^{\psi\psi}(\hat{\psi}, \hat{\chi})$ where $\hat{\chi}_0 = \hat{\chi}_{\psi_0}$, cf. in particular result (3.31). From the first of these forms we obtain

$$w_{\mathrm{p}}(\psi_0) = (\hat{\psi} - \psi_0)^{\mathrm{T}} \{ i^{\psi\psi}(\psi_0, \hat{\chi}_0) \}^{-1} (\hat{\psi} - \psi_0). \tag{3.33}$$

If we use as 'weighting matrix' $i^{\psi\psi}(\hat{\psi}, \hat{\chi})$, $j^{\psi\psi}(\psi_0, \hat{\chi}_0)$ or $j^{\psi\psi}(\hat{\psi}, \hat{\chi})$ the corresponding confidence regions are a set of similar ellipsoids centred on $\hat{\psi}$ because they are given by, for example,

$$\{ \psi; (\hat{\psi} - \psi)^{\mathrm{T}} \{ j^{\psi\psi}(\psi, \hat{\chi}_\psi) \}^{-1} (\hat{\psi} - \psi) \leqslant \chi^2_{d_\psi, \varepsilon} \}. \tag{3.34}$$

The development based on the score statistic is similar stemming from the asymptotic normality of $U_\psi(\psi_0, \chi_0)$ with zero mean and inverse covariance matrix $i^{\psi\psi}(\psi_0, \chi_0)$, so that, replacing χ_0 by a $\sqrt{n}$-consistent estimate, usually $\hat{\chi}_0$, we have

$$w_u(\psi_0) = U_\psi(\psi_0, \hat{\chi}_0)^T i^{\psi\psi}(\psi_0, \hat{\chi}_0) U_\psi(\psi_0, \hat{\chi}_0). \tag{3.35}$$

A particular computational advantage of this form is that χ has to be estimated only under the null hypothesis $\psi = \psi_0$.

To verify the first-order asymptotic properties of (3.33) and (3.35) it is necessary to show that when $\psi = \psi_0$, the statistics $w(\psi_0)$, $w_p(\psi_0)$ and $w_u(\psi_0)$ differ by $O_p(n^{-1/2})$ and that all have asymptotically a chi-squared distribution with d_ψ degrees of freedom. These results are set out in section 3.7.

The qualitative arguments in favour of $w(\psi_0)$, i.e. in favour of basing inference about ψ on the profile likelihood function to be given in section 4.2, remain essentially the same as when there are no nuisance parameters, although we shall see later that more refined arguments suggested some modification to the profile likelihood when nuisance parameters are present.

We now give some illustrations of these procedures.

Example 3.2 Normal mean

Let $Y_1, \ldots, Y_n$ be independently and identically distributed $N(\mu, \sigma^2)$ and let μ be the parameter of interest. The log-likelihood is, with $\tau = \sigma^2$,

$$\ell(\mu, \tau) = -\tfrac{1}{2} n \log \tau - \frac{\sum (Y_j - \mu)^2}{2\tau}. \tag{3.36}$$

It follows that $\hat{\tau}_\mu$, the maximum likelihood estimate of τ for given μ, is $\sum(Y_j - \mu)^2/n$; this is seen directly on differentiating (3.36), or alternatively from the exponential family form of the distribution, taken with the result that for prime exponential families the maximum likelihood estimates are achieved by equating canonical statistics to their expectation.

Consequently we have

$$\begin{aligned} \ell_p(\mu) &= \ell(\mu, \hat{\tau}_\mu) \\ &= -\tfrac{1}{2} n \log \frac{\sum (Y_j - \mu)^2}{n} - \tfrac{1}{2} n \\ &= -\tfrac{1}{2} n \log \left\{ \frac{\sum (Y_j - \bar{Y})^2 + n(\bar{Y} - \mu)^2}{n} \right\} - \tfrac{1}{2} n, \end{aligned} \tag{3.37}$$

so that, with $\hat{\mu} = \bar{Y}$,

$$\begin{aligned} w(\mu_0) &= 2\{\ell_P(\hat{\mu}) - \ell_P(\mu_0)\} \\ &= n \log\left\{1 + \frac{n(\bar{Y} - \mu_0)^2}{\sum(\bar{Y}_j - \bar{Y})^2}\right\} \\ &= n \log\left\{1 + \frac{t^2}{n-1}\right\}, \end{aligned} \tag{3.38}$$

where t denotes the standard Student t statistic with $n-1$ degrees of freedom.

There are a number of immediate generalizations of (3.37) and (3.38). Thus if the Y_j have a p-dimensional multivariate normal distribution with mean μ and covariance matrix Σ, then in (3.38) t^2 is replaced by Hotelling's T^2,

$$T^2 = (\bar{Y} - \mu_0)^{\mathrm{T}}\{S/(n-1)\}^{-1}(\bar{Y} - \mu_0),$$

where S is the matrix of sums and squares of products about the sample mean. In another generalization the Y_j are governed by a normal theory linear model and the parameter of interest is the vector of regression coefficients or a subvector thereof.

To calculate the statistics $w_u(\mu_0)$ and $w_p(\mu_0)$ for the single normal mean we evaluate the score function and information matrix,

$$U = \begin{pmatrix} \partial\ell/\partial\mu \\ \partial\ell/\partial\tau \end{pmatrix} = \begin{pmatrix} \sum(Y_j - \mu)/\tau \\ -\frac{1}{2}n/\tau + \sum(Y_j - \mu)^2/(2\tau^2) \end{pmatrix}, \tag{3.39}$$

$$i = \mathrm{diag}(n/\tau, n/(2\tau^2)). \tag{3.40}$$

Thus with $\hat{\mu} = \bar{Y}$, $\hat{\tau} = \sum(Y_j - \bar{Y})^2/n$ we have

$$w_u(\mu_0) = w_p(\mu_0) = (\bar{Y} - \mu_0)^2/(\tilde{\tau}/n),$$

where if i is evaluated at the maximum likelihood point then $\tilde{\tau} = \hat{\tau} = \sum(Y_j - \bar{Y})^2/n$, whereas if i is evaluated at the null hypothesis then $\tilde{\tau} = \hat{\tau}_0 = \sum(Y_j - \mu_0)^2/n$. If μ is replaced by the one-to-one function $h(\mu)$, then U and i are transformed respectively to

$$\begin{pmatrix} \sum(Y_j - \mu)/\{\tau h'(\mu)\} \\ -\frac{1}{2}n/\tau + \sum(Y_j - \mu)^2/\tau^2 \end{pmatrix}, \quad \mathrm{diag}\{n\tau^{-1}\{h'(\mu)\}^{-2}, n/(2\tau)\}.$$

Thus, provided the score and information are evaluated at the null hypothesis, w_u is unchanged. The simplest transformed version of

w_p is

$$\{h(\bar{Y}) - h(\mu_0)\}^2/[\{h'(\mu_0)\}^2\tau/n]. \qquad \square$$

Example 3.3 Non-standard normal parameters

The methods described above can be applied quite directly to less familiar problems connected with the normal distribution, such as inference about the parameter $\psi = \mu + \frac{1}{2}\sigma^2$. This arises naturally if it is required to estimate the mean of a log-normal distribution; if $Y = \log Z$ is $\mathrm{N}(\mu, \sigma^2)$, then $E(Z) = e^{\psi}$.

We write the log-likelihood (3.36) in terms of ψ and, say, τ, as

$$-\tfrac{1}{2}n\log\tau - \frac{\sum(Y_j - \psi + \frac{1}{2}\tau)^2}{2\tau} \tag{3.41}$$

so that for given ψ the maximum likelihood estimate of τ is

$$\hat{\tau}_\psi = 2\{1 + n^{-1}\textstyle\sum(Y_j - \psi)^2\}^{1/2} - 2 \tag{3.42}$$

and the profile likelihood for ψ is thus obtained by substitution in (3.41).

Of course in more complicated problems an explicit form for the profile likelihood may not be available.

Example 3.4 Generalized Pareto distribution

The generalized Pareto distribution has density and cumulative distribution function respectively

$$\sigma^{-1}(1 - \kappa y/\sigma)^{1/\kappa - 1}, \quad 1 - (1 - \kappa y/\sigma)^{1/\kappa}. \tag{3.43}$$

Here $\sigma > 0$ and κ are unknown parameters and the range of y is $\mathbb{R}_+ = \{y > 0\}$ if $\kappa < 0$ and is $\{0 < y < \sigma/\kappa = \tau\}$ if $\kappa > 0$. The case $\kappa = 0$, interpreted as a limit for $\kappa \to 0_-$, is the exponential density of mean σ. Given n independent observations the log-likelihood is simplest in terms of the parameters (τ, σ), with $\tau = \sigma/\kappa$, when

$$\ell(\tau, \sigma) = -n\log\sigma + (\tau/\sigma - 1)\textstyle\sum\log(1 - y_j/\tau). \tag{3.44}$$

Thus

$$U = \begin{pmatrix} U_\sigma \\ U_\tau \end{pmatrix} = \begin{pmatrix} -n/\sigma - \tau/\sigma^2\sum\log(1 - Y_j/\tau) \\ \sigma^{-1}\sum\log(1 - Y_j/\tau) + (\tau/\sigma - 1)\sum Y_j\tau^{-2}(1 - Y_j/\tau)^{-1} \end{pmatrix}. \tag{3.45}$$

Inference via profile log-likelihood is thus simplest for the parameter

τ because it follows immediately from (3.45) that, for given τ,

$$\hat{\sigma}_\tau = -n^{-1}\tau\sum\log(1 - Y_j/\tau).$$

Thus

$$\ell_{\mathrm{P}}(\tau) = -n\log\{-n^{-1}\tau\sum\log(1 - Y_j/\tau)\} - n - \tau\sum\log(1 - Y_j/\tau). \tag{3.46}$$

Care is needed in inference about τ because the range of y depends on τ if $\kappa > 0$; see section 3.8.

For parameters other than τ, explicit representation is not possible although the profile log-likelihood can be computed numerically.

□

Example 3.5 Some numerical examples
It is helpful in understanding the kind of conclusion, sometimes rather qualitative, that can be drawn from profile likelihood to see some particular cases. Figure 3.1 refers to problems in the analysis of survival data. For the original data and details of the analysis, see Cox and Oakes (1984, pp. 39, 42, 43, 138).

Figure 3.1(a) is a typical log-likelihood for the rate parameter of an exponential distribution, in fact derived from censored data. Note the asymmetry of the formal 95% likelihood-based confidence interval for ρ. Figures 3.1(b) and 3.1(c) refer to the same data, showing the formal 95% confidence region for the two parameters of a fitted Weibull distribution, the cumulative distribution function being taken in the form $1 - \exp\{-(\rho y)^\kappa\}$. Note that because of non-orthogonality $\hat{\rho}_\kappa$, the maximum likelihood estimate of ρ for given κ, varies quite strongly with κ; also possibly partly because of unsuitable parameterization, the formal 95% contour of constant log-likelihood is far from elliptical. Figure 3.1(c) shows the profile log-likelihood for κ; note that $\kappa = 1$, corresponding to an exponential distribution, is comfortably consistent with the data.

Finally, Fig. 3.1(d) refers to a much more complicated situation. In analyzing the possible effect on survival of a heart transplant, one possibility is that there is a transient effect on the hazard, involving a term proportional to $e^{-\gamma t}$, where t is time since transplant (days). A profile likelihood for γ is shown, together with the maximized log-likelihood without the transient term. The detailed shape of the profile log-likelihood is curious, but the key point for interpretation is that the total change of the log-likelihood over the whole range

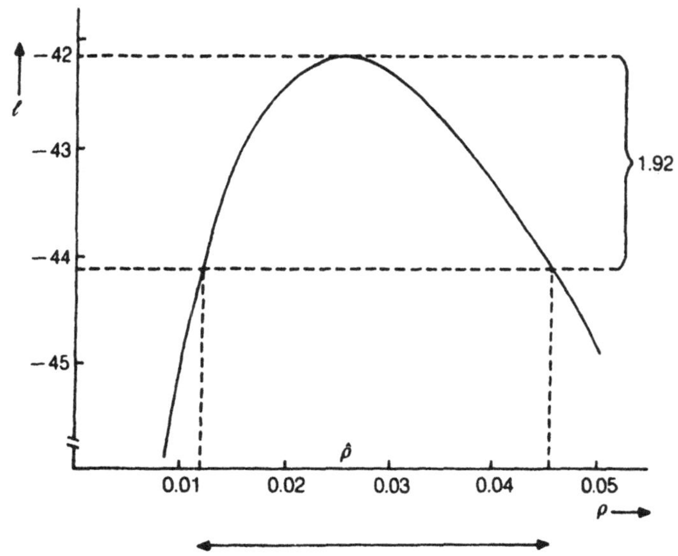
-42
-43
-44
-45
ℓ
1.92
ρ̂
0.01
0.02
0.03
0.04
0.05
ρ

(a)

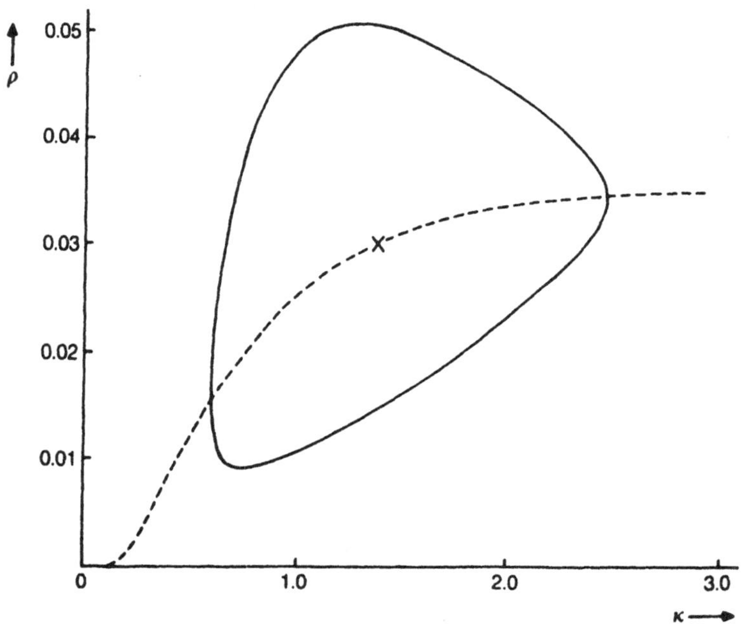
0.05
0.04
0.03
0.02
0.01
ρ
0
1.0
2.0
3.0
κ

(b)

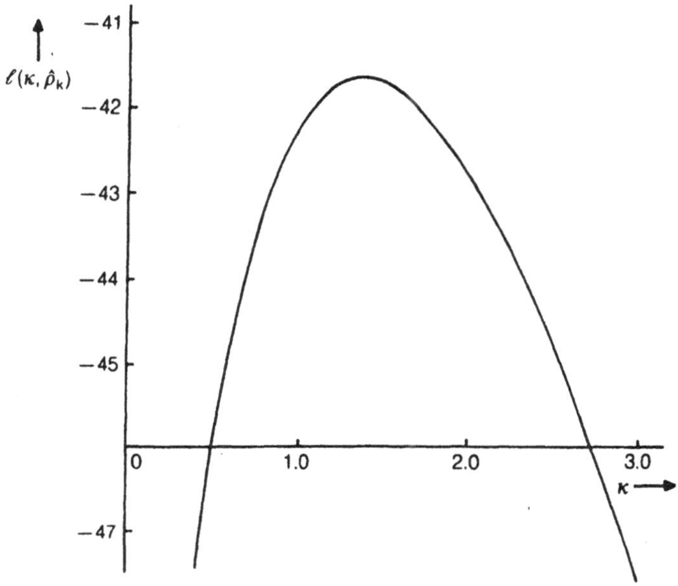

(c)

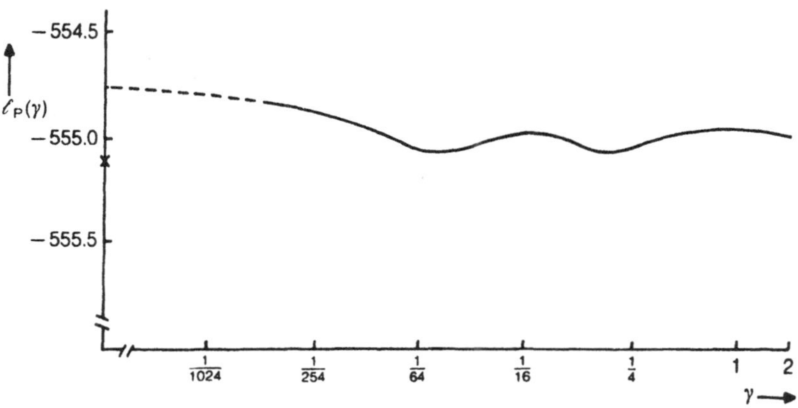

(d)

Figure 3.1 *Leukaemia data, 6-MP group. (a) Log-likelihood for rate parameter ρ of exponential distribution, maximum likelihood estimate and formal 95% likelihood-based confidence interval. (b) 95% formal contour of log-likelihood for parameters (ρ, κ) of Weibull distribution, and $\hat{\rho}_k$ as a function of κ and overall maximum likelihood point. (c) Profile likelihood for κ. (d) Heart transplant data. Profile log-likelihood for parameter γ of transient term in hazard. $\times$, transient term absent.*

is less than 0.5 and thus provides no evidence of a transient term of this form, even if the conventional chi-squared test is quite far from appropriate. This is an instance where inspection of the profile log-likelihood shows qualitatively that the data carry little information on the parameter under study. □

3.6 Effects of parameter orthogonality

In this section we assume that it is possible to make the parameter of interest ψ and the nuisance parameter, now to be denoted by λ, orthogonal. This is always possible if ψ is one-dimensional, cf. section 2.7. Any transformation from, say, (ψ, χ) to (ψ, λ) necessary to achieve this leaves the profile log likelihood, $\ell_P(\psi)$, invariant.

The first consequence of orthogonality follows from the block diagonal form of the matrices $i(\psi, \lambda)$ and $i^{-1}(\psi, \lambda)$. This implies that $\hat{\psi}$ and $\hat{\lambda}$ are asymptotically independent and that the asymptotic variance of $\hat{\psi}$ where λ is unknown is the same as that where λ known. A related aspect is that $\hat{\lambda}_\psi$, the maximum likelihood estimate of λ for specified ψ, varies only slowly with ψ in the neighbourhood of $\hat{\psi}$, and that there is a corresponding slow variation of $\hat{\psi}_\lambda$ with λ. More precisely if $\psi - \hat{\psi} = O_p(n^{-1/2})$, then $\hat{\lambda}_\psi - \hat{\lambda} = O_p(n^{-1})$. For a nonorthogonal nuisance parameter χ, we would have correspondingly only $\hat{\chi}_\psi - \hat{\chi} = O_p(n^{-1/2})$. When both parameter of interest and nuisance parameter are scalar, these results follow by expression of the log-likelihood in a $n^{-1/2}$ neighbourhood around $(\hat{\psi}, \hat{\chi})$ in the form

$$\ell(\psi, \chi) = \ell(\hat{\psi}, \hat{\chi}) - \tfrac{1}{2}\{\hat{j}_{\psi\psi}(\psi - \hat{\psi})^2 + 2\hat{j}_{\psi\chi}(\psi - \hat{\psi})(\chi - \hat{\chi}) + \hat{j}_{\chi\chi}(\chi - \hat{\chi})^2\} + O_p(n^{-1/2}).$$

It now follows that

$$\begin{aligned}\hat{\chi}_\psi - \hat{\chi} &= -\frac{\hat{j}_{\psi\chi}}{\hat{j}_{\chi\chi}}(\psi - \hat{\psi}) + O_p(n^{-1}) \\ &= -\frac{i_{\psi\chi}}{i_{\chi\chi}}(\psi - \hat{\psi}) + O_p(n^{-1}). \end{aligned} \tag{3.47}$$

Thus because $\psi - \hat{\psi} = O(n^{-1/2})$, $\hat{\chi}_\psi - \hat{\chi} = O_p(n^{-1/2})$ unless $i_{\psi\chi} = 0$, the orthogonal case, where the difference is $O_p(n^{-1})$. The proof for vector parameters is essentially the same. The orders of magnitude can be checked most explicitly by introducing standardized quantities in the usual way.

Note, in particular, that so far as asymptotic theory is concerned we can have $\hat{\chi}_\psi = \hat{\chi}$ independently of ψ only if χ and ψ are orthogonal. In this very special case we can write $\ell_P(\psi) = \ell(\psi, \hat{\lambda})$, whereas in the general orthogonal case

$$\ell_P(\psi) = \ell(\psi, \hat{\lambda}) + o_p(1), \tag{3.48}$$

so that for first-order theory we could use $\ell_P^*(\psi) = \ell(\psi, \hat{\lambda})$ instead of $\ell(\psi, \hat{\lambda}_\psi)$.

If, temporarily, we denote the true value of λ by λ_0, we have in the orthogonal case that in testing $\psi = \psi_0$

$$\begin{aligned} w(\psi_0) &= 2\{\ell_P(\hat{\psi}) - \ell_P(\psi_0)\} \\ &= 2\{\ell(\hat{\psi}, \hat{\lambda}) - \ell(\psi_0, \hat{\lambda}_0)\} \\ &= 2\{\ell(\hat{\psi}, \lambda_0) - \ell(\psi_0, \lambda_0)\} + o_p(1) \qquad (3.49) \\ &= 2\{\ell(\hat{\psi}_0, \lambda_0) - \ell(\psi_0, \lambda_0)\} + o_p(1) \qquad (3.50) \end{aligned}$$

where $\hat{\psi}_0 = \hat{\psi}_{\lambda_0}$. Here (3.49) follows by Taylor expansion and (3.50) from the result that $\hat{\psi} - \hat{\psi}_0$ is $O_p(n^{-1})$. The approximate equivalence of $w(\psi_0)$ to the statistic for known λ_0 gives some intuitive support to $w(\psi_0)$ as a test statistic.

Example 3.6 Some normal theory problems

The simplest illustration of the results concerning the dependence of $\hat{\psi}_\lambda$ on λ is obtained by writing the model for normal theory simple linear regression, in unfamiliar notation, in the two forms

$$E(Y_i) = \lambda + \psi(x_i - \bar{x}), \qquad E(Y_i) = \chi + \psi x_i$$

for $i = 1, \ldots, n$, where $\bar{x} = \sum x_i/n$, and we suppose that $\bar{x} \neq 0$. Then $\hat{\lambda}_\psi = \bar{Y} = \sum Y_i/n$, independently of ψ, whereas $\hat{\chi}_\psi = \bar{Y} - \psi\bar{x}$. Because $\psi - \hat{\psi} = O_p(n^{-1/2})$, $\hat{\chi}_\psi$ varies by $O_p(n^{-1/2})$ around $\hat{\chi}$. The notion of orthogonality used in the literature on experimental design specializes this. Thus in a balanced randomized block design, 'blocks' and 'treatments' are orthogonal in the sense that if the 'block' parameters are assigned particular values, e.g. zero, the least squares estimates of the treatment effects are unchanged.

A different normal theory illustration is provided by Example 3.2, a random sample from a normal distribution of mean μ and variance τ. Here μ and τ are orthogonal; note that μ is a component of the mean parameter and $1/\tau$ the complementary component of the canonical parameter, cf. Section 2.9. Here $\hat{\mu}_\tau = \hat{\mu} = \bar{Y}$, whereas

$$\hat{\tau}_\mu = n^{-1}\sum(Y_i - \mu)^2 = \hat{\tau} + (\bar{Y} - \mu)^2.$$

When we consider values of μ within $O(n^{-1/2})$ of $\bar{Y}$, $\hat{\tau}_\mu = \hat{\tau} + O(n^{-1})$.

□

3.7 Distribution theory: nuisance parameter case

We now consider the first order asymptotic distribution theory when nuisance parameters are present. For this it is convenient first to review some properties of the multivariate normal distribution.

Suppose that V is a $p \times 1$ random vector with a multivariate normal distribution of zero mean and covariance matrix Σ. It suffices to consider the non-singular case. Call $Q_V = V^{\mathrm{T}}\Sigma^{-1}V$ the **covariance form** associated with V. Then Q_V has the chi-squared distribution with p degrees of freedom, $Q_V \sim \chi^2_{\mathrm{p}}$. This can be seen with minimal calculation because

1. the covariance form is invariant under non-singular transformation of V;
2. the random variable V can be transformed to independent components of unit variance;
3. the chi-squared distribution then follows directly, with its obvious geometrical interpretation.

Now suppose that V is partitioned into two parts $V^{\mathrm{T}} = (V^{\mathrm{T}}_{(1)}, V^{\mathrm{T}}_{(2)})$ where $V_{(j)}$ is $p_j \times 1, p_1 + p_2 = p$. It follows directly from the above that $Q_{V_{(1)}} \sim \chi^2_{\mathrm{p}_1}$ but in addition

$$Q_{V_{(2)}\cdot V_{(1)}} = Q_V - Q_{V_{(1)}} \sim \chi^2_{\mathrm{p}_2} \tag{3.51}$$

independently of $Q_{V_{(1)}}$. This can be seen by extending the argument that $Q_V \sim \chi^2_{\mathrm{p}}$. Apply a linear transformation to V so that the first p_1 components are $V_{(1)}$ and the last p_2 components $V'_{(2)}$, say, are independent of $V_{(1)}$. Then by the invariance of the covariance form under non-singular transformation of V,

$$Q_V = Q_{V_{(1)}} + Q_{V'_{(2)}}, \tag{3.52}$$

so that $Q_{V'_{(2)}} = Q_{V_{(2)}\cdot V_{(1)}}$ and the stated properties of $Q_{V'_{(2)}}$ clearly hold. Note that because of the independence $Q_{V_{(2)}\cdot V_{(1)}}$ can be regarded as determining the conditional distribution of Q_V given $V_{(1)}$.

More generally, we can divide V into any number $s, s \leqslant p$, parts, $V^{(\mathrm{T})} = (V_{(1)}^{\mathrm{T}}, V_{(2)}^{\mathrm{T}}, \ldots, V_{(s)}^{\mathrm{T}})$ and

$$Q_{V_{(1)}}, Q_{V_{(2)} \cdot V^{(1)}}, \ldots, Q_{V_{(s)} \cdot V^{(s-1)}}$$

are independently $\chi^2_{p_1}, \chi^2_{p_2}, \ldots$, where $V^{(k)}$ denotes $(V_{(1)}, \ldots, V_{(k)})$.

For the application to the first-order theory of the distribution of the test statistic w, we use asymptotic versions of these normal theory results, based on random variables which are asymptotically normal with covariance matrix consistently estimated and using statistics that are asymptotically quadratic forms.

Now the statistic $w(\psi_0)$ of (3.30) can be written in the form

$$w(\psi_0) = 2\{\ell(\hat{\psi}, \hat{\chi}) - \ell(\psi_0, \chi)\} - 2\{\ell(\psi_0, \hat{\chi}_0) - \ell(\psi_0, \chi)\}, \quad (3.53)$$

i.e. as the difference of two statistics for testing hypotheses without nuisance parameters. By the discussion of Section 3.3, these are respectively equivalent asymptotically to the covariance form of the full score statistic U and to the score statistic of the part U_χ, i.e.

$$w(\psi_0) \sim Q_U - Q_{U_\chi} = Q_{U_\psi \cdot U_\chi} \quad (3.54)$$

which is thus asymptotically $\chi^2_{d_\psi}$, incidentally independently of U_χ and hence of $\hat{\chi}_0$.

Note also that if $d_\psi = 1$ the same argument shows that the directed likelihood

$$r(\psi_0) = \operatorname{sgn}(\hat{\psi} - \psi_0)\sqrt{w(\psi_0)} \quad (3.55)$$

has an asymptotic standard normal distribution.

A corollary of the above argument is that if for a general parameter $\theta = (\theta_1, \ldots, \theta_d)$ we consider a sequence of directed likelihood statistics $r_1, \ldots, r_d$ for $\theta_1 = \theta_{1,0}, \theta_2 = \theta_{2,0}$ given the value of $\theta_1, \theta_3 = \theta_{3,0}$ given the value of $\theta_{(2)} = (\theta_1, \theta_2), \ldots, \theta_d = \theta_{d0}$ given the value of $\theta_{(d-1)} = (\theta_1, \ldots, \theta_{d-1})$, then $r_1, \ldots, r_d$ are asymptotically independently standard normal.

Discussion of higher-order terms to be given in Chapters 5 and 6 will show that $w(\psi_0) \sim \chi^2_{d_\psi} + O_p(n^{-1})$.

The statistic $w_p(\psi_0)$ is based directly on the asymptotic covariance form of $\hat{\psi} - \psi_0$ and so is asymptotically $\chi^2_{d_\psi}$. To see that it is equivalent to $w(\psi_0)$, we can argue indirectly that both $w_p(\psi_0)$ and $w(\psi_0)$ are asymptotically the same covariance form, namely of d_ψ linear functions of $U(\psi_0)$. A similar straightforward argument can be used for $w_u(\psi_0)$.

The essential equivalence of these various statistics is sufficiently important that it is worth setting out the relations involved explicitly in terms of partitioned matrices. For this we write the linearized form of the maximum likelihood estimating equations in the form

$$\begin{pmatrix} i_{\psi\psi} & i_{\psi\chi} \\ i_{\chi\psi} & i_{\chi\chi} \end{pmatrix}\begin{pmatrix} \hat{\psi} - \psi_0 \\ \hat{\chi} - \chi \end{pmatrix} = \begin{pmatrix} U_\psi \\ U_\chi \end{pmatrix}, \quad \begin{pmatrix} \hat{\psi} - \psi_0 \\ \hat{\chi} - \chi \end{pmatrix} = \begin{pmatrix} i^{\psi\psi} & i^{\psi\chi} \\ i^{\chi\psi} & i^{\chi\chi} \end{pmatrix}\begin{pmatrix} U_\psi \\ U_\chi \end{pmatrix},$$

$$\hat{\chi}_0 - \chi = i^{\chi\chi} U_\chi,$$

$$\operatorname{cov}\begin{pmatrix} \hat{\psi} - \psi_0 \\ \hat{\chi} - \chi \end{pmatrix} = \begin{pmatrix} i^{\psi\psi} & i^{\psi\chi} \\ i^{\chi\psi} & i^{\chi\chi} \end{pmatrix}, \quad \operatorname{cov}\begin{pmatrix} U_\psi \\ U_\chi \end{pmatrix} = \begin{pmatrix} i_{\psi\psi} & i_{\psi\chi} \\ i_{\chi\psi} & i_{\chi\chi} \end{pmatrix}. \tag{3.56}$$

Further, to this order,

$$\begin{aligned} U_\psi(\psi_0, \hat{\chi}_0) &= U_\psi + (\partial U_\psi / \partial \chi)(\hat{\chi}_0 - \chi) \\ &= U_\psi - i_{\psi\chi}(\hat{\chi}_0 - \chi). \end{aligned} \tag{3.57}$$

We can now express the three statistics $w(\psi_0), w_{\mathrm{u}}(\psi_0), w_{\mathrm{p}}(\psi_0)$ in terms of the score vector U. Thus, always to first order,

$$w(\psi_0) = (U_\psi^{\mathrm{T}} U_\chi^{\mathrm{T}})\begin{pmatrix} i^{\psi\psi} & i^{\psi\chi} \\ i^{\chi\psi} & i^{\chi\chi} \end{pmatrix}\begin{pmatrix} U_\psi \\ U_\chi \end{pmatrix} - U_\chi^{\mathrm{T}} i_{\chi\chi}^{-1} U_\chi, \tag{3.58}$$

$$w_{\mathrm{u}}(\psi_0) = (U_\psi - i_{\psi\chi} i^{\chi\chi} U_\chi)^{\mathrm{T}} i^{\psi\psi} (U_\psi - i_{\psi\chi} i^{\chi\chi} U_\chi), \tag{3.59}$$

$$w_{\mathrm{p}}(\psi_0) = (i^{\psi\psi} U_\psi + i^{\psi\chi} U_\chi)^{\mathrm{T}} (i^{\psi\psi})^{-1} (i^{\psi\psi} U_\psi + i^{\psi\chi} U_\chi). \tag{3.60}$$

On simplifying these expressions via standard results for partitioned matrices, in particular

$$\begin{pmatrix} i^{\psi\psi} & i^{\psi\chi} \\ i^{\chi\psi} & i^{\chi\chi} \end{pmatrix} = \begin{pmatrix} i^{\psi\psi} & i^{\psi\chi} \\ i^{\chi\psi} & i^{\chi\psi} i_{\psi\psi}^{-1} i^{\psi\chi} \end{pmatrix} + \begin{pmatrix} 0 & 0 \\ 0 & i_{\chi\chi}^{-1} \end{pmatrix}, \tag{3.61}$$

the required identities follow. The identity (3.61) corresponds to the partition $Q_U = Q_{U_\psi \cdot U_\chi} + Q_{U_\chi}$.

Quite often a null hypothesis for the parameter $\theta = (\theta_1, \ldots, \theta_d)$ is expressed via constraints, either in the form

$$h_1(\theta) = \cdots = h_{d_0}(\theta) = 0 \tag{3.62}$$

or in the form of 'freedom' equations

$$\theta_j = \theta_j(\alpha_1, \ldots, \alpha_{d - d_0}). \tag{3.63}$$

In principle it is possible to calculate, for example, a likelihood ratio statistic by reparameterizing so that $\theta = (\psi, \chi)$ with the null hypothesis taking the form $\psi = \psi_0$. It is, however, sometimes more convenient to work directly from the forms (3.62) and (3.63).

For (3.62) a natural analogue of the maximum likelihood estimating statistic, w_p, can be obtained by first finding $\hat{\theta}$, the unrestricted maximum likelihood statistic and its asymptotic covariance matrix, for example in the form $\{j(\hat{\theta})\}^{-1}$. Then under the null hypothesis the vector

$$\hat{h} = \{h_1(\hat{\theta}), \ldots, h_{d_0}(\hat{\theta})\}^{\mathrm{T}}$$

is asymptotically normal with mean zero and covariance matrix

$$\hat{\Sigma}_h = \hat{\delta}\{j(\hat{\theta})\}^{-1}\hat{\delta}^{\mathrm{T}}, \tag{3.64}$$

where $\hat{\delta}^{\mathrm{T}} = \nabla_\theta \hat{h}^{\mathrm{T}}$ is the $d \times d_0$ matrix whose (r, s)th element is $\partial h_s / \partial \theta_r$ evaluated at $\theta = \hat{\theta}$ (cf. Barndorff-Nielsen and Cox, 1989, Section 2.7). This leads essentially to one form of the w_p statistic that would be obtained by reparameterizing so that $\psi = h$, i.e. to

$$\hat{h}^{\mathrm{T}}\hat{\Sigma}_h^{-1}\hat{h} \tag{3.65}$$

to be tested as chi-squared with d_0 degrees of freedom. The test is not exactly invariant under nonlinear transformations of the specification of the constraints.

Another approach involves maximization of the log-likelihood subject to the constraints introduced via Lagrange multipliers. The gradient vector $\nabla_\theta \ell$ evaluated at the constrained maximum $\tilde{\theta}$ will not vanish and a suitable squared norm corresponds to the score statistic introduced in (3.35).

The Lagrangian is $\ell(\theta) + \phi^{\mathrm{T}} h(\theta)$, where ϕ is a $d_0 \times 1$ vector and the maximum thus satisfies

$$\begin{aligned} \nabla_\theta \ell(\tilde{\theta}) + \tilde{\delta}^{\mathrm{T}}\tilde{\phi} &= 0; \\ h(\tilde{\theta}) &= 0, \end{aligned} \tag{3.66}$$

where $\tilde{\delta}^{\mathrm{T}} = \nabla_\theta h^{\mathrm{T}}$ evaluated at $\theta = \tilde{\theta}$. Thus a quadratic form in $\nabla_\theta \ell(\tilde{\theta})$ is asymptotically equivalent to one in $\tilde{\phi}$ and for that reason the test is sometimes called the Lagrange multiplier test.

It can be shown by the methods used above that $\tilde{\theta} - \theta_0$ and $\tilde{\phi}$ are $O_p(n^{-1/2})$ and therefore local linearization of (3.66) yields

$$\begin{pmatrix} i(\theta_0) & -\delta_0^{\mathrm{T}} \\ -\delta_0 & 0 \end{pmatrix} \begin{pmatrix} \tilde{\theta} - \theta_0 \\ \tilde{\phi} \end{pmatrix} = \begin{pmatrix} U(\theta_0) \\ 0 \end{pmatrix}, \tag{3.67}$$

where θ_0 is the true value of θ and $\delta_0^{\mathrm{T}} = \nabla_\theta h^{\mathrm{T}}$ evaluated at $\theta = \theta_0$. Thus

$$\begin{pmatrix} \tilde{\theta} - \theta_0 \\ \tilde{\phi} \end{pmatrix} = \begin{pmatrix} a & b^{\mathrm{T}} \\ b & c \end{pmatrix} \begin{pmatrix} U(\theta_0) \\ 0 \end{pmatrix}, \tag{3.68}$$

where a, b, c are component matrices of the inverse of the partitioned matrix in (3.67). Because $U(\theta_0)$ has covariance matrix $i(\theta_0)$ it follows that $\tilde{\theta}$ and $\tilde{\phi}$ are asymptotically normal with means θ_0 and 0 and with covariance matrix

$$\begin{pmatrix} ai(\theta_0)a^{\mathrm{T}} & ai(\theta_0)b^{\mathrm{T}} \\ bi(\theta_0)a^{\mathrm{T}} & bi(\theta_0)b^{\mathrm{T}} \end{pmatrix}.$$

It follows from the detailed form of a, b, c that

$$\begin{aligned} ai(\theta_0)b^{\mathrm{T}} &= 0, \\ bi(\theta_0)b^{\mathrm{T}} &= -c = [\delta_0\{i(\theta_0)\}^{-1}\delta_0^{\mathrm{T}}]^{-1}. \end{aligned} \tag{3.69}$$

Thus the standard quadratic statistic or covariance form associated with $\tilde{\phi}$ is

$$\tilde{\phi}^{\mathrm{T}}\delta_0\{i(\theta_0)\}^{-1}\delta_0^{\mathrm{T}}\tilde{\phi} \tag{3.70}$$

and this is asymptotically equivalent to

$$\{\nabla_\theta \ell(\tilde{\theta})\}^{\mathrm{T}}\{i(\theta_0)\}^{-1}\{\nabla_\theta \ell(\tilde{\theta})\} \tag{3.71}$$

by (3.69). Again $i(\theta_0)$ can be replaced by $i(\tilde{\theta})$ or $j(\tilde{\theta})$ and δ_0 by $\tilde{\delta}$. These statistics are under the null hypothesis asymptotically chi-squared with d_0 degrees of freedom. In the special case when $\theta = (\psi, \chi)$ and the null hypothesis expressed in the constraints (3.62) is $\psi = \psi_0$, we have $\tilde{\theta} = (\psi_0, \hat{\chi}_{\psi_0})$ and we recover the previous results for the score statistic, with $d_0 = d_\psi$.

A further application of these first-order asymptotic results is particularly relevant to tests of model adequacy. Suppose that $T_n = t_n(Y, \theta)$ is for large n asymptotically normal with mean zero for all θ and with variance of order n^{-1}. We consider the distribution of $\hat{T}_n = t_n(Y, \hat{\theta})$, where $\hat{\theta}$ is the maximum likelihood estimate or more generally any estimate equivalent to the maximum likelihood estimate to the first order of asymptotic theory. A quite general discussion is possible so long as the asymptotic joint distribution of T_n and $\hat{\theta}$ is multivariate normal. In this case $\hat{T}_n$ behaves like T_n adjusted for linear regression on $\hat{\theta}$.

To see this, we assume that $(\sqrt{n}T_n, \sqrt{n}(\hat{\theta} - \theta))$ has a limiting normal distribution of zero mean and non-degenerate covariance matrix. Also we assume that, corresponding to a first order Taylor expansion, we can write

$$\sqrt{n}\hat{T}_n = \sqrt{n}T_n + \sqrt{n}b^{\mathrm{T}}(\hat{\theta} - \theta) + o_{\mathrm{p}}(1), \tag{3.72}$$

where b is a $d_\theta \times 1$ vector; typically

$$b = \lim_{n \to \infty} b_n = E\{\nabla_\theta t_n(Y, \theta)\}. \tag{3.73}$$

Now $\hat{T}_n$ has asymptotic expectation zero and hence to any component of $\hat{\theta}$ we can add any multiple of $\hat{T}_n$ and still obtain an estimate of θ. But the asymptotic variances of the components of $\hat{\theta}$ are a minimum so that $\sqrt{n}\hat{T}_n$ and $\sqrt{n}(\hat{\theta} - \theta)$ are asymptotically independent. Thus if we rewrite (3.72) as

$$\sqrt{n}T_n = \sqrt{n}\hat{T}_n - \sqrt{n}b^{\mathrm{T}}(\hat{\theta} - \theta) + o_{\mathrm{p}}(1),$$

the two terms on the right-hand side are asymptotically independent so that asymptotically

$$\mathrm{var}(\sqrt{n}\hat{T}_n) = \mathrm{var}(\sqrt{n}T_n) - \mathrm{var}\{\sqrt{n}b^{\mathrm{T}}(\hat{\theta} - \theta)\},$$

i.e.

$$\mathrm{var}(\hat{T}_n) \sim \mathrm{var}(T_n) - b^{\mathrm{T}}i^{-1}(\theta)b, \tag{3.74}$$

where $i(\theta)$ is the information matrix for θ.

Example 3.7 Skewness of regression residuals

Consider the standardized sum of cubes of residuals from a normal theory linear regression model. For simplicity we take the case with a single explanatory variable, writing

$$E(Y_j) = \alpha + \beta z_j \qquad (j = 1, \ldots, n),$$

where, without loss of generality, we may suppose $\sum z_j = 0$. The vector of unknown parameters is $\theta^{\mathrm{T}} = (\alpha, \beta, \sigma^2)$, where σ^2 is the error variance. Thus $i(\theta) = \mathrm{diag}(n/\sigma^2, \sum z_j^2/\sigma^2, \frac{1}{2}n/\sigma^4)$.

Consider

$$\sqrt{n}T_n = n^{-1/2}\sum(Y_j - \alpha - \beta z_j)^3/\sigma^3, \tag{3.75}$$

Now a direct application of the central limit theorem shows that $\sqrt{n}T_n$ is asymptotically normal with zero mean and variance 15, the sixth moment of a standard normal variable. We have from (3.73) that $b^{\mathrm{T}} = (-3/\sigma, 0, 0)$ so that by (3.74) asymptotically

$$\mathrm{var}(\sqrt{n}\hat{T}_n) = (15 - 9)/n = 6/n. \tag{3.76}$$

Precisely the same result holds when there are a number of explanatory variables in the linear model, as is most easily seen by taking the explanatory variables in orthogonalized form. For the exact variances, see Anscombe (1961); the treatment here follows Pierce (1982). □

3.8 Non-standard cases

The discussion earlier in this chapter and throughout most of the book assumes regularity in the sense that local expansions can be carried out to whatever order is required, that the cumulants of log-likelihood derivatives are asymptotically proportional to a notional sample size (amount of information), that expectations (whether integrals or infinite sums) can be differentiated under the expectation sign with respect to parameters and that distinct parameter values represent distinct distributions. In multi-parameter problems it may well happen that the problem is irregular only as regards certain components.

We shall not attempt a full treatment of the difficult issues that arise when some or all of these conditions fail.

One set of problems is typified in extreme form by independent values from rectangular distributions over say $(0, \theta)$ or over $(\theta, \theta + 1)$ or over (θ_1, θ_2). More generally, there is a fixed distribution $g(y)$ over the interval $(0, 1)$ converted into a scale or location or scale and location family. For example, in the density $\theta^{-1} g(y/\theta)$ over $(0, \theta)$, with $g(1) > 0$, generalizing Example 2.17, it is easy to show that the largest observation differs from θ by $O_p(n^{-1})$ and that the limiting distribution of the largest value is exponential. Failure of the standard results stems from the impossibility, noted following (2.34), of differentiating under the expectation sign, leading in particular to $E\{U(\theta)\} \neq 0$. See also *Further results and exercises* 3.16.

In general when the support of a distribution depends on an unknown parameter some but not all of the information is contributed by the extreme observations. The following very informal discussion illustrates the ideas involved.

Example 3.8 A general family with parameter-dependent support
Suppose that $Y_1, \ldots, Y_n$ are independent and identically distributed with density $g(y; \theta)$ $(0 < y < \theta)$, The log-likelihood is $\sum \log g(y_j; \theta)$

provided $m = \max(y_1, \ldots, y_n) < \theta$. Now the density of M is

$$ng(m;\theta)\{G(m;\theta)\}^{n-1}, \tag{3.77}$$

where G is the distribution function corresponding to g. Suppose that locally near θ

$$g(m;\theta) \sim g_1(\theta - m)^a, \quad G(m;\theta) = 1 - g_1(\theta - m)^{a+1}/(a+1), \tag{3.78}$$

where we shall be interested in $a \geqslant 0$. Then the density of M for large n is

$$ng_1(\theta - m)^a \exp\{-ng_1(\theta - m)^{a+1}/(a+1)\}, \tag{3.79}$$

a familiar derivation of the Weibull distribution as one of the extreme value laws; it follows that $M - \theta = O_p(n^{-1/(a+1)})$. In particular, for the rectangular distribution on $(0, \theta)$ in which $a = 0$, $M - \theta = O_p(n^{-1})$, whereas if $a = 1$, when the density, while zero at the upper terminal, has positive slope there, $M - \theta = O_p(n^{-1/2})$ and if $a > 1$, $M - \theta = o_p(n^{-1/2})$. We can now divide the information about θ into that arising from M and that arising from the conditional log-likelihood given $M = m$, namely

$$\sum{}' \log g(y_j;\theta) - (n-1)\log G(m;\theta) \quad (y_j \leqslant m), \tag{3.80}$$

where the sum is over the observations other than the largest. Now this in general is a well-behaved log-likelihood and, unless it is vacuous, as it is in the rectangular case, will give a (partial) maximum likelihood estimate of θ with error $O_p(n^{-1/2})$ in the usual way.

The general conclusion is that if $0 \leqslant a \leqslant 1$ one should concentrate on M, if $1 < a$ one should analyse (3.80), but that if $a = 1$ a delicate balance of behaviour is to be expected. If there are additional parameters the argument is essentially the same. For example, consider the three-parameter Weibull distribution with density in the form

$$\kappa\sigma^{\kappa}(y - \alpha)^{\kappa - 1} \exp[-\{\rho(y - \alpha)\}^{\kappa}] \quad (y > \alpha), \tag{3.81}$$

the three parameters being (ρ, κ, α) and the lower terminal being the point of irregularity. The earlier discussion suggests that if $0 < \kappa < 2$ most of the information about κ is contained in the smallest observation, whereas if $\kappa > 2$ the problem can be treated as regular in all three component parameters. In the former case the error in $\hat{\alpha}$ is negligible compared with that in $(\hat{\rho}, \hat{\kappa})$ and therefore to a first approximation α can be regarded as known in inference about (ρ, κ). □

The second form of irregular behaviour, the so-called **change-point problem**, illustrates discontinuous behaviour in a particularly striking form. In the simplest version of the problem $Y_1, \dots, Y_n$ are independent random variables, $Y_1, \dots, Y_\psi$ having a known density $f_0(y)$ and $Y_{\psi+1}, \dots, Y_n$ a different and known density $f_1(y)$. There are more realistic versions in which f_0 and f_1 contain unknown parameters. The likelihood corresponding to any specified ψ can be written down and a maximum likelihood estimate $\hat{\psi}$ computed but the study of its asymptotic distribution calls for arguments quite different from those for regular problems. This can be seen most directly when f_0 and f_1 are (1, 1) exponential family distributions with canonical statistic y and known canonical parameters ϕ_0 and ϕ_1. Then the log likelihood is

$$\phi_0 \sum_{j=1}^{\psi} y_j + \phi_1 \sum_{j=\psi+1}^{n} y_j - \psi k(\phi_0) - (n-\psi)k(\phi_1)$$
$$= \left\{\phi_1 \sum_{j=1}^{n} y_j - nk(\phi_1)\right\} + (\phi_0 - \phi_1) \sum_{j=1}^{\psi} \left\{y_j - \frac{k(\phi_0) - k(\phi_1)}{\phi_0 - \phi_1}\right\}. \tag{3.82}$$

Thus the maximum likelihood estimate of ψ is determined by the extreme value of a random walk the direction of whose drift changes sign on passing the point ψ. Note that the increments in the random walk can be written $y_j - \eta(\phi')$, where $\eta(\phi')$ is the mean parameter corresponding to an intermediate value of ϕ which will often the approximately $\frac{1}{2}(\phi_0 + \phi_1)$.

A third possibility is that the parameter space has an unusual topology so that Taylor expansion in a neighbourhood of the true parameter value is not possible. One example, arising fairly infrequently in applications, is where the parameter space is discrete; for example suppose that the parameter θ must be an integer as when it represents a population size or atomic number. Another possibility is that θ lies on the boundary of the parameter space.

If the problem can be embedded in a regular problem, in which, for example, θ can take any real value, inference about θ is fairly simple by modification of that based on the distribution of the ordinary maximum likelihood estimate. For example, if $\hat{\theta}$ is asymptotically normal with mean θ and variance $i^{-1}(\theta)$, then a $1-\varepsilon$ confidence interval in the restricted problem in which θ is an integer consists of all those integers not 'rejected' at level ε via the statistic w. This set may be null. If it is necessary to use extremely small ε in order to obtain a non-degenerate confidence interval, then the formulation is suspect. An example where θ may lie on a boundary is provided by the estimation from one observation (or a random sample) of the mean vector μ of a multivariate normal distribution of known covariance matrix, when the parameter space is constrained, say to the positive orthant. The specification of a $1-\varepsilon$ confidence region (possibly void) for μ via that for the unconstrained problem is straightforward, even though the study of the maximum likelihood point estimate, found by definition to lie in the restricted parameter space, raise difficult issues.

A fourth source of irregular behaviour arises in testing a null hypothesis $\psi = \psi_0$ in the presence of a nuisance parameter χ, when under $\psi = \psi_0$ the dimension of χ is reduced, in the extreme case that we shall consider χ being undefined when $\psi = \psi_0$. A simple example is provided by two random variables (Y_1, Y_2) independently normally distributed with mean $(\psi \cos \chi, \psi \sin \chi)$ and unit variance, it being required to test $\psi = 0$ on the basis of n independent and identically distributed copies; under $\psi = 0$, χ is not defined. A more complicated example is provided by the densities

$$(1-\psi)e^{-y} + \psi\chi e^{-\chi y}, \quad (1-\psi)\chi_1 e^{-\chi_1 y} + \psi\chi_2 e^{-\chi_2 y}. \qquad (3.83)$$

In the first, again under $\psi = 0$, χ is undefined, whereas in the second the null hypothesis that there is a single component can be expressed as $\psi = 0$, χ_1 arbitrary, χ_2 undefined; as $\psi = 1$, χ_2 arbitrary, χ_1 undefined; or as $\chi_1 = \chi_2$ arbitrary, ψ undefined. We shall sketch a brief treatment of the simplest problem with one-dimensional χ.

One approach to testing $\psi = 0$ versus local alternatives $\psi > 0$ with χ known, $\chi = \chi'$ say, is via the standardized score statistic

$$t_{\mathrm{u}}(\chi') = \frac{1}{\sqrt{n}} \sum \left[\frac{\partial \log p(y_j; \psi, \chi')}{\partial \psi} \right]_{\psi=0} \{v(\chi')\}^{-\frac{1}{2}}, \qquad (3.84)$$

where

$$v(\chi') = \operatorname{var}\left[\frac{\partial \log p(Y_j; \psi, \chi')}{\partial \psi} \right]_{\psi=0}.$$

Under $\psi = 0$, $T_{\mathrm{u}}(\chi')$ is asymptotically standard normal. Unfortunately χ' will rarely be known. There are various possibilities. One is to choose a value of χ' in the middle of some plausible range and to hope that the outcome is not sensitive to the choice made. Another is to average $T_u(\chi')$ with respect to some formal prior distribution over χ'. A third is to take the supremum of $T_u(\chi')$ over χ' in its parameter space, possibly restricted a priori on the basis of supplementary information about plausible values of χ. One motivation is that one tests simultaneously against all possible values of χ', takes the most significant outcome and then adjusts the significance level for selection bias. For this, we need the distribution under $\psi = 0$ of

$$\sup_{\chi'} T_u(\chi'). \tag{3.85}$$

The marginal distribution of $T_u(\chi')$ for each fixed χ' is asymptotically standard normal so that it is natural to approximate $T_u(\cdot)$ as a Gaussian stochastic process described by its covariance kernel $\gamma(\chi', \chi'') = \text{cov}\{T_u(\chi'), T_u(\chi'')\}$, which can be calculated, at least in sufficiently simple cases. The theory of extremes of Gaussian processes (Cramér and Leadbetter, 1967, p. 285) can then be used to show that asymptotically when $\psi = 0$

$$P\left\{\sup_{a \leqslant \chi' \leqslant b} T_u(\chi') > c\right\} \leqslant \Phi(-c) + \phi(-c)\int_a^b \left\{-\frac{\partial^2 \gamma(\chi', \chi'')}{\partial^2 \chi'^2}\right\}_{\chi'=\chi''}^{1/2} d\chi'. \tag{3.86}$$

Simulation may well be the best route for significance evaluations in many practical cases. The analysis here follows Davies (1977; 1987).

The distinction between regular and irregular problems is quite sensitive to the specification. For example suppose that $X_1, \ldots, X_n$ are independently distributed with the density $\rho e^{-\rho(x-\alpha)}(x \geqslant \alpha)$. Then estimation of α, whether or not ρ is known, is an irregular problem. Now suppose that the X_j are perturbed by errors of observation independently normally distributed with zero mean and known variance σ_0^2 leading to observations $Y_1, \ldots, Y_n$. The density of Y_j is

$$\rho\Phi(y - \alpha - \sigma_0^2\rho)\exp\{-\rho(y - \alpha) + \tfrac{1}{2}\sigma_0^2\rho^2\},$$

with support $(-\infty, \infty)$, estimation in which is regular. There is, of course, non-uniformity of behaviour as $\sigma_0^2 \to 0$. For given values of (n, σ_0^2) the estimation problem may be close to irregular behaviour, or adequately covered by regular theory with precision determined by the information matrix, or may be in some intermediate regime. This situation is a simplified version of the econometric problem of the estimation of 'frontiers'.

Further results and exercises

3.1 Let Y_{it} $(i = 1,2; t = 1,\ldots,r)$ be independently normally distributed with known variance σ_0^2 and with $E(Y_{1t}) = \theta, E(Y_{2t}) = c\theta^2$, where c is a known constant. Obtain a reduction by sufficiency to means $(\bar{Y}_{1\cdot}, \bar{Y}_{2\cdot})$ showing membership of the (2, 1) exponential family and then examine the form taken by the statistics $w(\theta_0)$, $w_u(\theta_0)$, $w_p(\theta_0)$ for testing the hypothesis $\theta = \theta_0$.

[Section 3.2]

3.2 The number of local maxima of the log-likelihood function for estimating the location parameter μ from n independent observations with density $\pi^{-1}\{1 + (y - \mu)^2\}^{-1}$ that are different from the global maximum has for large n a Poisson distribution of mean π^{-1}. The proof hinges on showing that with probability tending to one each observation differing from μ by more than $2n - \sqrt{n}$ in absolute value is associated with a 'false' local maximum, i.e. the 'false' maxima are associated with the sparse tails of the scatter and not in the central portion.

[Reeds, 1985]

3.3 Examine the form of the statistics $w(\theta_0)$, $w_u(\theta_0)$, $w_p(\theta_0)$ for testing the hypothesis $\theta = \theta_0$ on the basis of n independent and identically distributed observations from the Cauchy density $\pi^{-1}\{1 + (y - \theta)^2\}^{-1}$. Consider in particular the cases $n = 1, 2$. Design and implement a computer simulation comparison of these procedures, say for $n = 10$, 20, noting the result of the previous exercise.

[Sections 3.2 and 3.3]

3.4 In the balanced normal theory components of variance model, Exercise 1.2, suppose that the parameter of interest is the proportion of 'groups' with positive mean $\mu + \xi_i$. Show that the parameter of interest can be taken as $\psi = \mu/\sqrt{\tau_\xi}$. Obtain its profile log-likelihood.

[Section 3.4]

3.5 Let Y have a Poisson distribution with mean $\psi + \chi$, and independently Z have a Poisson distribution with mean χ, the two corresponding respectively to observations on signal plus noise and on noise alone. Show that asymptotic theory can be applied to this problem provided ψ and χ are large, for example by regarding both as sums of a large number of components. Obtain the profile log-likelihood for ψ and compare the inference from it with that obtained by regarding $Y - Z - \psi$ as normally distributed with zero mean and variance estimated by $\sqrt{(Y+Z)}$.

[Section 3.4]

3.6 Based on observations $Y_1, \ldots, Y_n$ independently normally distributed with unknown mean and variance, obtain the profile log-likelihood for $P(Y > a)$, where a is an arbitrary constant, and compare it with the 'exact' answer from the non-central t distribution.

[Section 3.4]

3.7 The lifetimes $Y_1, \ldots, Y_n$ of n independent individuals are observed subject to non-informative right censoring. That is, there are values $c_1, \ldots, c_n$ such that if $Y_j > c_j$ only c_j is observed, whereas if $Y_j \leqslant c_j$, Y_j is observed. Suppose that, given $c_1, \ldots, c_n$, the random variables $Y_1, \ldots, Y_n$ are independently identically distributed with density $f(y; \theta)$. The data are (t_j, d_j), where $d_j = 0$ (censored), when $t_j = c_j$ and $d_j = 1$ (not censored), when $t_j = y_j$. Obtain the likelihood function when (a) $c_1, \ldots, c_n$ are arbitrary constants; (b) $c_1, \ldots, c_n$ are values of a vector random variable C with (joint) density $f_C(c; \gamma)$, where the parameter spaces of Y and C are independent. In case (b) find also the profile log-likelihood for θ. Examine the special case where Y_j is exponentially distributed with density $\theta e^{-\theta y}$ $(y \geqslant 0)$. What is the modification to (a) and (b) if in addition the censoring times of uncensored individuals are observed?

[Section 3.5]

3.8 Sometimes explicit forms are available for maximum likelihood fitting of a 'full' model with a large number of parameters, e.g. a saturated model, but it is in fact required to fit a 'reduced' model with a smaller number of parameters for which no explicit solution is available. Suppose that the full model is parameterized by $\theta = (\psi, \chi)$ and that the reduced model has $\chi = \chi_0$. If the information matrix is partitioned in the usual way show that

$$\tilde{\psi}_r = \hat{\psi}_f + i_{\psi\psi}^{-1} i_{\psi\chi}(\hat{\chi}_f - \chi_0),$$

where the suffices r, f refer to the reduced and full models, differs from the maximum likelihood estimate $\hat{\psi}_r$ by $O_p(n^{-1})$, and that

$i_{\psi\psi}, i_{\psi\chi}$ can be replaced by consistent estimates if, for instance, they involve unknown parameters.

For example, suppose that in the full model (Y_1, Y_2) are independently normally distributed with mean $(\psi, c\psi^2 + \chi)$ and known variance σ_0^2, so that $\chi = 0$ recovers the (2, 1) exponential family model of Example 1.5. Given n independent copies of (Y_1, Y_2), show that

$$\hat{\psi}_{\mathrm{f}} = \bar{Y}_1, \hat{\chi}_{\mathrm{f}} = \bar{Y}_2 - c\bar{Y}_1^2,$$

$$\hat{\psi}_{\mathrm{r}} = \bar{Y}_1 + \frac{2c\bar{Y}}{1 + \psi c^2 \bar{Y}_1^2}(-c\bar{Y}_1^2 + \bar{Y}_2).$$

Prove that asymptotically $\mathrm{var}(\hat{\psi}_{\mathrm{r}}) = (\sigma_0^2/n)(1 + \psi c^2\theta^2)^{-1}$.

[Cox and Wermuth, 1990]

3.9 In the simplest form of the seemingly unrelated regression models of Zellner (1962), there are two component response variables $(Y^{(1)}, Y^{(2)})$ bivariate normal with constant covariance matrix Σ and with $Y^{(1)}$ having linear regression on one explanatory variable and $Y^{(2)}$ linear regression on another explanatory variable. For n independent pairs this corresponds to the reduced model $\chi = 0$ obtained from the full model

$$E(Y) = \begin{bmatrix} x_{11} & x_{12} \\ \vdots & \vdots \\ x_{n1} & x_{n2} \end{bmatrix} \begin{bmatrix} \psi_1 & \chi_2 \\ \chi_1 & \psi_2 \end{bmatrix},$$

where Y is the $n \times 2$ matrix of response variables, so that in the full model a multivariate linear model holds in which the maximum likelihood estimates are obtained by separate least squares analysis.

Show that

$$\hat{\psi}_{\mathrm{r}} = \hat{\psi}_{\mathrm{f}} + \begin{bmatrix} \hat{\sigma}_{11}\sum x_{t1}^2 & \hat{\sigma}_{12}\sum x_{t1}x_{t2} \\ \hat{\sigma}_{12}\sum x_{t1}x_{t2} & \hat{\sigma}_{22}\sum x_{t2}^2 \end{bmatrix}^{-1}$$

$$\cdot\begin{bmatrix} \hat{\sigma}_{11}\sum x_{1t}x_{t2} & \hat{\sigma}_{12}\sum x_{t1}^2 \\ \hat{\sigma}_{12}\sum x_{t2}^2 & \hat{\sigma}_{22}\sum x_{t1}x_{t2} \end{bmatrix} \begin{bmatrix} \hat{\chi}_{1f} \\ \hat{\chi}_{2f} \end{bmatrix}$$

where $\hat{\sigma}_{ij}$, elements of $\hat{\Sigma}$, are obtained from the residual line of an analysis of covariance table.

[Zellner, 1962; Cox and Wermuth, 1990]

3.10 Extend the discussion of Example 3.7 to standardized fourth moments of residuals.

[Section 3.7]

3.11 Let $Y_1, \ldots, Y_n$ be independent binary random variables satisfying the linear logistic model

$$\log\{P(Y_j = 1)/P(Y_j = 0)\} = \alpha + \beta x_j + \gamma z_j,$$

where x_j is binary and z_j is an arbitrary scalar. Compare the merits, including computational simplicity, of the likelihood ratio and score tests of the hypothesis $\gamma = 0$, with (α, β) as unknown nuisance parameters. Give also the form of the test statistic when z_j is a vector.

[Section 3.7]

3.12 For some unknown ψ the independent random variables $Y_1^\psi, \ldots, Y_n^\psi$ follow a normal theory linear model of constant variance σ^2 and with $E(Y^\psi) = x\beta$. Develop a score test of the hypothesis $\psi = 1$.

[Atkinson and Lawrance, 1989; Section 3.7]

3.13 If $f_0(.)$ and $f_1(.)$ are two densities with the same support, a family of densities can be formed with

$$f_\psi(y) = c(\psi)\{f_0(y)\}^{1-\psi}\{f_1(y)\}^\psi,$$

where $c(\psi)$ is a normalizing constant. Develop a score statistic for the hypothesis $\psi = 0$ based on n independent and identically distributed observations. Examine further the situation in which $f_0(.)$ and $f_1(.)$ contain nuisance parameters.

[Cox, 1961; Atkinson, 1970; Section 3.7]

3.14 When there is a scalar parameter, θ, to be estimated from independent observations show that the variance of the maximum likelihood estimate can be estimated by

$$[\sum\{\partial \log p(Y_j; \theta)/\partial\theta\}^2/i^2(\theta)]_{\theta=\hat{\theta}}.$$

This is sometimes called the **sandwich estimate**. Show that, while for an exponential model this generally conflicts with sufficiency, the variance is correctly estimated for a broader family than the initial one, provided always that the original maximum likelihood estimate continues to estimate the target parameter. As an example consider the estimation of the mean of the Poisson distribution. The results generalize to vector parameters.

[Royall, 1986]

3.15 Show that under weak regularity conditions REML estimators are asymptotically efficient. Note, however, that this is a situation in which direct use of the form of likelihood on which REML is based is to be preferred to point inference and limits based on the asymptotic normal distribution of those estimates.

[Cressie and Lahiri, 1993; *Further results and exercises*, 2.10]

3.16 For n independent observations from the uniform distribution on $(0, \theta)$, obtain the distribution of the maximum observation and show that it is asymptotically exponential. Obtain exact and asymptotic confidence limits for θ from these two distributions and compare these with those obtained from naive use of the w statistic and its asymptotic chi-squared distribution.

[Section 3.8]

3.17 Suppose that the (1, 1) exponential family density $a(y)\exp\{s\phi - k(\phi)\}$ is reparameterized by the transformation

$$\psi(\phi) = \int_{\phi_0}^{\phi} \{k''(\omega)\}^{\delta} d\omega$$

for fixed δ. Show that the following special cases arise: $\delta = 0$, canonical parameter; $\delta = 1/3$, normal likelihood in that the expected third derivative of the likelihood function vanishes; $\delta = 1/2$, maximum likelihood estimate has constant variance; $\delta = 2/3$, zero asymptotic skewness of maximum likelihood estimate; $\delta = 1$, mean parameter η. Examine the special case of the exponential distribution. Show that the results can be extended to curved exponential families.

[Wedderburn, unpubl.; Sprott, 1973; Hougaard, 1982]

3.18 Show that the expression

$$Y_{is} = \mu + \xi_i + \eta_{is} + \alpha_{20}\xi_i^2 + \alpha_{11}\xi_i\eta_{is} + \alpha_{02}\eta_{is}^2$$

for $i = 1, \ldots, m$, $s = 1, \ldots, r$, in which $\{\xi_i\}$, $\{\eta_{is}\}$ are independently normally distributed with zero mean and variances $\sigma_\xi^2, \sigma_\eta^2$ and the αs are small, represents a departure from a normal theory components of variance model in which α_{20} and α_{02} give the skewness of the between and within components and α_{11} determines the variation with x of the conditional standard deviation of Y_{is} given $\xi_i = x$. Show further that if, for some power λ, Y_{is}^{λ} obeys a normal theory model then approximately, if the variation about μ is relatively small, the quadratic representation applies with $\alpha_{20} = \alpha_{02} = \frac{1}{2}\alpha_{11}$. Assuming $\alpha_{20} = \alpha_{02} = 0$ in the general model, obtain the score statistic for testing $\alpha_{11} = 0$.

[Solomon and Cox, 1992]

3.19 Consider a family of parametric models, for a d-dimensional parameter θ and indexed by a quantity v. Let $\mathcal{M}_v$ denote the model specified by v and let u_v and j_v be the score vector and the observed information calculated under $\mathcal{M}_v$. Furthermore, let $b_v(\theta)$ denote a positive definite non-random $d \times d$ matrix and define $w_v(\theta)$ by

$$w_v(\theta) = b_v(\theta)^{-\frac{1}{2}} j_v(\theta) b_v^{\mathrm{T}}(\theta)^{-\frac{1}{2}},$$

where $b_v(\theta)^{1/2}$ denotes the left Cholesky square root of $b_v(\theta)$.

Suppose that $b_v(\theta)$ can be chosen so that, as v tends to some limit v_0 (possibly $v_0 = \infty$), we have:

(i) $w_v(\theta) \xrightarrow{\sim} w_0(\theta)$;

(ii) $\sup_{\theta' \in N_v(\theta,c)} (\theta, c)|b_v(\theta)^{-\frac{1}{2}} - I_d| \xrightarrow{\sim} 0$;

(iii) $\sup_{\theta' \in N_v(\theta,c)} (\theta, c)|b_v(\theta)^{-\frac{1}{2}}\{j_v(\theta') - j_v(\theta)\} b_v^{\mathrm{T}}(\theta)^{-\frac{1}{2}}| \xrightarrow{\sim} 0.$

Here $w_0(\theta)$ denotes a positive definitive random matrix, $\xrightarrow{\sim}$ means weak convergence and $N_v(\theta, c) = \{\theta' : |b_v^{\mathrm{T}}(\theta)^{\frac{1}{2}}(\theta' - \theta)| \leqslant c\}$. Show that if the convergences are locally uniform,in a suitably defined sense, then

$$j_v(\theta)^{-\frac{1}{2}} u_v(\theta) \xrightarrow{\sim} N(0, I_d)$$

as $v \to v_0$, with similar results for the maximum likelihood estimator, the directed likelihood and the likelihood ratio.

The result extends straightforwardly to the case where v may be depend on θ and vice versa, with an estimate for θ being inserted eventually.

[Sweeting, 1980; 1992a; 1992b]

3.20 Apply the technique of the previous note to the following settings: (a) exponential disperson models

$$a(\lambda, y)\exp[\lambda\{\theta^{\mathrm{T}}y - k(\theta)\}]$$

with $\nu = \lambda$, i.e. small dispersion asymptotics; (b) a birth process observed over a time interval $[0, t_0]$ and having initial population size n_0 and birth rate λ, and $\nu = n_0(e^{\lambda t_0} - 1)$ (note that both n_0 and t_0 may be fixed provided $\lambda \to \infty$).

[Jørgensen, 1987; Sweeting, 1992b)

Bibliographic notes

Most of the work in this chapter has a long history going back to the introduction of the likelihood ratio test by Neyman and Pearson (1928) and the chi-squared distributional results of Wilks (1938). Bartlett (1953a; 1953b) studied systematically the use of the score function to obtain tests and confidence intervals, including higher-order terms. See Rao (1973) and the books of Lehmann (1983; 1986) and Cox and Hinkley (1974) for methodological treatments. A fairly comprehensive discussion from the viewpoint of martingale theory is available in Barndorff-Nielsen and Sørensen (1993).

For some examples of the application of profile likelihood, see Box and Cox (1964), Kalbfleisch and Sprott (1970), Jørgensen (1982), Cox and Oakes (1984, ch. 3), Brown and Sundberg (1987) and Davison and Smith (1990). A method for computing confidence intervals based on a special version of the Wald statistic has been given by Venzon and Moolgavkar (1988). Properties similar to those of profile likelihood hold in many cases if instead of the maximum likelihood estimate for the nuisance parameter given the interest parameter some other estimate is used. For some discussion of this, see Gong and Samaniego (1981) and also Pfanzagl (1985; 1990).

Difficulties with the Wald statistic were stressed by Mantel (1987) and Væth (1985).

Silvey (1961) established the connection between maximum likelihood estimation and martingales.

For reviews of various kinds of non-regular problem, see Smith (1989) and Cheng and Iles (1987); a range of interesting examples is given by Romano and Siegel (1986). For change-point problems, see Hinkley and Hinkley (1970) and Cobb (1978). The discussion of problems in which the nuisance parameter dimension collapses at the null hypothesis follows Davies (1977; 1987).

For a quite different, mathematically powerful approach to maximum likelihood estimation and associated issues, emphasizing decision-theoretic ideas, see Le Cam (1986; 1990a; 1990b) and Le Cam and Yang (1990). Sen and Singer (1993) develop the theorems underlying first-order asymptotic theory together with some applications.

Patefield (1977) gave the important relation (3.31) connecting the curvature of the profile log-likelihood and the variance of the maximum likelihood estimate.

CHAPTER 4

Higher-order theory: preliminaries and motivations

4.1 Introduction

In the previous chapter the emphasis has been on the equivalence to first order of the likelihood ratio statistic, the directed likelihood, the score statistic and the statistic based on the maximum likelihood estimate, the last two statistics having a number of variants. These statistics often, but not always, give virtually identical answers in applications and it is therefore necessary to consider three rather different higher-order aspects: arguments for choosing among the different procedures; approximations to the distribution theory; and modifications to the procedures, in particular to achieve better behaviour and especially when the number of nuisance parameters is appreciable.

We now deal with these issues, in the present chapter discussing relatively qualitative aspects and in Chapters 5–8 developing the rather intricate general theory involved.

4.2 Choice between different statistics: qualitative considerations

We have seen in Chapter 3 that, even if attention is restricted to statistics computed from the log-likelihood function, there is a considerable variety of test statistics for a null hypothesis $\theta = \theta_0$ differing by $O_p(n^{-1/2})$ and hence equivalent to the first order of asymptotic theory. Additional such statistics could be found by replacing the log-likelihood function by some other function having the same first two derivatives to the right order of magnitude. We shall here keep to statistics based on the log-likelihood function and examine rather qualitative arguments concerning the relative merits of the various statistics. There are essentially five such considerations, as follows.

1. As we have already seen, the test, w_p, based on the maximum likelihood estimate, and the test based on the score function using information evaluated at the maximum likelihood point are not invariant under one-to-one transformations of the parameter. The likelihood ratio statistic $w(\theta_0)$ and the score statistic $w_u(\theta_0)$ do have the right invariance. This is easily seen from general transformation formulae. Now invariance considerations have to be used with caution, but in the present context of confidence regions for a parameter point, they do seem compelling. The conclusion is that either w_p should not be used or that, if it is used, attention to an appropriate parameterization is needed. In full exponential family problems the score statistic compares the canonical statistic with its expectation and is therefore equivalent to w_p, with the mean parameterization as the preferred one.
2. In problems with no nuisance parameters there is some appeal in making the inference directly from the log-likelihood function, pointing towards the use of w or w_u or often their signed forms r, r_u. This brings some conformity with the Bayesian approach where, whatever the prior density for the unknown parameter, two observed points y', y'' with proportional likelihood functions yield identical posterior distributions. When there are just two possible values θ', θ'' of θ given a priori there seems no need to do more than give the value of the likelihood ratio $L(\theta'; y)/L(\theta''; y)$, which can be interpreted as a factor in a hypothetical application of Bayes's theorem. In more complicated cases, however, calibration of $w(\theta_0)$ or $r(\theta_0)$ by considering their distribution under $\theta = \theta_0$ seems desirable and then the advantages of $w(\theta_0)$ are rather less clear.
3. Suppose that the log-likelihood function is anomalous in that

   (a) it has several local maxima of roughly equal magnitude; or
   (b) the likelihood is zero for some values in the parameter space; or
   (c) the log-likelihood approaches its supremum as, say, $\theta \rightarrow \infty$ with no local maximum.

   In all these cases w and, where applicable, its signed version give qualitatively sensible answers, but the other statistics discussed above do not. Thus in (a) confidence regions consist of subregions surrounding the largest local maxima, in (b) values of zero likelihood are automatically excluded, whereas in (c) confidence regions consist of all sufficiently large values of the parameter. The coverage probabilities calculated from the chi-squared approximation will in general not be correct, but at least the general form of the solutions is reasonable. The same is not true for the other procedures. For example, in case (b) it is possible that confidence intervals computed from the statistic w_p include parameter values logically inconsistent with the data.

Example 4.1 The index of a binomial distribution
As a somewhat artificial example of (c) above, let Y have a binomial distribution with parameter $\frac{1}{2}$ and unknown index θ. The log-likelihood function is

$$l(\theta) = \log \theta! - \log(\theta - y)! - \theta \log 2 \qquad (\theta = y, y+1, \ldots).$$

Now if $y = 0$ it is hardly reasonable to expect a good answer from asymptotic theory, but it is reassuring if at least qualitatively sensible answers are recovered. The 'exact' upper $1 - \varepsilon$ confidence limit for θ is $[-\log \varepsilon / \log 2] + 1$, where $[.]$ the denotes integer part. The maximum likelihood estimate is $\hat{\theta} = 0$, so that $w(\theta_0) = 2\theta_0 \log 2$ and the formal limit from (3.10) is thus $[\frac{1}{2}\chi^2_{1,\varepsilon}/\log 2] + 1$. At $\varepsilon = 0.05$, exact and approximate limits are 5 and 3, both procedures suggesting that small values of θ are supported. At $y = 0$ there is no local maximum of the likelihood so that w_u and w_p cannot be applied. Cormack (1992) discusses a related, more realistic, example connected with mark-recapture sampling. □

4. For estimation purposes w_p, with information evaluated at $\hat{\theta}$, has computational advantages, because it is necessary to locate only the position of the maximum of the log-likelihood function and the second derivatives at the maximum. For testing a specific null hypothesis in complex problems with nuisance parameters there are compensating computational advantages to the score test in that some quantities have to be found only under the null hypothesis.
5. An important aspect of any analysis lies in presenting the conclusions in a way that is both amenable to direct interpretation and useful for subsequent reanalysis. For the former, summarization via a formal point estimate (usually the maximum likelihood estimate itself), an estimate of error and a pivotal distribution has many advantages, although, as noted above, care is needed

over choice of parameterization. When the parameter dimension d_θ is small, recording of the log-likelihood function, or the summarization of key features, is good, especially if the log-likelihood is of unusual form. If the data currently being analysed are later to be combined with other related data, availability of the log-likelihood function will enable composite conclusions to be drawn via the testing of mutual consistency of parameters in different sets of data and, subject to that consistency, addition of log-likelihood functions across independent sets of data.

Sometimes estimates of parameters from separate sets of data, corresponding, say, to independent studies, are treated as response variables for a further stage of analysis in which variation between studies is to be explained. If that further analysis is to be linear, for example to involve the fitting of a linear model by least squares, it will usually be reasonable to require the first-stage estimates to be nearly unbiased. That is, if the first-stage estimates are maximum likelihood estimates, it will be reasonable to correct them for bias; see Chapter 5.

The traditional approach to the comparison of alternative test procedures is via the study of power functions. The limitations of this approach are partly that it is cumbersome and does not lead to results of any generality, except for local power, and partly that, as soon as we leave simple exponential family problems for which there are uniformly most powerful procedures, the optimal test depends on the particular alternative against which high power is required and in the great majority of cases we do not believe it realistic in applications to specify this with any precision. Therefore we do not pursue this line of argument further; see bibliographic notes to Chapter 4 for some key references.

There are two broad stages to the choice of an inference procedure for θ, the selection of a test statistic which in complex problems largely determines the 'shape' of the resulting solution and the resolution of a distributional problem that calibrates the confidence regions according to the confidence coefficient involved. It is an advantage, but not a matter of principle, if this distributional problem has, if not an 'exact' solution, then a simple reasonably accurate approximation. Here we shall see in section 4.4 that a simple adjustment to the chi-squared approximation leads, for the likelihood ratio statistic, to an accurate approximation to the distribution required, and a slightly more complicated adjustment deals with the signed version. Similar adjustments to the other statistics tend to be rather more complicated or non-existent.

As examples consider normal theory nonlinear regression one of the simplest instances of which involves random variables $Y_1, \ldots, Y_n$ independently normally distributed with constant variance σ^2 and with

$$E(Y_j) = \chi e^{\psi x_j} \qquad (j = 1, \ldots, n),$$

where $(x_1, \ldots, x_n)$ are given constants not all the same. A rather similar model for independent binary random variables has

$$P(Y_j = 1) = \psi e^{\chi_1 + \chi_2 x_j}/(1 + e^{\chi_1 + \chi_2 x_j})$$

where it is known that $\chi_2 > 0$ and the parameter space for ψ, the limiting probability as $x_j \to \infty$, is $0 < \psi < 1$. Given large amounts of data, informal methods for estimating ψ can be developed via 'common-sense' arguments but the efficiency of these is quite unclear, and, moreover, without general theory as a guide it is not possible to see what functions of the data should be calculated for efficient inference about ψ; that is, the 'shape' of a good solution is unclear, at least offhand. In some simpler problems, for example many of these connected with the full exponential family, the statistic of primary interest may be clearer, either on some general grounds or via consideration of sufficiency. Then interest focuses on the distributional properties of the resulting statistic and there are many examples throughout the book.

We return to these issues in Chapter 6, where further arguments supporting use of the likelihood ratio statistic or its signed version, i.e. the directed likelihood, emerge.

4.3 The reasons for modifying simple likelihood-based methods

The procedures described in Chapter 3, especially the use of profile likelihoods, are very useful for parametric problems with a modest number of nuisance parameters. If, however, the dimension of the nuisance parameter is quite large, the procedures may in extreme cases break down, giving quite misleading answers. This suggests the need for a modification to the procedures, and for this we need the higher-order asymptotic arguments to be developed in Chapters 6–8. For the moment we illustrate the nature of the difficulty in fairly general terms. As with other applications of higher-order asymptotics, there are two broad considerations involved. If the modification induced by the higher-order terms is numerically small, there is some reassurance that the initial asymptotic argument is applicable in the context considered. If, on the other hand, the modification is substantial, the modified version will usually be a major improvement; in critical cases, however, the failure of the first-order asymptotic argument is a warning that checking by simulation or other methods may be wise.

Central to the discussions is the notion that the score statistic is unbiased, i.e. that in particular

$$E\{U_{\psi}(\psi, \chi); \psi, \chi\} = 0. \tag{4.1}$$

If $\chi = \chi_0$ is known, the equation in ψ

$$U_{\psi}(\psi, \chi_0) = 0 \tag{4.2}$$

yields the maximum likelihood estimate of ψ. If the amount of information about ψ is large, (4.2) establishes that $U_{\psi}(\psi, \chi_0)$ will with high probability be small for the true ψ and hence the basis is laid for a local linearization of (4.2) around the true value. If χ_0 in (4.2) is replaced by a sufficiently close estimate, (4.1) will hold approximately and the argument still applies. If, however, the dimension of χ is large, or, for any reason, the amount of information about certain components of χ is not great, it may happen that even if χ_0 is replaced by the best available estimate $\tilde{\chi}$,

$$n^{-1}E\{U_{\psi}(\psi, \tilde{\chi}); \psi, \chi\} \neq 0, \tag{4.3}$$

not being zero even approximately and then the local linearization around the true ψ is not sensible.

The issues involved are best seen by examining in detail an important special case, the difficulties of which were appreciated in the earliest work on analysis of variance; the relation to maximum likelihood theory was discussed carefully by Neyman and Scott (1948).

Example 4.2 Neyman–Scott problem

Suppose that the parameter of interest is the error variance τ in a normal theory linear regression problem $E(Y) = x\beta$, where Y is $n \times 1$ and x is $n \times d_{\beta}$ of rank $d_{\beta} < n$. Now the maximum likelihood estimate of τ is SSD/n, where SSD is the residual sum of squares, whereas the corresponding unbiased estimate is $SSD/(n-d_\beta)$. It follows that if in a notional limiting operation as $n\to\infty$, $d_\beta \sim nd_\beta^*$, for some fixed $d_\beta^*\in(0,1)$, then $\hat{\tau}$ tends in probability to $(1-d_\beta^*)\tau$. Thus the profile likelihood is with high probability sharply peaked around the 'wrong' value and therefore, in particular, direct use of the profile likelihood function or one of its equivalents is misleading.

In the analysis of variance of experiments with many factors, and in particular in the analysis of fractional factorial experiments, the degrees of freedom available for estimating error may be quite small relative to the total sample size, i.e. d_β^* will be close to one and so the relative systematic error in the use of $\hat{\tau}$ very large. Neyman and Scott examined the special case of m pairs of observations, each pair of observations having a different mean, i.e. they took the special case of the linear model with observations Y_{js} $(j=1,\dots,m; s=1,2)$ with $E(Y_{js})=\mu_j$ all random variables being independently normally distributed with variance τ. Here $n=2m$, $d_\beta=m$, $d_\beta^*=\frac{1}{2}$.

The log-likelihood is

$$-\tfrac{1}{2}n\log\tau-\textstyle\sum_j\left\{\dfrac{(Y_{j1}-\mu_j)^2+(Y_{j2}-\mu_j)^2}{2\tau}\right\} \tag{4.4}$$

and the details $\hat{\mu}_j=\bar{Y}_{j+}$, $\hat{\tau}=\frac{1}{2}n^{-1}\sum_j(Y_{j2}-Y_{j1})^2\to\frac{1}{2}\tau$ in probability are easily verified directly.

Differentiation of (4.4) with respect to τ gives for the τ-component of the score function

$$U_\tau=-\frac{n}{2\tau}+\frac{1}{2\tau^2}\textstyle\sum_j\{(Y_{j1}-\mu_j)^2+(Y_{j2}-\mu_j)^2\}. \tag{4.5}$$

When the unknown parameters μ_j are used on the right-hand side of (4.5), $E(U_\tau)=0$, as is clear from the general discussion. If, however, as in forming the maximum likelihood estimating equations, μ_j is replaced by $\hat{\mu}_j$, then $E(U_\tau)=-\frac{1}{2}m/\tau$, with a similar result for the general linear model. Thus with μ_j replaced by $\hat{\mu}_j$, U_τ has a distribution not centred around zero even approximately. □

Before discussing the general implications of Example 4.2, we illustrate another aspect of the difficulties that can be produced by a large number of nuisance parameters via a problem that is in a sense 'opposite' to that of Example 4.2.

Example 4.3 Weighted mean
Suppose that random variables $Y_{js}(j=1,\ldots,m; s=1,\ldots,r_j)$ are independently normally distributed with

$$E(Y_{js})=\mu, \operatorname{var}(Y_{js})=\tau_j$$

and that μ is the parameter of interest. Reduction by sufficiency yields the sample means $\bar{Y}_{j+}$ and sums of squares $SSD_j=\sum(Y_{js}-\bar{Y}_{j+})^2$; in some rough sense μ is to be estimated via a weighted mean of the $\bar{Y}_{j+}$, the weights depending on the sample sizes r_j and on the variance. Similar results hold if the constant mean μ is replaced by a linear model for the 'group' means.

Note that if the τ_j are known, $\tau_j=\tau_{j0}$, say, reduction by sufficiency leads to

$$\frac{\sum \bar{Y}_{j.}(\tau_{j0}/r_j)^{-1}}{\sum(\tau_{j0}/r_j)^{-1}}, \tag{4.6}$$

say, which is normally distributed with mean μ and variance $1/\sum(\tau_{j0}/r_j)^{-1}$ leading to very simple 'exact' confidence limits for μ.

The log-likelihood in general is

$$-\tfrac{1}{2}\sum r_j\log\tau_j-\tfrac{1}{2}\sum_j\sum_s(Y_{js}-\mu)^2/\tau_j. \tag{4.7}$$

The μ-component of the score vector is thus

$$U_\mu=\sum_j\sum_s(Y_{js}-\mu)/\tau_j=\sum_j r_j\bar{Y}_{j+}/\tau_j-\mu\sum r_j/\tau_j. \tag{4.8}$$

Now if τ_j is replaced by a function of SSD_j, which is therefore independent of $\bar{Y}_{j+}$, then $E(U_\mu)=0$. Indeed, for more general choices to replace τ_j, the expectation remains approximately or exactly zero, so that bias in the first step in the study of maximum likelihood (and other) estimates encountered in Example 4.2 does not arise here.

The maximum likelihood estimate of μ is, from (4.8), $\sum\bar{Y}_{j+}(\hat{\tau}_j/r_j)^{-1}/\sum(\hat{\tau}_j/r_j)^{-1}$, where by (4.7)

$$\hat{\tau}_j=\frac{\sum(Y_{js}-\hat{\mu})^2}{r_j}=\frac{SSD_j+r_j(\bar{Y}_{j+}-\hat{\mu})^2}{r_j}. \tag{4.9}$$

That is, the maximum likelihood estimate of τ_j 'recovers' the information contained in the contrast of $\bar{Y}_{j+}$ with $\hat{\mu}$. It is easily shown that on replacing τ_j in the score U_μ by $\hat{\tau}_\mu$ the property $E(U_\mu)=0$ is preserved. The formal information matrix for $(\mu, \tau_1, \ldots, \tau_m)$ calculated from (4.7) is diagonal and if the usual relation between information and asymptotic variance of maximum likelihood estimates holds we would have

$$\operatorname{var}(\hat{\mu}) = 1/\{\sum(\tau_j/r_j)^{-1}\} \tag{4.10}$$

and $\hat{\mu}$ and (4.6), the estimate when the τ_j are known, would be equivalent.

Now in the asymptotic argument in which m is fixed and the r_j increase, this conclusion is correct. A more challenging, and often more relevant, asymptotic argument fixes the sequence $\{r_j\}$ to one having suitable stability and allows $m \to \infty$. The equation defining $\hat{\mu}$ can be written more conveniently as

$$\sum \frac{r_j^2(\bar{Y}_{j+} - \hat{\mu})}{SSD_j + r_j(\bar{Y}_{j+} - \hat{\mu})^2} = 0,$$

and it is instructive, again following Neyman and Scott (1948), to analyse the more general equation

$$\sum \frac{a_j(\bar{Y}_{j+} - \tilde{\mu})}{SSD_j + r_j(\bar{Y}_{j+} - \tilde{\mu})^2} = 0, \tag{4.11}$$

which it is convenient to write as $\sum B_j(\tilde{\mu}) = 0$. Local linearization about the true value μ is justified via $E\{\sum B_j(\mu)\} = 0$, $\operatorname{var}(m^{-1}\sum B_j(\mu)\} \to 0$ and it follows that $\tilde{\mu} - \mu$ is asymptotically normal with zero mean and variance

$$\frac{\operatorname{var}\{\sum B_j(\mu)\}}{[E\{\sum B_j'(\mu)\}]^2} = \frac{\sum a_j^2/\{r_j^2(r_j - 2)\tau_j\}}{\{\sum a_j/(r_j\tau_j)\}^2}, \tag{4.12}$$

where we suppose that $a_j = 0$ if $r_j = 1, 2$ and that the sums in the numerator and denominator of (4.12) are asymptotically proportional to m. The maximum likelihood estimate, with $a_j = r_j^2$, is thus asymptotically normal with variance

$$\frac{\sum r_j^2/\{(r_j - 2)\tau_j\}}{(\sum r_j/\tau_j)^2} \tag{4.13}$$

exceeding the formal variance given by the information matrix. Note also that the choice $a_j = r_j(r_j - 2)$ minimizes (4.12) and produces the asymptotic variance $\{\sum(r_j - 2)/\tau_j\}^{-1}$ smaller than that of the maximum likelihood estimate, but still larger than $\{\sum(r_j/\tau_j)\}^{-1}$ from the information matrix. It is unclear whether an estimating equation of a form different from (4.11) can improve on these results for all $\{\tau_j\}$. Of course for particular τ_j, e.g. constant τ_j, improvement is trivially possible, for example by the choice of fixed weights appropriate for the given $\{\tau_j\}$, as determined by (4.6).

For a review of this problem in a broader setting, see Cox (1982a). □

Examples 4.2 and 4.3 illustrate important general issues. For Example 4.2, the estimation of residual variance, a widely accepted 'exact' solution is available in which a residual sum of squares is divided by the residual degrees of freedom rather than by total sample size. The problem here is to incorporate this solution into a generally applicable modification of likelihood-based procedures, so that other similar problems not amenable to 'exact' discussion can be addressed. Example 4.3 does not have a recognized 'exact' solution; it raises similar and in some ways more difficult issues.

A more detailed discussion is deferred to section 8.5. One broad approach to problems with many similar nuisance parameters is to dismiss them as misspecified. Thus in Example 4.2, the m unknown means could be represented via a small number of unknown regression parameters or, if they are not rationally distinguishable, specified as random variables with an appropriately parameterized distribution. Similar comments apply to the variances in Example 4.3. We thereby achieve a formulation with a much smaller number of unknown parameters and to which the difficulties sketched above do not apply. While such formulations are often physically appropriate and hence attractive, they do not, in our view, provide a full answer to the difficulties under discussion. The second broad approach is to find an exact or approximate factorization of the likelihood so that one factor contains all or most of the information about the parameter of interest largely free of contamination by nuisance parameters, and to show that likelihood-like methods based solely on the factor that has been isolated have the desired features. This notion is developed much further in Chapter 8.

We now give 'random-effects' versions of Examples 4.2 and 4.3.

Example 4.4 Neyman–Scott problem

Suppose that in the formulation of Example 4.2, the means $\mu_1, \ldots, \mu_m$ are independently normally distributed with mean ν and variance

ω. The pairs $(Y_{j1}, Y_{j2})^{\mathrm{T}}$ are now independently bivariate normal with mean $(\nu, \nu)^{\mathrm{T}}$ and covariance matrix

$$\begin{bmatrix} \omega+\tau & \omega \\ \omega & \omega+\tau \end{bmatrix}. \tag{4.14}$$

It follows either via the bivariate normal form or by direct calculation that the log-likelihood is

$$-\tfrac{1}{2}m\log(\omega+\tfrac{1}{2}\tau)-\frac{\sum(\bar{Y}_{j+}-\nu)^2}{2\omega+\tau}-\tfrac{1}{2}m\log(2\tau)-\frac{\sum(Y_{j2}-Y_{j1})^2}{4\tau}. \tag{4.15}$$

There is now a simple reduction by sufficiency to three statistics $\{\bar{Y}_{++}, \sum(\bar{Y}_{j+}-\bar{Y}_{++})^2, \sum(Y_{j2}-Y_{j1})^2\}$ or equivalently, in analysis of variance notation,

$$\bar{Y}_{++}, MS_{\mathrm{b}} = SSD_{\mathrm{b}}/(m-1) = 2\sum(\bar{Y}_{j+}-\bar{Y}_{++})^2/(m-1),$$
$$MS_{\mathrm{w}} = SSD_{\mathrm{w}}/m = \tfrac{1}{2}\sum(Y_{j2}-Y_{j1})^2/m.$$

The maximum likelihood estimate of τ is

$$\begin{cases} MS_{\mathrm{w}} & (MS_{\mathrm{w}} \leqslant MS_{\mathrm{b}}), \\ (SSD_{\mathrm{w}}+SSD_{\mathrm{b}})/(n-1) & (MS_{\mathrm{w}} > MS_{\mathrm{b}}), \end{cases} \tag{4.16}$$

the complication arising because the parameter space is $\nu\in\mathbb{R}$, $\tau, \omega\in\mathbb{R}^+$, so that if $MS_{\mathrm{b}} < MS_{\mathrm{w}}$ then the maximum likelihood is achieved on the boundary $\omega = 0$. Provided that the complications induced on the boundary can be ignored, the usual properties of likelihood-based procedures hold; some 'exact' distributional results are, of course, available. □

Example 4.5 Weighted mean

Similarly, in Example 4.3 if variation in the τ_j cannot be represented in systematic form, it may be reasonable to take $\tau_1, \ldots, \tau_m$ as values of independent and identically distributed random variables. The inverse gamma distribution

$$p_0(\tau) = (\tfrac{1}{2}d_0\tau'_0)^{\frac{1}{2}d_0}\tau^{-\frac{1}{2}d_0-1}\exp(-\tfrac{1}{2}d_0\tau'_0/\tau)/\Gamma(\tfrac{1}{2}d_0) \tag{4.17}$$

is a convenient form, where d_0 is an effective degrees of freedom and the prior mean is $\tau_0 = \tau'_0 d_0/(d_0-2)$. For one sample of size r the likelihood is thus

$$\int_0^\infty \frac{1}{(2\pi\tau)^{\frac{1}{2}r}} \exp\left\{-\frac{\sum(Y_j-\mu)^2}{2\tau}\right\} p_0(\tau)d\tau. \tag{4.18}$$

The full log-likelihood is

$$-\tfrac{1}{2}\sum(r_j+d_0)\log\left\{1+\frac{r_j(\bar{Y}_{j+}-\mu)^2}{SSD_j+d_0\tau'_0}\right\}$$

and the maximum likelihood estimating equation is

$$\sum \frac{r_j(r_j+d_0)(\bar{Y}_{j+}-\tilde{\mu})}{SSD_j+r_j(\bar{Y}_{j+}-\tilde{\mu})^2+d_0\tau'_0}=0. \tag{4.19}$$

The quantities d_0, τ'_0 would typically be unknown and thus have to be replaced by estimates obtained from the marginal distribution of $(SSD_1, \ldots, SSD_m)$. Note that $d_0=0$ reproduces the maximum likelihood estimating equation and that the formal choice $d_0=-2$, $\tau'_0=0$ recovers the estimator with minimum asymptotic variance in the family (4.11). □

The phenomena illustrated in Examples 4.2–4.5 are not specific to the normal distribution and we therefore give some further examples for non-normal distributions illustrating the above issues and some new points.

Example 4.6 Matched binary pairs
Let (Y_{j1}, Y_{j2}) $(j=1,\ldots,m)$ be independent pairs of independent binary (0, 1) random variables with $P(Y_{j1}=1)=e^{\chi_j}/(1+e^{\chi_j})$, $P(Y_{j2}=1)=e^{\chi_j+\psi}/(1+e^{\chi_j+\psi})$ so that the parameter of interest ψ is the logistic difference between the probabilities. We have that the log-likelihood is

$$\sum\chi_j(Y_{j1}+Y_{j2})+\psi\sum Y_{j2}-\sum\log(1+e^{\chi_j})-\sum\log(1+e^{\chi_j+\psi}).$$

Now

$$\frac{\partial\ell_j}{\partial\chi_j}=Y_{j1}+Y_{j2}+\frac{1}{1+e^{\chi_j}}+\frac{1}{1+e^{\chi_j+\psi}}-2 \tag{4.20}$$

and for fixed ψ it follows that the maximum likelihood estimate of χ_j is

$$\hat{\chi}_{j\psi} = \begin{cases} -\infty & (Y_{j1} + Y_{j2} = 0), \\ -\frac{1}{2}\psi & (Y_{j1} + Y_{j2} = 1), \\ \infty & (Y_{j1} + Y_{j2} = 2). \end{cases}$$

Further,

$$\frac{\partial \ell}{\partial \psi} = \sum Y_{j2} - \sum \frac{e^{\chi_j + \psi}}{1 + e^{\chi_j + \psi}}. \tag{4.21}$$

On inserting $\hat{\chi}_{j\psi}$ for χ_j it follows that the equation defining $\hat{\psi}$ is

$$\sum Y_{j2} + R_0 + \frac{R_1}{1 + e^{\frac{1}{2}\hat{\psi}}} - n = 0,$$

i.e.

$$1 + e^{\frac{1}{2}\hat{\psi}} = R_1/(n - \sum Y_{j2} - R_0),$$

where R_0, R_1 denote the numbers of pairs for which $Y_{j1} + Y_{j2} = 0$ or 1.

As $m \to \infty$ the right-hand side tends in probability to

$$\begin{aligned} &\sum \{P(Y_{j1} = 1)P(Y_{j2} = 0) + P(Y_{j1} = 0)P(Y_{j2} = 1)\}/ \\ &\cdot \sum \{1 - P(Y_{j2} = 1) - P(Y_{j1} = Y_{j2} = 0)\} \end{aligned} \tag{4.22}$$

and a simple calculation shows that this is $1 + e^{\psi}$, so that $\hat{\psi} \to 2\psi$ in probability.

In this example conditioning on the statistic $Y_{j1} + Y_{j2}$ $(j = 1, \ldots, m)$, which for each fixed ψ is sufficient for $\chi_1, \ldots, \chi_m$, yields a conditional log-likelihood which is essentially that of the binomial distribution of R_{01}, the number of pairs with $y_{j0} = 0$, $y_{j1} = 1$ given $R_{01} + R_{10}$, which has parameter $e^{\psi}/(1 + e^{\psi})$ and from which estimation of ψ follows routinely.

This example is of practical importance both in its own right, for instance in connection with case-control studies in epidemiology, and because it illustrates in simple terms a phenomenon arising quite widely in applications with a considerable number of poorly estimated nuisance parameters; see *Further results and exercises* 4.5. □

4.4 Higher-order theory: the key questions

The above discussion pinpoints four issues in connection with higher-order asymptotic theory:

1. When there are no nuisance parameters, which of the various procedures equivalent to the first order is to be preferred?
2. Is there an effective general way of assessing the accuracy of and possibly improving the approximations to the distributions involved?
3. When nuisance parameters are present, is it possible to modify the profile log-likelihood to achieve properties closer to those applying in the absence of nuisance parameters?
4. The distributional question corresponding to issue 2 arises out of any answer to question 3.

These issues will be addressed in Chapters 6, 7 and 8 using the mathematical tools set out in Chapter 5. Here we outline some further considerations largely in rather qualitative terms.

Over issue 1, we have already in section 4.2 indicated some advantages of the likelihood ratio statistic $w(\theta_0)$ or its signed form $r(\theta_0)$, the directed likelihood, when there are no nuisance parameters. These arguments are consolidated by the following more technical argument. In 'exact' discussions of, for example, those curved exponential family problems in which there is an ancillary statistic A^*, i.e. a component of the minimal sufficient statistic with a distribution not depending on θ, probability calculations for inference are made conditionally on the observed value of A^*. This strongly suggests that if in higher-order asymptotic theory a statistic A has a distribution not depending on θ to an appropriate order of approximation, i.e. is approximately ancillary, then inference should be conditional on the observed value of A. The simplest way to achieve such conditioning is to modify any test statistic by a function of A in order to achieve independence from A and then to use the unconditional distribution of the resulting statistic. It turns out that to a higher order of asymptotics not only is $w(\theta)$ independent of A but also $w_u(\theta)$ and $w_p(\theta)$, when modified as above, become equivalent to $w(\theta)$, thus consolidating its position as the preferred quadratic statistic.

If the parameter ψ of interest is multi-dimensional, confidence regions for ψ can be obtained from the various essentially quadratic statistics described above. Wherever possible, however, it will be good to aim for scalar parameters of interest, essentially because these are more likely to correspond to sharply formulated scientific questions. Of course, in analysing a complex problem there will be several or many issues of concern, and ideally these are to be broken into largely separate features, each with its scalar parameter of concern. In obtaining confidence intervals for a scalar parameter ψ from the likelihood ratio statistic there are two broad approaches. One is to form likelihood-based regions, i.e. ones in which all parameter points included have higher (profile) likelihood than those excluded; these are formed directly from $w(\psi)$. A second and in many ways much more informative approach is to specify at level $1-\varepsilon$ a $1-\frac{1}{2}\varepsilon$ upper confidence level and a $1-\frac{1}{2}\varepsilon$ lower confidence level. For this we study the signed version $r(\psi)$.

Question 2 above concerns the distribution of $w(\psi_0)$. Here a simple adjustment applies. It can be shown that typically, under repeated sampling with sample size n,

$$E\{w(\theta_0)\} = d_\theta\{1 + b(\theta_0)/n\} + O(n^{-2}) \tag{4.23}$$

and the adjusted statistic

$$w'(\theta_0) = w(\theta_0)/\{1 + b(\theta_0)/n\} \tag{4.24}$$

not only has expectation close to d_θ, the mean of the appropriate chi-squared distribution, but remarkably the distribution function is for continuous random variables correspondingly improved,

$$P\{w'(\theta_0) \leqslant x\} = P\{\chi^2_{d_0} \leqslant x\}\{1 + O(n^{-2})\}. \tag{4.25}$$

Here $\{1 + b(\theta_0)/n\}^{-1}$ is called a **Bartlett adjustment**.

A corresponding result does not in general hold for the alternative statistics $w_u(\psi_0)$ and $w_p(\psi_0)$, although of course often some improvement in distributional approximation would be produced by simple rescaling.

The corresponding result for the signed likelihood ratio statistic, $r(\psi_0)$, which is of more direct practical relevance in most applications, involves in addition a correction for skewness.

Example 4.7 Proportionality of covariance matrices (Eriksen, 1987) As a first numerical example, we outline some results connected with a relatively complex problem in normal theory multivariate analysis. Some more details are given in Chapter 6. Suppose that samples are available from m multivariate normal distributions with $p \times p$ covariance matrices $\Sigma_1, \ldots, \Sigma_m$. By sufficiency, we can replace the data by m matrices $S_1, \ldots, S_m$ having Wishart distributions with expectations $\Sigma_1, \ldots, \Sigma_m$ and degrees of freedom $d_1, \ldots, d_m$. There are at least two null hypotheses of possible interest. First consider.

$$\Sigma_j = \kappa_j \Sigma_1 \qquad (j = 2, \ldots, m), \tag{4.26}$$

where $\kappa_2, \ldots, \kappa_m$ are unknown scalars and Σ_1 is an unknown covariance matrix; this is the hypothesis of proportionality of covariance matrices. Secondly, assuming (4.26), we may consider the null hypothesis

$$\kappa_2 = \cdots = \kappa_m = 1 \tag{4.27}$$

of equality of covariance matrices.

Likelihood ratio statistics can be found for (4.26) and for (4.27) and the Bartlett adjustments found either by direct evaluation of expectations or preferably by the indirect methods set out in section 6.5. Eriksen (1987) found by simulation the actual tail probabilities corresponding to the chi-squared approximation to the likelihood ratio statistic with and without adjustment. It turns out that the Bartlett factor for (4.27) is close to one, suggesting that the performance of the unadjusted statistic should be satisfactory, as is indeed the case. On the other hand, the Bartlett factor for (4.26) is substantial, the performance of the unadjusted statistic in small samples very poor and appreciably improved by adjustment. Table 4.1 gives a few specimen values. For example, in the last column, where the unadjusted likelihood ratio statistic is tested as χ^2_{10}, 11.5% of simulated values exceed the nominal 1% point. Because of the very small sample sizes Table 4.1 provides an exceptionally severe test of the adequacy of the distributional approximations. □

Table 4.1 *Percentage tail areas by simulation of unadjusted and adjusted likelihood ratio statistics for testing proportionality of covariance matrices*

Dimensionality, p	2	3	2	3
Degrees of freedom of S_j	2, 2, 2, 2, 2	3, 3, 3	3, 3, 4, 4, 5	4, 6, 8
Degrees of freedom of w statistic	8	10	8	10
Bartlett adjustment	2.32	2.37	1.48	1.47
10% level, unadj.	64.5	76.2	32.5	38.3
adj.	7.7	10.9	9.7	11.0
5% level, unadj.	54.1	67.4	22.0	27.1
adj.	3.5	6.2	4.8	5.8
1% level, unadj.	32.9	48.5	8.9	11.5
adj.	0.7	1.7	0.9	1.2

We now turn to problems involving nuisance parameters. Question 3 at the beginning of this section is the most difficult and can be tackled from a number of viewpoints:

(a) In problems with an 'exact' solution, this is typically achieved either by conditioning or marginalizing. It is therefore natural to look for some process of approximate conditioning or marginalization to produce an adjusted likelihood function that will have properties closer to those for problems without nuisance parameters.
(b) More generally, it is appealing to look for an adjusted likelihood function that is as close as possible to the function that would be used were the value of nuisance parameters known.
(c) In inference for a component canonical parameter in an exponential family, a standard approach with good properties is to condition on the remaining components of the canonical statistic, and the mean value vector of these is orthogonal to the parameter of interest. This suggests a general procedure wherein a new likelihood would be obtained conditional on the maximum likelihood estimates of the orthogonalized nuisance parameters.
(d) We have in section 4.3 already commented on the desirability of arranging that the linearizing expansions on which asymptotic theory is based can justifiably be made around the true value and that this may call for some adjustment to the likelihood function, especially if the number of nuisance parameters is large.
(e) The gradient of any adjusted log-likelihood defines a correspondingly adjusted score function and it is sensible to require that as far as is feasible this adjusted score function should have the properties of a simple score function in the absence of nuisance parameters.
(f) While most of the discussion that follows is based on asymptotic arguments in which the dimensionality of the nuisance parameter is notionally fixed as the number n of observations, or amount of information, increases, it is important that in many of the most fruitful applications the number of nuisance parameters can be quite large. Ideally, we need a double asymptotic theory in which the number of nuisance parameters tends to infinity at a suitable rate. In the absence of this we must look cautiously at terms which are specified to be small only to a given order in $n^{-1/2}$.

A conceptually different line of argument which provides a simple formal motivation for the type of amendment to the likelihood to be analysed in more detail in Chapter 8 is provided by a Bayesian analysis.

Let ψ be the parameter of interest and λ the nuisance parameter taken to be orthogonal to ψ. Let the prior density of ψ and λ be $\pi(\psi, \lambda)$. Then the posterior density of ψ is proportional to

$$\int \exp\{\ell(\psi, \lambda)\}\pi(\psi, \lambda)d\lambda. \tag{4.28}$$

Consider this at a fixed value of ψ; as a function of λ, the log-likelihood $\ell(\psi, \lambda)$ takes its maximum at $\lambda = \hat{\lambda}_\psi$, and, expanding about this point by Laplace's method (Barndorff-Nielsen and Cox, 1989, section 6.2), (4.28) is approximately

$$(2\pi)^{\frac{1}{2}d_\lambda}\pi(\psi, \hat{\lambda}_\psi)\exp\{\ell(\psi, \hat{\lambda}_\psi)\}/[\det\{j_{\lambda\lambda}(\psi, \hat{\lambda}_\psi)\}]^{1/2}, \tag{4.29}$$

where $j_{\lambda\lambda}(\psi, \hat{\lambda}_\psi)$ is the observed information matrix for λ at a fixed value of ψ evaluated at $\lambda = \hat{\lambda}_\psi$. Now as ψ varies in its range of interest within $O(n^{-1/2})$ of $\hat{\psi}$, $\hat{\lambda}_\psi$ varies by $O_p(n^{-1})$ and therefore so, too, does the contribution from the prior density, whereas, because of its dependence on ψ, the determinantal factor varies by $O(n^{-1/2})$ and the part depending on the likelihood by $O(1)$. It follows that with a relative error that is $O(n^{-1})$ inference about ψ is that based on an effective log-likelihood of

$$\begin{aligned} \ell_A(\psi) &= \ell(\psi, \hat{\lambda}_\psi) - \tfrac{1}{2}\log\det j_{\lambda\lambda}(\psi, \hat{\lambda}_\psi) \\ &= \ell_P(\psi) - \tfrac{1}{2}\log\det j_{\lambda\lambda}(\psi, \hat{\lambda}_\psi), \end{aligned} \tag{4.30}$$

where $\ell_P(\psi)$ is the profile log likelihood. We call $\ell_A(\psi)$ an **adjusted log likelihood**. No further assumption about the prior densities is required.

Note that (4.30) is a penalized version of the profile log-likelihood. Values of ψ that are relatively informative about the nuisance parameter λ have their value of the profile log-likelihood reduced on that account. It will turn out in Chapter 8 that (4.30) and adaptations thereof provide adjusted versions of the log-likelihood with the desired properties.

While (4.30) is in many ways the simplest version of adjusted profile likelihood, it has two disadvantages. First it requires, at least in the version (4.30), that the nuisance parameter can be expressed in a form orthogonal to ψ. Second, the right-hand side of (4.30) depends on the choice of λ; for example, in a simple scalar case the transformation from λ to $\log \lambda$, while preserving orthogonality to ψ, would change the correction term (4.30). In terms of asymptotic theory the correction term in (4.30) is of order $n^{-1/2}$, whereas the non-invariance affects terms of order n^{-1} and so is negligible to the order considered, again provided the number of nuisance parameters is modest. Other things being equal, however, exact invariance is to be preferred and is indeed achieved in the rather more complicated versions to be developed in Chapter 8. It can be seen that no change is needed to achieve exact invariance if $\hat{\lambda}_\psi$ is independent of ψ, and this applies to some of our examples, notably the important Example 4.9 below.

We now give some simple examples to illustrate (4.30).

Example 4.8 Exponential regression

Let $Y_1, \ldots, Y_n$ be independently exponentially distributed with $E(Y_j) = \lambda \exp(-\psi x_j)$, where ψ is the parameter of interest and $\sum x_j = 0$. The log-likelihood is

$$\ell(\psi, \lambda) = -n \log \lambda - \lambda^{-1} \sum Y_j \exp(\psi x_j).$$

Orthogonality of ψ and λ can be verified directly and we have

$$\hat{\lambda}_\psi = n^{-1} \sum Y_j \exp(\psi x_j),$$
$$j_{\lambda\lambda}(\psi, \lambda) = 2\lambda^{-3} \sum Y_j \exp(\psi x_j) - n\lambda^{-2},$$
$$j_{\lambda\lambda}(\psi, \hat{\lambda}_\psi) = n\hat{\lambda}_\psi^{-2}.$$

Further, the profile log-likelihood for ψ is, ignoring constants,

$$\ell_{\mathrm{P}}(\psi) = \ell(\psi, \hat{\lambda}_\psi) = -n \log \hat{\lambda}_\psi \tag{4.31}$$

and therefore

$$\ell_{\mathrm{A}}(\psi) = -(n-1) \log \hat{\lambda}_\psi. \tag{4.32}$$

Thus the adjustment in this instance is very minor, in part because only a single nuisance parameter is involved. Further comparisons and modifications of this example will be discussed later. □

Example 4.9 Normal theory linear model

Consider the normal theory linear model (1.1), supposing the variance τ to be the parameter of interest. As noted in Example 2.27, τ is orthogonal to the parameters β in the model for $E(Y)$. The log-likelihood function, using the notation standard for the linear model, is

$$\ell(\tau, \beta) = -\tfrac{1}{2}n \log \tau - \frac{SSD + (\hat{\beta} - \beta)^{\mathrm{T}}(x^{\mathrm{T}}x)(\hat{\beta} - \beta)}{2\tau},$$

where $SSD = (Y - x\hat{\beta})^{\mathrm{T}}(Y - x\hat{\beta})$ is the residual sum of squares and $\hat{\beta}$ is the least squares estimate, satisfying $x^{\mathrm{T}}(Y - x\hat{\beta}) = 0$. Here $\hat{\beta}_\tau = \hat{\beta}$ for all τ and the profile likelihood for τ is

$$\ell_{\mathrm{P}}(\tau) = \ell(\tau, \hat{\beta}) = -\tfrac{1}{2}n \log \tau - \tfrac{1}{2}SSD/\tau. \tag{4.33}$$

Now the information matrix for β for given τ is the $d_\beta \times d_\beta$ matrix $(x^{\mathrm{T}}x)/\tau$ having determinant $\det(x^{\mathrm{T}}x)\tau^{-d_\beta}$. Thus, except for constants,

$$\begin{aligned} \ell_{\mathrm{A}}(\tau) &= \ell_{\mathrm{P}}(\tau) + \tfrac{1}{2}d_\beta \log \tau \\ &= -\tfrac{1}{2}(n - d_\beta)\log \tau - \tfrac{1}{2}SSD/\tau. \end{aligned} \tag{4.34}$$

An important feature is that, whereas (4.33) has its maximum at $\hat{\tau} = SSD/n$, (4.34) has its maximum at $SSD/(n - d_\beta)$. Thus, at least when d_β is appreciable, $\ell_{\mathrm{A}}(\tau)$ is a qualitatively better guide to inference about τ than is $\ell_{\mathrm{P}}(\tau)$.

Further, suppose that we were to agree that the information about τ is wholly (or largely) contained in SSD, a random variable such that SSD/τ has the chi-squared distribution with $n - d_\beta$ degrees of freedom. Then we would consider the log-likelihood derived by considering SSD on its own as the only data and this is exactly $\ell_{\mathrm{A}}(\tau)$.

The example thus provides a particularly clear example of the need for and advantages of adjusting the profile log-likelihood function. □

Example 4.10 Pairs of exponentially distributed random variables
We can give a further illustration of these ideas via two special cases in which $(Y_{j1}, Y_{j2})(j = 1, \ldots, m)$ are m pairs of independently exponentially distributed random variables. First let the rate parameters, i.e. reciprocals of means, be $(\lambda_j\psi, \lambda_j/\psi)$, so that the parameter of interest is the square root of the ratio of the rates. A direct calculation shows that $\hat{\lambda}_{j\psi} = 2(\psi Y_{j1} + Y_{j2}/\psi)^{-1}$ and that the profile log-likelihood is

$$\ell_{\mathrm{P}}(\psi) = -2\sum \log(\psi Y_{j1} + Y_{j2}/\psi).$$

An alternative approach is to replace the data by $R_j = Y_{j2}/Y_{j1}$. It is easily shown that the density of R_j, which is proportional to that of the standard F statistic with (2, 2) degrees of freedom, yields exactly the log-likelihood function $\ell_P(\psi)$. This does not prove an efficiency property of $\ell_P(\psi)$ but does show that it is the ordinary log-likelihood of a set of independent and identically distributed random variables and that it therefore has the usual asymptotic properties.

Next suppose that the rate parameters are $(\lambda_j\psi, \psi/\lambda_j)$, so that now, somewhat artificially, it is the product of rates that is of interest. Orthogonality of λ_j and ψ can be proved and a direct calculation yields

$$\ell_P(\psi) = -2m\log\psi - 2\psi^{-1}\sum\sqrt{(Y_{j1}Y_{j2})},$$
$$\ell_A(\psi) = \ell_P(\psi) + \tfrac{1}{2}m\log\psi. \tag{4.35}$$

The properties of these log-likelihood functions can be partly studied via the associated maximum likelihood estimates of ψ which are respectively

$$\hat{\psi}_P = m^{-1}\sum\sqrt{(Y_{j1}Y_{j2})},$$
$$\hat{\psi}_A = \tfrac{4}{3}\hat{\psi}_P. \tag{4.36}$$

Some idea of the behaviour of these estimates can be gained by noting that $E(\hat{\psi}_P) = \frac{1}{4}\pi\psi$, $E(\hat{\psi}_A) = \frac{1}{3}\pi\psi$ so that, in particular, if $m \to \infty$ both estimates are inconsistent but $\hat{\psi}_A$ has substantially smaller bias than $\hat{\psi}_P$. It can be shown that if the data consist of groups of four observations, two with parameter $\lambda_j\psi$ and two with parameter ψ/λ_j, then the corresponding expectations are 0.884ψ and 1.010ψ. □

As mentioned earlier, in general ℓ_A is not exactly invariant under monotonic transformations of λ, i.e. transformations which preserve orthogonality and leave the parameter of interest unchanged. For ℓ_P, however, exact invariance does hold; see Chapter 8 for further discussion.

Further results and exercises

4.1 Show that in the regression transformation model in which Y_j^ψ is normally distributed with mean and variance

$$\lambda_1^\psi + \psi\lambda_1^\psi \sum_{s=2}^{p} x_{js}\lambda_s,\ \lambda_1^{2\psi-2}\psi^2\lambda_0,\ \text{with } x_{j1} \equiv 1, \sum_j x_{js} = 0 \quad (s = 2, \ldots, p)$$

the transformation parameter ψ is approximately orthogonal to

$\lambda_0, \ldots, \lambda_p$. Note that λ_1 and $\lambda_0^{1/2}$ have the dimensions of Y and that $\lambda_2, \ldots, \lambda_q$ are dimensionless. Because of approximate orthogonality $\{\text{var}(\hat{\psi})\}^{-1} \doteq E(-\partial^2 \ell / \partial \psi^2)$ and hence show that this is approximately

$$n\left\{\frac{7}{4}CV_e^2 + \frac{3}{2}CV_\Delta^2 + \frac{1}{4}\frac{CV_\Delta^4}{CV_e^2}(1 + c_\Delta)\right\},$$

where

$$CV_\Delta^2 = n^{-1}\sum \Delta_j^2 = n^{-1}\sum\left(\sum_{s=2}^{p} x_{js}\lambda_s\right)^2, n^{-1}\sum \Delta_c^4 = CV_\Delta^4(1 + c_\Delta)$$

and $CV_e^2 = \text{var}(Y_j)/\{E(Y_j)\}^2$. Interpret the various terms, noting that for the single sample problem $CV_\Delta^2 = 0$.

[Box and Cox, 1964; Cox and Reid, 1987a]

4.2 As a model for a three-dimensional contingency table (with three random response variables) independent individuals are classified by three criteria A, B, C having respectively m_1, m_2, m_3 levels. Let θ_{ijk}^{ABC} be the probability of falling in level $(A = i, B = j, C = k)$, $\sum \theta_{ijk} = 1$ so that the dimensionality of the parameter space is $m_1 m_2 m_3 - 1$. Log-linear models for θ_{ijk} specify a representation in terms of factorial effects of various orders; for instance, a model with no three-factor interaction has

$$\log \theta_{ijk}^{ABC} = \mu + \alpha_i^A + \alpha_j^B + \alpha_k^C + \beta_{jk}^{BC} + \beta_{ki}^{CA} + \beta_{ij}^{AB},$$

with constraints to remove parameter redundancy. From the classification of n independent individuals, an unconditional likelihood ratio statistic can be obtained for testing the null hypothesis of no three-factor interaction by comparison with the 'saturated' model in which the θ_{ijk}^{ABC} are arbitrary. Discuss the dangers of this unless n is very large.

[Section 4.3]

4.3 A computer package under development has an unknown number ψ of 'bugs'. In testing, the time for any particular bug to be detected is exponentially distributed with rate parameter ρ independently for all bugs. On detection the bug is removed with probability one. After a time t_0 on test, n bugs have been detected at times $t_1 \leqslant \cdots \leqslant t_n < t_0$. Obtain the profile likelihood for ψ, discuss the general shapes that it may take and the associated difficulties of inference about ψ.

[Joe and Reid, 1985; Section 4.3]

4.4 Discuss how in the weighted means problem of Example 4.3 consistency with the assumption that all observations have the same expected value should be tested.

[Section 4.3]

4.5 In the Rasch model Y_{jk} $(j = 1, \ldots, m; k = 1, \ldots, r)$ are independent binary random variables with

$$P(Y_{jk} = 1) = e^{\chi_j + \psi_k}/(1 + e^{\chi_j + \psi_k})$$

where, without loss of generality, we may take $\sum \psi_k = 0$. This model is exponential. Show that the row and column totals of the observed array are the canonical statistics. (For example, j may index subjects, k test items in a battery of true–false questions administered to m subjects; m might be very large and r moderate). Suppose that m is large and r fixed. Then the unconditional maximum likelihood estimates of the $\{\psi_k\}$ are inconsistent, whereas the conditional maximum likelihood estimates have the usual properties.

[Section 4.4]

4.6 For the first paired exponential variable problem of Example 4.10 involving the ratio of means, obtain a parametric empirical random effects formulation in which the nuisance parameters $\{\lambda_j\}$ are given a parametric distribution of suitably convenient form.

[Section 4.4]

Bibliographic notes

The higher-order asymptotic theory of procedures based on likelihood can be approached from a number of viewpoints. Fisher (1925) emphasized the role of ancillary statistics in recovering the (expected) information lost by concentrating on the maximum likelihood estimate; Rao's (1962) notion of second-order efficiency is a development from this.

The higher-order asymptotic distribution theory associated with Edgeworth expansions and saddlepoint (tilted) expansions is summarized by Barndorff-Nielsen and Cox (1989). Peers (1971) and Pfanzagl (1985) applied Edgeworth expansions to compare the powers of tests equivalent to the first order of asymptotic theory; see also Amari (1985), the collections of papers by Akahira (1991; 1992) and the results on local power summarized and extended by Chandra and Mukerjee (1985). The present treatment is an elaboration of that reviewed by Cox (1988). For higher-order asymptotic inference connected with exponential families, see Pierce and Peters (1992).

Bartlett (1936a; 1936b; 1937) gave an early discussion of conditional and marginal likelihoods. Welch and Peers (1963) discussed the relation between unconditional confidence limits and a Bayesian approach.

The need for conditional inference in dealing with large numbers of parameters was discussed in detail by Andersen (1970; 1971; 1973).

CHAPTER 5

Some tools of higher-order theory

5.1 Introduction

A number of the basic tools of the theory of higher-order asymptotic inference, in particular Edgeworth expansions, saddlepoint expansions and Laplace's method, are discussed by Barndorff-Nielsen and Cox (1989) and these will not be reviewed here. Another important technique is the determination of stochastic expansions of likelihood quantities such as the maximum likelihood estimator and the log-likelihood ratio statistic. The basic tool for this is Taylor expansion but for general understanding of the details, some of which initially look rather forbidding, it is often valuable to bring certain key differential geometric concepts and techniques to bear. In particular, consideration of invariance and of more general properties under transformation of the parameters is important.

From one point of view it would be desirable to carry out the calculations in a manner that is 'geometric' or 'invariant' throughout. However, the development of a suitable 'invariant statistical calculus' has only recently begun and is far from complete. Furthermore, such an approach might make the non-geometrical parts of the techniques less conspicuous, and those parts are important in many other contexts where the geometrical viewpoint is not, or at least less, relevant. Also, in specific applications calculations in particular parameterizations may be called for. We shall therefore, in the main, proceed by straightforward Taylor expansion followed by rearrangement of terms according to asymptotic order and in order to bring in geometrical aspects. A general technique for invariant, and hence geometrical, Taylor expansion will, however, be described at the end of this chapter. The geometrical study of the expansions could be brought much further than we shall do, but that would require a more extensive background in differential geometry.

There are two main types of expansion, **observed** and **expected/observed**.

In observed expansions the coefficients consist of mixed derivatives of the log model function, while the stochastic parts consist entirely of the components of the score vector. This type is both 'closer to the data' and more geometrical in nature, but in most applications it requires specification of an ancillary statistic.

Expected/observed expansions employ moments of log-likelihood derivatives for the coefficients and such derivatives centred at their mean values for the stochastic parts, and they are more generally applicable.

In many important cases it is, however, possible to 'translate' from the one type of expansion to the other.

Example 5.1 Exponential distribution
For Example 2.10 we can give in different parameterizations a number of the likelihood-based functions connected with the exponential distribution. No ancillary statistic is involved. In particular, with mean μ and rate $\rho = \mu^{-1}$ the log-likelihood from n independent and identically distributed observations is

$$\ell(\mu; \hat{\mu}) = -n\log\mu - n\hat{\mu}/\mu \quad \text{or} \quad \ell(\rho; \hat{\rho}) = n\log\rho - n\rho/\hat{\rho}$$

and the score function is

$$u(\mu; \hat{\mu}) = -n/\mu + n\hat{\mu}/\mu^2 \quad \text{or} \quad u(\rho; \hat{\rho}) = n/\rho - n/\hat{\rho}.$$

Observed expansions employ mixed partial derivatives with respect to μ and $\hat{\mu}$ or ρ and $\hat{\rho}$.

We shall use this example below for simple illustrative purposes. □

As before, we assume given a statistical model with model function $p(y; \theta)$ and log-likelihood function $\ell = \ell(\theta)$. Generic coordinates of the parameter θ, assumed d-dimensional, will be denoted by $\theta^r, \theta^s, \ldots$. Throughout, a circumflex over the functional symbol for a combinant, $q = q(x; \theta)$ say, indicates that $\hat{\theta}$ has been substituted for θ, i.e. $\hat{q} = q(x; \hat{\theta})$.

As a convenient notation for differentiation, we write $f_{/u}$ for the partial derivative with respect to u of a function $f(u, v)$ of two variables u and v. Here f, u and v may, in fact, be of dimension more than one; if their dimensions are denoted by m, p and q, respectively, we interpret $f_{/u}$ as the $p \times m$ matrix of the ordinary partial derivatives in question, i.e. if the coordinates of f and u are f^r and u^a we have that $f_{/u}$ is the matrix with (a, r)th element $\partial f^r/\partial u^a$. Often the latter symbol will be abbreviated to $f^r_{/a}$, and if $A = a_1 \ldots a_k$ is a block of possible values of the index a we write $f^r_{/A} = f^r_{/a_1 \ldots a_k}$ for the kth-order partial derivative $\partial^k f^r/\partial u^{a_1} \ldots \partial u^{a_k}$, etc. In the special case where f is a log-likelihood function ℓ we adhere, however, to our usual notation by dropping the / symbol, thus writing, for instance, ℓ_r instead of $\ell_{/r} = (\partial \ell/\partial \theta^r)$. Underlying many of the calculations to follow is the asumption (cf. section 2.5) of the existence of an exact or approximate ancillary statistic a so that $(\hat{\theta}, a)$, where $\hat{\theta}$ is the maximum likelihood estimator, constitutes a sufficient statistic. Under this assumption the log-likelihood function ℓ can, in its dependence on the data, without loss of generality be considered as a function of $(\hat{\theta}, a)$, and we indicate this by writing $\ell(\theta; \hat{\theta}, a)$.

5.2 Log-likelihood derivatives and mixed log model derivatives

We shall use ν to indicate joint moments of log-likelihood derivatives. More specifically if $R_1 = r_{11} \ldots r_{1n_1}, \ldots, R_m = r_{m1} \ldots r_{mn_m}$ are sets of coordinate indices we let

$$\nu_{R_1, \ldots, R_m} = E(\ell_{R_1} \ldots \ell_{R_m}). \tag{5.1}$$

The corresponding joint cumulants are denoted by

$$\lambda_{R_1, \ldots, R_m} = K\{\ell_{R_1}, \ldots, \ell_{R_M}\}. \tag{5.2}$$

For definitions and results on joint, or multivariate, moments and cumulants, see Barndorff-Nielsen and Cox (1989) or McCullagh (1987).

Lifting of indices by the expected formation i^{rs} produces quantities that we denote as follows

$$\nu^{R_1, \ldots, R_m} = i^{r_{11}s_{11}} \ldots i^{r_{mn_m}s_{mn_m}} \nu_{s_{11} \ldots s_{1n_1}, \ldots, s_{m1} \ldots s_{mn_m}} \tag{5.3}$$

and

$$\lambda^{R_1, \ldots, R_m} = i^{r_{11}s_{11}} \ldots i^{r_{mn_m}s_{mn_m}} \lambda_{s_{11} \ldots s_{1n_1}, \ldots, s_{m1} \ldots s_{mn_m}}. \tag{5.4}$$

Here, as elsewhere, we use the summation convention that if an index occurs twice in a term, once in lower or covariant position and once in upper or contravariant position, then summation over that index is understood.

Let ω be an alternative parameterization of the parametric model, i.e. $\omega = (\omega^1, \ldots, \omega^d)$ is a one-to-one smooth transformation of θ. We denote generic coordinates of ω by ω^a, ω^b, etc. and in an unambiguous, if somewhat illogical, notation, the corresponding likelihood derivatives by ℓ_a, ℓ_b, etc.

The score vector ℓ_r transforms as a covariant vector in the sense of differential geometry, i.e.

$$\ell_a = \ell_r \theta^r_{/a}.$$

The mean of ℓ_r is exactly zero,

$$\nu_r = E(\ell_r; \theta) = 0, \tag{5.5}$$

and the variance, or expected information,

$$i_{rs} = \nu_{r,s} = E(\ell_r \ell_s; \theta),$$

is a covariant metric tensor, i.e. it satisfies the transformation law

$$i_{ab} = i_{rs} \theta^r_{/a} \theta^s_{/b},$$

where $i_{ab} = E(\ell_a \ell_b; \omega)$.

Differentiation of (5.5) with respect to θ^s yields

$$\nu_{r,s} + \nu_{rs} = 0,$$

where $\nu_{rs} = E\{\ell_{rs}; \theta\}$. This provides the well-known alternative formula for expected information i as the mean value of the observed information j, defined as minus the matrix of second order derivatives of ℓ, i.e.

$$j_{rs} = -\ell_{rs}.$$

The inverse matrices i^{-1} and j^{-1} are called the **expected formation** and the **observed formation**, respectively.

On further differentiation one obtains

$$\nu_{r,s,t} + \nu_{rs,t}[3] + \nu_{rst} = 0, \tag{5.6}$$

$$\nu_{r,s,t,u} + \nu_{rs,t,u}[6] + \nu_{rs,tu}[3] + \nu_{rst,u}[4] + \nu_{rstu} = 0, \tag{5.7}$$

etc., the general form being

$$\sum_{\pi=1}^{|R|} \sum_{R/\pi} \nu_{R_1, \ldots, R_\pi} = 0. \tag{5.8}$$

Here $R = r_1 \ldots r_m$ is an arbitrary index set of length $|R| = m$ and the inner sum in (5.8) is over all partitions of R into π blocks $R_1, \ldots, R_\pi$. In (5.6) and (5.7) the notation [3], etc. indicates a sum over the indicated number of terms obtained by rearranging the subscripts.

Remarkably, the exactly analogous formula holds for the joint cumulants:

$$\sum_{\pi=1}^{|R|} \sum_{R/\pi} \lambda_{R_1,\ldots,R_\pi} = 0. \tag{5.9}$$

This follows from the fact, noted by Skovgaard (1986), that the cumulants λ of log-likelihood derivatives obey the same rule under differentiation with respect to the parameter coordinates as do the moments ν; that in turn may be proved by applying the general formulae relating moments to cumulants. These formulae are given, for instance, by Barndorff-Nielsen and Cox (1989, section 5.5). We refer to (5.8) and (5.9) as the **expected balance relations**. They are also referred to as the **Bartlett identities**.

Being a second-order derivative of $-\ell$, observed information follows the transformation rule

$$j_{ab} = j_{rs}\theta^r_{/a}\theta^s_{/b} - \ell_r\theta^r_{/ab}.$$

Thus, when the observed information is evaluated at $\hat{\theta}$, we have

$$\hat{j}_{ab} = \hat{j}_{rs}\hat{\theta}^r_{/a}\hat{\theta}^s_{/b}, \tag{5.10}$$

showing that $\hat{j}$, but not j, behaves as a metric tensor; see section 2.5.

Given an ancillary statistic a, it is often appropriate to consider differentiation of $\ell = \ell(\theta; \hat{\theta}, a)$ with respect to components of $\hat{\theta}$ as well as of θ, holding a fixed. Let

$$\ell_{r_1\ldots r_m;s_1\ldots s_n}(\theta;\hat{\theta},a) = \frac{\partial}{\partial\theta^{r_1}}\cdots\frac{\partial}{\partial\theta^{r_m}}\frac{\partial}{\partial\hat{\theta}^{s_1}}\cdots\frac{\partial}{\partial\hat{\theta}^{s_n}}\ell(\theta;\hat{\theta},a). \tag{5.11}$$

Also, write

$$\not{\ell}_{r_1\ldots r_m;s_1\ldots s_n} = \ell_{r_1\ldots r_m;s_1\ldots s_n}(\theta;\theta,a), \tag{5.12}$$

i.e. the quantities $\not{\ell}$ are obtained from (5.11) by substituting θ for $\hat{\theta}$ after any differentiation. We refer to the quantities (5.11) and (5.12) as **mixed log model derivatives**.

The likelihood equation $\hat{\ell}_r = 0$ is

$$\ell_r(\hat{\theta};\hat{\theta},a) = 0$$

identically in $\hat{\theta}$. Replacing $\hat{\theta}$ by θ the equation becomes

$$\not{\ell}_r = \ell_r(\theta; \theta, a) = 0$$

and on repeated differentiation of this we find

$$\not{\ell}_R + \sum_{R/2} \not{\ell}_{R_1;R_2} = 0, \tag{5.13}$$

the sum being over all partitions of $R = r_1 \ldots r_m$ into two blocks. Equation (5.13) is closely analogous to (5.8) and (5.9) and is termed the **observed balance relation**.

In particular, we have

$$\not{\ell}_{rs} + \not{\ell}_{r;s} = 0 \tag{5.14}$$

or, equivalently,

$$\left[\frac{\partial^2 \ell(\theta; \hat{\theta}, a)}{\partial \theta^r \partial \hat{\theta}^s}\right]_{\hat{\theta}=\theta} = \not{\ell}_{r;s} = \not{j}_{rs} = \left[-\frac{\partial^2 \ell(\theta; \hat{\theta}, a)}{\partial \theta^r \partial \theta^s}\right]_{\hat{\theta}=\theta},$$

where $\not{j}_{rs} = -\not{\ell}_{rs}$. In view of the first equality one sees that $\not{j}_{rs}$ and $\not{\ell}_{r;s}$ are metric tensors. The relation (5.14) and its expected analogue $\lambda_{r,s} = \lambda_{rs}$ are of crucial importance in the development of 'observed geometry' and 'expected geometry', cf. respectively Barndorff-Nielsen (1986b; 1988a).

A simple instance of (5.14) is provided by the formula $\ell(\mu; \hat{\mu}) = -n\log\mu - n\hat{\mu}/\mu$ of Example 5.1, no ancillary being needed. Direct differentiation gives

$$\partial^2\ell/\partial\mu\partial\hat{\mu} = n/\mu^2, \quad -\partial^2\ell/\partial\mu^2 = -n/\mu^2 + 2n\hat{\mu}/\mu^3$$

these being equal at $\mu = \hat{\mu}$.

Suppose the parameter vector θ is partitioned as $\theta = (\psi, \chi)$. We may then express the determinants of the observed and expected information matrices as

$$|j| = |j_{\psi\psi}||j^{\chi\chi}|^{-1}, \tag{5.15}$$

$$|i| = |i_{\psi\psi}||i^{\chi\chi}|^{-1}. \tag{5.16}$$

5.3 Expansion of likelihood quantities: expected/observed case

The first step in deriving stochastic as well as non-stochastic expansions of likelihood quantities is, most commonly, to make a Taylor expansion in $\hat{\theta}$ around θ.

We begin by expanding the generic component ℓ_r of the score vector, or rather its 'estimated value' $\hat{\ell}_r = 0$, in $\hat{\theta}$ around θ to give

$$0 = \hat{\ell}_r = \ell_r + \ell_{rs}(\hat{\theta}-\theta)^s + \tfrac{1}{2}\ell_{rst}(\hat{\theta}-\theta)^s(\hat{\theta}-\theta)^t \\ + \tfrac{1}{6}\ell_{rstu}(\hat{\theta}-\theta)^s(\hat{\theta}-\theta)^t(\hat{\theta}-\theta)^u + \cdots, \qquad (5.17)$$

where $(\hat{\theta}-\theta)^r = \hat{\theta}^r - \theta^r$, etc.

Next we invert this into an asymptotic expansion for $\hat{\theta}-\theta$. With j^{rs} denoting the observed formation, i.e. the inverse matrix of the observed information matrix $j_{rs} = -\ell_{rs}$, we find, from (5.17),

$$(\hat{\theta}-\theta)^r = j^{rs}\ell_s + \tfrac{1}{2}j^{rs}\ell_{stu}(\hat{\theta}-\theta)^t(\hat{\theta}-\theta)^u \\ + \tfrac{1}{6}j^{rs}\ell_{stuv}(\hat{\theta}-\theta)^t(\hat{\theta}-\theta)^u(\hat{\theta}-\theta)^v + \cdots, \qquad (5.18)$$

and iterating this leads to

$$(\hat{\theta}-\theta)^r = j^{rs}\ell_s + \tfrac{1}{2}j^{rs}j^{tu}j^{vw}\ell_{stv}\ell_u\ell_w \\ + \tfrac{1}{6}j^{rs}j^{tu}j^{vw}j^{xy}(\ell_{suyw} + 3\ell_{swp}j^{pq}\ell_{quy})\ell_t\ell_v\ell_x + \cdots. \qquad (5.19)$$

Note, incidentally, that j^{rs} will generally exist and be positive definite only in some neighbourhood of $\hat{\theta}$ and that outside the full exponential models it usually does not transform as a tensor. To get around this difficulty we write

$$j = i\{I - i^{-1}(i-j)\}, \\ j^{-1} = \{I - i^{-1}(i-j)\}^{-1}i^{-1}, \qquad (5.20)$$

where I denotes the identity matrix, and on expansion of the inverse we obtain

$$j^{-1} = i^{-1} + i^{-1}(i-j)i^{-1} + i^{-1}(i-j)i^{-1}(i-j)i^{-1} + \cdots, \qquad (5.21)$$

or in coordinate form

$$j^{rs} = i^{rs} + i^{rt}i^{su}(i_{tu} - j_{tu}) + i^{rt}i^{su}i^{vw}(i_{tv} - j_{tv})(i_{uw} - j_{uw}) + \cdots. \qquad (5.22)$$

Introducing the general notation

$$H_R = \ell_R - \nu_R \qquad (5.23)$$

for any set $R = r_1 \ldots r_m$ of coordinate indices, we may rewrite (5.22) as

$$j^{rs} = i^{rs} + i^{rt}i^{su}H_{tu} + i^{rt}i^{su}i^{vw}H_{tv}H_{uw} + \cdots. \qquad (5.24)$$

Now, inserting (5.24) in (5.19) and collecting terms of the same asymptotic order we obtain finally

$$\begin{aligned}(\hat{\theta}-\theta)^r = {} & i^{rs}\ell_s \blacktriangledown + \tfrac{1}{2}v^{rst}\ell_s\ell_t + i^{rs}i^{tu}H_{st}\ell_u \\ & \blacktriangledown + \tfrac{1}{6}(v^{rstu} + 3v^{rsv}i_{vw}v^{wtu})\ell_s\ell_t\ell_u \\ & + v^{rsu}i^{tv}H_{uv}\ell_s\ell_t + \tfrac{1}{2}i^{rs}v^{tuv}H_{st}\ell_u\ell_v \\ & + \tfrac{1}{2}i^{rs}i^{tu}i^{vw}H_{stv}\ell_u\ell_w \\ & + i^{rs}i^{tu}i^{vw}H_{st}H_{uv}\ell_w \blacktriangledown + \cdots . \end{aligned} \tag{5.25}$$

Here, and in the following, the symbol $\blacktriangledown$ indicates a drop in asymptotic magnitude of order $n^{-1/2}$ under ordinary repeated sampling. For instance, while $i^{rs}\ell_s$ is of order $n^{-1/2}$ in probability the terms $\frac{1}{2}v^{rst}\ell_s\ell_t$ and $i^{rs}i^{tu}H_{st}\ell_u$ are of order n^{-1}.

An expansion for the mean of the maximum likelihood estimator is readily obtained by taking termwise expectations in (5.25). The first correction to the leading term θ^r is of order n^{-1} and is given by

$$E(\hat{\theta}-\theta)^r = \tfrac{1}{2}i^{rs}i^{tu}(v_{stu} + 2v_{st,u}) + O(n^{-3/2}). \tag{5.26}$$

By subtracting an estimate of the right-hand side from $\hat{\theta}^r$ a bias-corrected version of the maximum likelihood estimate is obtained.

Example 5.2 Exponential distribution

About the simplest illustration of (5.26) is provided by the exponential distribution, again considered in two parametrizations, the mean μ and the rate $\rho = \mu^{-1}$. Here the two terms required to evaluate the bias in terms of μ are

$$v_{111} = E\left(\frac{\partial^3 \ell}{\partial \mu^3}\right) = \frac{4n}{\mu^3},$$

$$v_{11,1} = E\left(\frac{\partial^2 \ell}{\partial \mu^2}\frac{\partial \ell}{\partial \mu}\right) = -\frac{2n}{\mu^3}$$

so that the bias is zero, as indeed will happen in general when the parameter of interest is the mean parameter of a prime exponential family model. In terms of ρ the corresponding vs are respectively $2n\rho^{-3}$ and 0, so that with $i^{-1} = \rho^2/n$ we have from (5.26) that

$$E(\hat{\rho}-\rho) = \rho/n + O(n^{-3/2}),$$

suggesting the bias-adjusted estimate $(1-n^{-1})\hat{\rho}$. In fact exact calculation with the gamma distribution of $\hat{\mu} = \hat{\rho}^{-1}$ shows that

$$E(\hat{\rho}-\rho) = \rho/(n-1). \qquad \square$$

As the next likelihood quantity we consider the log-likelihood difference $\ell(\hat{\theta})-\ell(\theta)$. We find by Taylor expansion around θ that

$$\begin{aligned}\ell(\hat{\theta})-\ell(\theta)=&\ \ell_r(\hat{\theta}-\theta)^r+\tfrac{1}{2}\ell_{rs}(\hat{\theta}-\theta)^r(\hat{\theta}-\theta)^s\\&+\tfrac{1}{6}\ell_{rst}(\hat{\theta}-\theta)^r(\hat{\theta}-\theta)^s(\hat{\theta}-\theta)^t\\&+\tfrac{1}{24}\ell_{rstu}(\hat{\theta}-\theta)^r(\hat{\theta}-\theta)^s(\hat{\theta}-\theta)^t(\hat{\theta}-\theta)^u+\cdots.\end{aligned}\tag{5.27}$$

Further, we express ℓ_{rs}, ℓ_{rst} and ℓ_{rstu} as $v_{rs}+H_{rs}$, $v_{rst}+H_{rst}$ and $v_{rstu}+H_{rstu}$, where the quantities H are defined by (5.23), insert these expressions as well as the expansion (5.25) for $\hat{\theta}-\theta$ in (5.27) and rearrange terms according to their asymptotic order. A considerable number of terms are found to cancel (in this connection, see section 5.5) and the resulting expansion is

$$\begin{aligned}\ell(\hat{\theta})-\ell(\theta)=&\ \tfrac{1}{2}i^{rs}\ell_r\ell_s\blacktriangledown+\tfrac{1}{6}(v^{rst}\ell_r\ell_s\ell_t+3i^{rs}i^{tu}H_{rt}\ell_s\ell_u)\\&\blacktriangledown+\tfrac{1}{24}(v^{rstu}+3v^{rsv}i_{vw}v^{wtu})\ell_r\ell_s\ell_t\ell_u\\&+\tfrac{1}{2}i^{rs}v^{tuv}H_{rt}\ell_s\ell_u\ell_w\\&+\tfrac{1}{2}i^{rs}i^{tu}i^{vw}H_{rt}H_{sv}\ell_u\ell_w+\tfrac{1}{6}i^{rs}i^{tu}i^{vw}H_{rtv}\ell_s\ell_u\ell_w\blacktriangledown+\cdots.\end{aligned}\tag{5.28}$$

An important general fact is relevant here and at a number of places in the following. When any parameterization invariant quantity, such as $\ell(\hat{\theta})-\ell(\theta)$, is expanded in an asymptotic series then the entire term of first asymptotic order, the entire term of second asymptotic order, etc., must each be invariant, at least to that order. More generally, if the quantity expanded is a geometric quantity in the sense of following a coordinate transformation law, for instance if the quantity is a tensor, then each of the entire asymptotic terms must follow the same transformation law, to that order. Note, however, that component terms in an entire term will not in general individually follow the transformation law.

We may illustrate these points by considering expansion to order $O_p(n^{-1})$ of $\ell(\hat{\theta})-\ell(\theta)$. In (5.28) the leading term is half the quadratic score statistic which is invariant and, as we shall now verify, the term of order $n^{-1/2}$ is also exactly invariant. Letting $\ell^r=i^{rs}\ell_s$ and using our standard index notation relating to change of parameterization from $\theta=[\theta^r]$ to $\omega=[\omega^a]$ we find, in view of the facts that ℓ^r is a contravariant vector and ℓ_{rs}, v_{rs} and v_{rst} transform as do second- and third-order partial derivatives,

$$\begin{aligned}
v^{rst}\ell_r\ell_s\ell_t + 3i^{rs}i^{tu}H_{rt}\ell_s\ell_u &= v_{rst}\ell^r\ell^s\ell^t + 3H_{rs}\ell^r\ell^s \\
&= (v_{abc}\omega^{abc}_{/rst} + v_{ab}\omega^{ab}_{/rst})\theta^{rst}_{/def}\ell^d\ell^e\ell^f \\
&\quad + 3(H_{ab}\omega^{ab}_{/rs} + \ell_a\omega^a_{/rs})\theta^{rs}_{/de}\ell^d\ell^e \\
&= v_{abc}\ell^a\ell^b\ell^c + 3H_{ab}\ell^a\ell^b + v_{ab}\omega^a_{/r}\omega^b_{/st}\theta^{rst}_{/def}\ell^d\ell^e\ell^f[3] \\
&\quad + 3\omega^a_{/rs}\theta^{rs}_{/de}\ell_a\ell^d\ell^e \\
&= v^{abc}\ell_a\ell_b\ell_c + 3i^{ab}i^{cd}H_{ac}\ell_b\ell_d,
\end{aligned}$$

where in the last step we have used the fact that $v_{ab} = -i_{ab}$, and where

$$\begin{aligned}
\omega^{ab}_{/rs} &= \omega^a_{/r}\omega^b_{/s} \\
\omega^{ab}_{/rst} &= \omega^a_{/r}\omega^b_{/st} + \omega^a_{/rs}\omega^b_{/t} + \omega^a_{/rt}\omega^b_{/s}
\end{aligned}$$

etc. Note that, while the entire term is invariant, this is not true of the two individual terms $v^{rst}\ell_r\ell_s\ell_t$ and $i^{rs}i^{tu}H_{rt}\ell_s\ell_u$.

Note also that if the last term of the entire n^{-1}-order term of (5.28) were included with the entire $n^{-\frac{1}{2}}$-order term, yielding

$$i^{rs}i^{tu}i^{vw}\ell_{rtv}\ell_s\ell_u\ell_w + 3i^{rs}i^{tu}H_{rt}\ell_s\ell_u$$

we would still have a 'natural' $n^{-1/2}$-order expression which is not, however, exactly invariant but only invariant to order $n^{-1/2}$.

The entire n^{-1}-order term of (5.28) can likewise be shown to be invariant.

Returning to (5.28), we find that the mean value of the log-likelihood ratio statistic $w = w(\theta)$ has an asymptotic expansion of the form

$$E(w) = d\blacktriangledown\blacktriangledown + R\blacktriangledown\blacktriangledown + \cdots \tag{5.29}$$

where R, the **expected Bartlett adjustment**, is of order n^{-1} and is given by

$$R = \tfrac{1}{12}\{K_{rstu}i^{rs}i^{tu} + K_{rstuvw}i^{rs}i^{tu}i^{vw}\} \tag{5.30}$$

with

$$K_{rstu} = 3v_{rstu} + 12(v_{rt,su} + v_{r,t,su} + v_{r,stu}), \tag{5.31}$$

$$\begin{aligned}
K_{rstuvw} &= 3v_{rst}v_{uvw} + 6v_{rtv}v_{suw} + 12\{v_{r,st}(v_{uvw} + v_{uv,w}) + v_{r,tv}v_{su,w}\} \\
&\quad + 4v_{r,t,v}v_{suw} + 24v_{r,tv}v_{suw}.
\end{aligned} \tag{5.32}$$

We have here used relations such as

$$\begin{aligned}
v_{rst} + v_{r,st} &= v_{st/r}, \\
v_{rst,uv} &= (v_{ruv} - v_{uv/r})v_{st},
\end{aligned}$$

the latter being valid to order $O(n^{-1})$ only. Note the apparently fortuitous cancellation of many terms evidenced by the relatively simple form (5.30). An alternative derivation of R, which shows some light on why those cancellations occur, will be indicated at the end of section 5.5.

The quantities K_{rstu} and K_{rstuvw} are covariant tensors and they may be meaningfully written as combinations of certain other tensors. More specifically we have

$$K_{rstu} = 3t_{rstu} + 12t_{rs;tu}, \tag{5.33}$$

$$K_{rstuvw} = 3t_{rst}t_{uvw} + 2t_{rtv}t_{suw}, \tag{5.34}$$

where t_{rst}, t_{rstu} and $t_{rs;tu}$ are tensors given by

$$t_{rst} = v_{r,s,t}, \tag{5.35}$$

$$t_{rs;tu} = v_{rs,tu} + v_{rs,t,u} - v_{rs,v}(v_{w,tu} + v_{w,t,u})i^{vw}, \tag{5.36}$$

$$t_{rstu} = v_{rstu} + v_{rst,u}[4] + \tfrac{1}{2}v_{rs,v}(v_{tu,w} + 2v_{t,u,w})i^{vw}[6]. \tag{5.37}$$

An expansion for the more general log-likelihood ratio

$$w = 2\{\ell(\hat{\theta}) - \ell(\tilde{\theta})\},$$

where $\hat{\theta} = (\hat{\psi}, \hat{\chi})$ and $\tilde{\theta} = \hat{\theta}_\psi = (\psi, \hat{\chi}_\psi)$, is readily obtained by writing $\ell(\hat{\theta}) - \ell(\tilde{\theta})$ as the difference between $\ell(\hat{\theta}) - \ell(\theta)$ and $\ell(\tilde{\theta}) - \ell(\theta)$ and applying (5.28) to both these terms. If χ is chosen to be expected orthogonal to ψ at the true parameter point, then the leading term of the expansion simplifies but no major simplification of the higher-order terms occurs.

Writing d_χ for the dimension of χ, we have, by (5.29),

$$E_\theta[2\{\ell(\tilde{\theta}) - \ell(\theta)\}] = d_\chi + R_\psi + \cdots,$$

where R_ψ is of order n^{-1} and is determined by the formulae (5.30)–(5.32) as applied to the model given by the fixed value of ψ. Hence $E(w)$ is of the form

$$E(w) = (d - d_\chi) + (R - R_\psi) + \cdots. \tag{5.38}$$

In particular, then, the quantities R follow the same rule of calculation by differencing as do the degrees of freedom.

We go on to consider higher-order expansions for the set of directed likelihood statistics $r_1, \ldots, r_d$ discussed in section 3.7.

Initially we assume that the expected information matrix is a diagonal matrix at the true parameter value θ. We then find, by means of (5.28) and dropping the summation convention for the time being,

$$r_k^2 = i^{kk}\ell_k\ell_k \blacktriangledown + \tfrac{1}{3}v^{kkk}\ell_k\ell_k\ell_k + \sum_{r>k} v^{rkk}\ell_r\ell_k\ell_k + \sum_{r,s>k} v^{rsk}\ell_r\ell_s\ell_k$$
$$+ i^{kk}i^{kk}H_{kk}\ell_k\ell_k + 2\sum_{r>k} i^{rr}i^{kk}H_{rk}\ell_r\ell_k \blacktriangledown + \cdots \tag{5.39}$$

and from this we obtain, by writing $r_k = \sqrt{i^{kk}}\ell_k + \varepsilon_k$, inserting in (5.39) and solving for ε_k,

$$r_k = \sqrt{i^{kk}}\ell_k \blacktriangledown + \sqrt{i_{kk}}\Bigg(\tfrac{1}{6}v^{kkk}\ell_k\ell_k + \tfrac{1}{2}\sum_{r>k} v^{rkk}\ell_k\ell_k + \sum_{r,s>k} v^{rsk}\ell_r\ell_s$$
$$+ \tfrac{1}{2}i^{kk}i^{kk}H_{kk}\ell_k + \sum_{r>k} i^{rr}i^{kk}H_{rk}\ell_r\Bigg) \blacktriangledown + \cdots. \tag{5.40}$$

In the single-parameter case, i.e. $d = 1$, this expansion for r may be rewritten as

$$r = i^{-\frac{1}{2}}\ell_\theta[1 + \tfrac{1}{6}i^{-2}(3iH_2 + v_3\ell_\theta) + \tfrac{1}{72}i^{-3}\{27iH_2^2$$
$$+ 30v_3H_2\ell_\theta + 12H_3\ell_\theta + (3v_4 + 8v_3^2 i^{-1})\ell_\theta^2\} + \cdots] + \cdots. \tag{5.41}$$

where, in contrast to (5.40), we have also included the second correction term.

For multivariate parameters, formula (5.41) may be generalized to $r = i^{-\frac{1}{2}}s$, where $r = (r_1, \ldots, r_d)^{\mathrm{T}}$, $s = (s_1, \ldots, s_d)^{\mathrm{T}}$, $i^{-1/2}$ denotes an arbitrary matrix square root of the expected formation i^{-1} and

$$s_r = \ell_r + \tfrac{1}{6}(3H_{rs}\ell^s + v_{rst}\ell^s\ell^t) + \tfrac{1}{72}(27H_{rs}H_{tu}i^{st}\ell^u$$
$$+ 30H_{rs}v_{tuv}i^{st}\ell^u\ell^v + 12H_{rst}\ell^s\ell^t + 3v_{rstu}\ell^s\ell^t\ell^u$$
$$+ 8v_{rst}v_{uvw}i^{tu}\ell^s\ell^v\ell^w) + \cdots, \tag{5.42}$$

in which the expected information $i = [i_{rs}]$ is not assumed to be diagonal. Defining r in this way we have

$$\ell(\hat{\theta}) - \ell(\theta) = r_t r_u i^{tu} \tag{5.43}$$

with error $O(n^{-3/2})$ under ordinary repeated sampling.

It follows further from formula (5.42) that

$$E(s_r) = \tfrac{1}{6}(3v_{rs,t} + v_{rst})i^{st} + O(n^{-3/2})$$
$$= -\tfrac{1}{6}v_{r,s,t}i^{st} + O(n^{-3/2}), \tag{5.44}$$

$$\begin{aligned}\operatorname{cov}(s_r, s_s) = i_{rs} + \tfrac{1}{72}(18\nu_{rstu} + 72\nu_{rs,tu} \\ + 72\nu_{r,t,su} + 12\nu_{rtu}\nu_{svw}t^{vw} \\ + 16\nu_{r,s,t}\nu_{u,v,w}i^{vw})i^{tu} + O(n^{-2})\end{aligned} \tag{5.45}$$

and that all higher-order cumulants of $s_1, \ldots, s_d$ are of asymptotic order $O(n^{-3/2})$ or smaller. From these expressions asymptotic formulae for $E(r_t)$ and $\operatorname{cov}(r_t, r_u)$ may be derived.

5.4 Expansion of likelihood quantities: observed case

Our starting point is again the Taylor expansion (5.17) of the score ℓ_r and its inverted form (5.19). But whereas in the previous section we substituted, into (5.19), the expansion (5.22) of the observed formation j^{rs} in terms of j_{rs}, i_{rs} and i^{rs}, we will here employ an expansion of j^{rs} in terms of mixed log model derivatives and the score vector. To derive this we first note that, in analogy with (5.22), we have

$$j^{rs} = \not{\jmath}^{rs} + \not{\jmath}^{rt}\not{\jmath}^{su}(\not{\jmath}_{tu} - j_{tu}) + \not{\jmath}^{rt}\not{\jmath}^{su}\not{\jmath}^{vw}(\not{\jmath}_{tv} - j_{tv})(\not{\jmath}_{uw} - j_{uw}) + \cdots \tag{5.46}$$

where

$$\not{\jmath}_{rs} = -\not{\ell}_{rs} = -[\partial^2\ell(\theta; \hat{\theta}, a)/\partial\theta^r\partial\theta^s]_{\theta=\hat{\theta}}$$

is defined in section 5.2. Next, we expand the first few log-likelihood derivatives in $\hat{\theta}$ around θ as

$$\ell_{rs} = \not{\ell}_{rs} + \not{\ell}_{rs;t}(\hat{\theta}^t - \theta^t) + \tfrac{1}{2}\not{\ell}_{rs;tu}(\hat{\theta}^t - \theta^t)(\hat{\theta}^u - \theta^u) + \cdots, \tag{5.47}$$

$$\ell_{rst} = \not{\ell}_{rst} + \not{\ell}_{rst;u}(\hat{\theta}^u - \theta^u) + \cdots, \tag{5.48}$$

$$\ell_{rstu} = \not{\ell}_{rstu} + \cdots; \tag{5.49}$$

see the notation for mixed log model derivatives introduced in section 5.2.

We combine these expressions with formula (5.19) and write

$$\ell^r = \not{\jmath}^{rs}\ell_s, \tag{5.50}$$

noting, however, that in the previous section ℓ^r stood instead for $i^{rs}\ell_s$. We find, again using the symbol ▼ as defined in the previous section,

$$j_{rs} = \not{\jmath}_{rs}\blacktriangledown - \not{\ell}_{rs;t}l^t\blacktriangledown - \tfrac{1}{2}(\not{\ell}_{rs;tu} - \not{\ell}_{rs;v}\not{\ell}_{w;tu}\not{\jmath}^{vw}[3])l^tl^u\blacktriangledown + \cdots, \tag{5.51}$$

$$j^{rs} = \not{j}^{rs}\blacktriangledown + \not{j}^{rt}\not{j}^{su}\not{\ell}_{tu;v}l^v\blacktriangledown + \not{j}^{rt}\not{j}^{su}\{\tfrac{1}{2}(\not{\ell}_{tu;vw} - \not{\ell}_{tu;x}\not{\ell}_{y;vw}\not{j}^{xy}[3] + \not{\ell}_{tx;v}\not{\ell}_{uy;w}\not{j}^{uw}\}l^v l^w\blacktriangledown + \cdots, \tag{5.52}$$

$$(\hat{\theta}-\theta)^r = \ell^r\blacktriangledown - \tfrac{1}{2}\not{j}^{rs}\not{\ell}_{s;tu}l^t l^u - \tfrac{1}{6}\not{j}^{rs}(\not{\ell}_{s;tuv} - \ell_{s;tx}\not{\ell}_{y;uv}\not{j}^{xy}[3])\ell^t\ell^u\ell^v\blacktriangledown + \cdots. \tag{5.53}$$

Furthermore, in parallel with (5.26), (5.28) and (5.30) we find

$$E(\hat{\theta}^r - \theta^r) = -\tfrac{1}{2}\not{\ell}_{s;tu}\not{j}^{rs}j^{tu}\blacktriangledown + \cdots, \tag{5.54}$$

$$\ell(\hat{\theta}) - \ell(\theta) = \tfrac{1}{2}\not{j}_{rs}l^r l^s\blacktriangledown + \tfrac{1}{6}\not{t}_{rst}\ell^r\ell^s\ell^t\blacktriangledown + \tfrac{1}{24}\{\not{t}_{rstu} + 6\not{t}_{rs;tu} + 3\not{t}_{rsv}\not{t}_{tuw}\not{j}^{vw}[3]\}l^r l^s l^t l^u\blacktriangledown + \cdots \tag{5.55}$$

$$\not{R} = \tfrac{1}{12}\{\not{K}_{rstu}\not{j}^{rs}\not{j}^{tu} + \not{K}_{rstuvw}\not{j}^{rs}\not{j}^{tu}\not{j}^{vw}\} \tag{5.56}$$

where $\not{R}$, the **observed Bartlett adjustment**, satisfies $E(w) = d\blacktriangledown\blacktriangledown + \not{R}\blacktriangledown\blacktriangledown + \cdots$ and where

$$\not{K}_{rstu} = 3\not{t}_{rstu} + 12\not{t}_{rt;su}, \tag{5.57}$$

$$\not{K}_{rstuvw} = 3\not{t}_{rst}\not{t}_{uvw} + 2\not{t}_{rtv}\not{t}_{suw}. \tag{5.58}$$

The quantities $\not{t}_{rst}, \not{t}_{rstu}$ and $\not{t}_{rs;tu}$ are tensors and their definitions are given by

$$\not{t}_{rst} = -(\not{\ell}_{rst} + \not{\ell}_{rs;t}[3]), \tag{5.59}$$

$$\not{t}_{rs;tu} = \not{\ell}_{rs;tu} - \not{\ell}_{rs;v}\not{\ell}_{w;tu}\not{j}^{vw}, \tag{5.60}$$

$$\not{t}_{rstu} = \not{\ell}_{rstu} + \not{\ell}_{rst;u}[4] + \tfrac{1}{2}\not{\ell}_{rs;v}(\not{t}_{tuw} + \not{\ell}_{w;tu})\not{j}^{vw}[6]. \tag{5.61}$$

Note the analogy of (5.56)–(5.61) with the formulae (5.30), (5.33)–(5.37).

The derivation of the above formulae is, on the whole, and particularly as concerns (5.55) and (5.56), more straightforward than that of the corresponding formulae in the expected/observed case. In this connection, see the end of section 5.5.

For later use we note here two further expansions. The key to these are some formulae for the differentiation of the inverse and of the determinant of a matrix function. The formulae, which are easy to derive by applying index notation, are established in section 5.7, and applying those we find

$$\log\{|\hat{j}||\not{j}|/|\not{\ell}_{;}|^2\} = -\not{t}_{rst}\not{j}^{rs}\ell^t\blacktriangledown + \tfrac{1}{2}\{\not{t}_{rstu} + 3\not{t}_{rs;tu} + \not{t}_{rt;su} + \not{t}_{ru;st} + (\not{t}_{rsv}\not{t}_{tuw} + \not{t}_{rtv}\not{t}_{suw})\not{j}^{vw}\}\not{j}^{rs}\ell^t\ell^u\blacktriangledown + \cdots, \tag{5.62}$$

$$\log\{|\hat{j}|/|\mathring{j}|\} = (\ell_{st;r} + \ell_{r;st})\mathring{j}^{rs}\ell^{t} \blacktriangledown + \tfrac{1}{2}\{\ell_{rstu} + 3\ell_{rs;tu} + \ell_{rt;su} + \ell_{ru;st} + (\ell_{rsv}\ell_{tuw} + \ell_{rtv}\ell_{suw})\mathring{j}^{vw}\}\ell^{t}\ell^{u} \blacktriangledown + \cdots. \tag{5.63}$$

Here, as elsewhere, $\hat{j}, \mathring{j}$ and $\ell_{;}$ are the matrices whose elements are given by

$$\hat{j}_{rs} = [-\partial^2\ell(\theta;\hat{\theta},a)/\partial\theta^r\partial\theta^s]_{\theta=\hat{\theta}},$$
$$\mathring{j}_{rs} = [-\partial^2\ell(\theta;\hat{\theta},a)/\partial\theta^r\partial\theta^s]_{\theta=\hat{\theta}},$$
$$\ell_{r;s} = \partial^2\ell(\theta;\hat{\theta},a)/\partial\theta^r\partial\hat{\theta}^s.$$

By analogy with (5.42) we have

$$s_r = \mathring{j}_{rs}\ell^s \blacktriangledown + \tfrac{1}{6}\ell_{rst}\ell^s\ell^t \blacktriangledown + \tfrac{1}{72}(3\ell_{rstu} + 18\ell_{rs;tu} + 8\ell_{rsv}\ell_{tuw}\mathring{j}^{vw})\ell^s\ell^t\ell^u \blacktriangledown + \cdots. \tag{5.64}$$

To derive (5.64) we introduce the power series of s_r in terms of l^r, which we write as

$$s_r = c_{rs}\ell^s + \tfrac{1}{2}c_{rst}\ell^s\ell^t + \frac{1}{3!}c_{rstu}\ell^s\ell^t\ell^u + \cdots. \tag{5.65}$$

From this we form a power-series representation of $\frac{1}{2}\mathring{j}^{rs}s_r s_s$ and we then determine the coefficients c of (5.65) so that the power series representation agrees with the representation (5.55) of $\ell(\hat{\theta}) - \ell(\theta)$, thus obtaining (5.64).

A bias-adjusted version of s_r is given by

$$s_r^{\dagger} = s_r + \tfrac{1}{6}\ell_{rst}\mathring{j}^{st}. \tag{5.66}$$

5.5 Translation between observed and expected expressions

In this section we derive formulae which give asymptotic expressions for certain types of mixed log model derivative in terms of moments or cumulants of log-likelihood derivatives, and vice versa.

Given a parametric statistical model with d-dimensional parameter θ and probability density function $p(y;\theta)$, the **observed** and **expected likelihood yokes** are defined on $\Omega \times \Omega$, where Ω is the domain of variation of θ, by

$$g(\theta;\theta') = \ell(\theta;\theta',a) - \ell(\theta';\theta',a), \tag{5.67}$$

$$h(\theta;\theta') = E\{\ell(\theta) - \ell(\theta');\theta'\}, \tag{5.68}$$

respectively. Inherent in definition (5.67) is the assumption that the log-likelihood function ℓ can be represented as $l(\theta; \hat{\theta}, a)$ with a ancillary, cf. the end of section 5.1. For a fixed value of a we may then also think of ℓ as a mapping from $\Omega \times \Omega$ onto the reals. Such a mapping will later be called a yoke provided it satisfies two conditions (see further in section 5.6), which are met in particular by the functions defined by (5.67) and (5.68).

We shall also consider the **conditional expected likelihood yoke**

$$h(\theta; \theta' \mid a) = E\{\ell(\theta) - \ell(\theta') \mid a; \theta'\} \tag{5.69}$$

i.e. the expected likelihood yoke in the conditional model given a.

In wide generality we have, under ordinary repeated sampling with sample size n, that $g(\theta; \hat{\theta})$, $h(\theta; \hat{\theta})$ and $h(\theta; \hat{\theta} \mid a)$ are of order $O(1)$ and that with error $O_p(n^{-\frac{1}{2}})$

$$\begin{aligned} g(\theta; \hat{\theta}) &\doteq h(\theta; \hat{\theta}) \\ &\doteq h(\theta; \hat{\theta} \mid a). \end{aligned} \tag{5.70}$$

Example 5.3 Prime exponential models

For a prime exponential family (2.105) the log-likelihood function is, in the canonical parameterization,

$$l(\phi) = \phi^{\mathrm{T}} t - k(\phi)$$

and hence the observed and expected likelihood yokes g and h are

$$\begin{aligned} g(\phi; \hat{\phi}) &= (\phi - \hat{\phi})^{\mathrm{T}} s - k(\phi) + k(\hat{\phi}), \\ h(\phi; \phi') &= (\phi - \phi')^{\mathrm{T}} \eta(\phi') - k(\phi) + k(\phi'), \end{aligned}$$

respectively, where $\eta(\phi) = E(S; \phi)$ is the mean of the canonical statistic S. Since for prime exponential models we have $\eta(\hat{\phi}) = s$, there is in this case no difference between the two kinds of yoke. □

Example 5.4 von Mises distribution

The von Mises distribution of precision λ has density

$$\{2\pi I_0(\lambda)\}^{-1} \exp\{\lambda \cos(y - \phi)\},$$

where $I_0(\lambda)$ is the Bessel function of zero order and imaginary argument arising as a normalizing constant.

Thus if the precision λ is known, then the log-likelihood is of the form

$$l(\phi) = \lambda \cos(\hat{\phi} - \phi)$$

and the observed and expected likelihood yokes are

$$g(\phi;\hat{\phi}) = \lambda\{\cos(\hat{\phi}-\phi)-1\},$$
$$h(\phi;\phi') = \{\lambda I_0'(\lambda)/I_0(\lambda)\}\{\cos(\hat{\phi}-\phi)-1\}.$$

Here g and h happen to be proportional. □

As usual we denote generic coordinates of $\theta = (\theta^1,\ldots,\theta^d)$ by the letters $r, s, \ldots$, and for $m, n = 1, 2, \ldots$ we let $R_m = r_1 \cdots r_m$ and $S_n = s_1 \cdots s_n$ be sets of indices. Furthermore, R_0 and S_0 both denote an empty set. For a real-valued function $f(\theta;\theta')$ and for $m, n = 0, 1, 2, \ldots$ with $m + n > 0$ we use the notation

$$f_{R_m;S_n}(\theta;\theta') = \frac{\partial}{\partial\theta^{r_1}}\cdots\frac{\partial}{\partial\theta^{r_m}}\frac{\partial}{\partial\theta'^{s_1}}\cdots\frac{\partial}{\partial\theta'^{s_n}} f(\theta;\theta'), \tag{5.71}$$

$$\not{f}_{R_m;S_n}(\theta) = f_{R_m;S_n}(\theta;\theta). \tag{5.72}$$

This notation extends that for mixed log model derivatives introduced by (5.11) and (5.12).

In particular, from the definition (5.68) of the expected likelihood yoke one finds that $\not{h}_{r;s}$ and $-\not{h}_{rs;}$ are equal and, in fact, equal to the (r,s)th element of the expected information matrix.

If we differentiate the relations (5.70) with respect to coordinates of θ and $\hat{\theta}$ and then substitute θ for $\hat{\theta}$, we obtain a system of relations each holding to order $O_p(n^{-\frac{1}{2}})$. For instance, we find that, to the order claimed,

$$\not{g}_{R_m;S_n}(\theta) \doteq \not{h}_{R_m;S_n}(\theta). \tag{5.73}$$

We now, in turn, consider the right-hand side and the left-hand side of (5.73).

For $m, n = 1, 2, \ldots$ one has

$$\not{h}_{R_m;S_n}(\theta) = \nu_{R_m;S_n}(\theta), \tag{5.74}$$

where $\nu_{R_m;S_n}(\theta)$ is defined by

$$\nu_{R_m;S_n}(\theta) = \sum_{k=1}^{n}\sum_{S_n/k} \nu_{R_m,S_{n1},\ldots,S_{nk}}(\theta). \tag{5.75}$$

Here S_n/k under the summation sign has the sense defined in connection with formula (5.8).

Let

$$\lambda_{R_m;S_n}(\theta) = \sum_{k=1}^{n}\sum_{S_n/k} \lambda_{R_m,S_{n1},\ldots,S_{nk}}(\theta). \tag{5.76}$$

Using the formulae relating cumulants and moments (Barndorff-Nielsen and Cox, 1989, section 5.5) and the balance relations (5.8) and (5.9) it can be shown that

$$\lambda_{R_m;S_n}(\theta) = \nu_{R_m;S_n}(\theta). \tag{5.77}$$

Consequently, one may replace ν with λ in the expression on the right hand side of (5.74).

Furthermore, for $m = 1, 2, \ldots,$ we have

$$\not{h}_{R_m;}(\theta) = \nu_{R_m;}(\theta) = \nu_{R_m}(\theta) = \lambda_{R_m}(\theta) = \lambda_{R_m;}(\theta), \tag{5.78}$$

$$\not{h}_{;S_m}(\theta) = \sum_{\mu=1}^{m-1} (m-\mu) \sum_{S_m/\mu} \nu_{S_{m1},\ldots,S_{m\mu}}(\theta), \tag{5.79}$$

with the convention that the right-hand side of (5.79) is to be interpreted as 0 for $m = 1$.

Next, suppressing in the notation the dependence of the observed likelihood yoke (5.67) on the ancillary statistic a we find that, for $m, n = 1, 2, \ldots,$

$$\not{g}_{R_m;S_n}(\theta) = \not{l}_{R_m;S_n}(\theta), \tag{5.80}$$

$$\not{g}_{R_m;}(\theta) = \not{l}_{R_m;}(\theta), \tag{5.81}$$

$$\not{g}_{;S_n}(\theta) = -\sum_{S_n/2} \not{l}_{S_{n2};S_{n1}}(\theta). \tag{5.82}$$

From the above formulae it follows that

$$\not{\ell}_{R_m;S_n}(\theta) \doteq \lambda_{R_m;S_n}(\theta) = \nu_{R_m;S_n}(\theta), \tag{5.83}$$

$$\not{\ell}_{R_m;}(\theta) \doteq \lambda_{R_m;}(\theta) = \lambda_{R_m}(\theta) = \nu_{R_m}(\theta) = \nu_{R_m;}(\theta), \tag{5.84}$$

$$-\sum_{S_n/2} \not{\ell}_{S_{n2};S_{n1}}(\theta) \doteq \sum_{k=1}^{n-1} (n-k) \sum_{S_n/k} \lambda_{S_{n1},\ldots,S_{nk}}(\theta), \tag{5.85}$$

where the error of the approximation is $O_p(n^{-1/2})$.

In particular, the lowest-order cases of the general relations (5.83) and (5.84) are

$$\not{\ell}_{r;s} \doteq \lambda_{r;s} = \lambda_{r,s},$$
$$\not{\ell}_{r_1r_2;} \doteq \lambda_{r_1r_2},$$
$$\not{\ell}_{r;s_1s_2} \doteq \lambda_{r;s_1s_2} = \lambda_{r,s_1s_2} + \lambda_{r,s_1,s_2},$$
$$\not{\ell}_{r_1r_2;s} \doteq \lambda_{r_1r_2;s} = \lambda_{r_1r_2,s},$$
$$\not{\ell}_{r_1r_2r_3;} \doteq \lambda_{r_1r_2r_3},$$

$$
\begin{aligned}
\not\ell_{r;s_1s_2s_3} &\doteq \lambda_{r;s_1s_2s_3} \\
&= \lambda_{r,s_1s_2s_3} + \lambda_{r_1,s_1s_2,s_3}[3] + \lambda_{r,s_1,s_2,s_3} \\
\not\ell_{r_1r_2;s_1s_2} &\doteq \lambda_{r_1r_2;s_1s_2} = \lambda_{r_1r_2,s_1s_2} + \lambda_{r_1r_2,s_1,s_2}, \\
\not\ell_{r_1r_2r_3;s} &\doteq \lambda_{r_1r_2r_3;s} = \lambda_{r_1r_2r_3,s}, \\
\not\ell_{r_1r_2r_3r_4;} &\doteq \lambda_{r_1r_2r_3r_4}, \\
\not\ell_{r;s_1s_2s_3s_4} &\doteq \lambda_{r;s_1s_2s_3s_4} \\
&= \lambda_{r,s_1s_2s_3s_4} + \lambda_{r,s_1,s_2s_3s_4}[4] + \lambda_{r,s_1s_2,s_3s_4}[3] \\
&\quad + \lambda_{r,s_1,s_2,s_3s_4}[6] + \lambda_{r,s_1,s_2,s_3,s_4}, \\
\not\ell_{r_1r_2;s_1s_2s_3} &\doteq \lambda_{r_1r_2;s_1s_2s_3} \\
&= \lambda_{r_1r_2,s_1s_2s_3} + \lambda_{r_1r_2,s_1s_2,s_3}[3] + \lambda_{r_1r_2,s_1,s_2,s_3}, \\
\not\ell_{r_1r_2r_3;s_1s_2} &\doteq \lambda_{r_1r_2r_3;s_1s_2} = \lambda_{r_1r_2r_3,s_1s_2} + \lambda_{r_1r_2r_3,s_1,s_2}, \\
\not\ell_{r_1r_2r_3r_4;s} &\doteq \lambda_{r_1r_2r_3r_4;s} = \lambda_{r_1r_2r_3r_4,s}, \\
\not\ell_{r_1r_2r_3r_4r_5;s} &\doteq \lambda_{r_1r_2r_3r_4r_5,s}.
\end{aligned}
$$

Note that because of (5.73), (5.74) and (5.77) λ may be replaced by ν in the formulae above.

If in the formula (5.56) for the observed Bartlett adjustment $\not R$ we re-express the various terms, as given by (5.57)–(5.61), using the above 'dictionary' we arrive at the expected Bartlett adjustment R as presented by (5.30)–(5.32).

On account of the second relation in (5.70) all cumulant combinations $\lambda_{R_m;S_n}$ will to order $O(n^{-1/2})$ be the same whether calculated unconditionally or conditionally on a, and in this sense they are **stable**; cf. *Further results and exercises* 5.5.

5.6 Yokes and invariant calculations

In the previous section we have already met two key examples of the concept of yoke, the observed likelihood yoke and the expected likelihood yoke.

Yokes are of interest for several reasons. One is that they lead in a natural way to a definition of invariant Taylor expansions. Further, they induce geometric quantities of both statistical and mathematical import, in particular some key tensors. While yokes are not needed for the statistical development in the rest of the book we give a preliminary account because of the potential interest for future work.

To define a yoke, let Ω be subset of $\mathbb{R}^d$ and let Ω' denote a copy of Ω. In statistical applications Ω will be the parameter space and Ω' will often have the role of the domain of variation of the maximum likelihood estimator. We shall consider functions $g = g(\theta, \theta')$ defined on the product space $\Omega \times \Omega'$ and we employ the notation for mixed derivatives introduced by (5.71) and (5.72).

The function g is said to be a **yoke** if it statisfies the following two conditions for every $\theta \in \Omega$

1. $\not{g}_r = g_r(\theta, \theta) = 0$;
2. the matrix $[\not{g}_{rs}] = [g_{rs}(\theta, \theta)]$ is negative definite.

The observed likelihood yoke and the expected likelihood yoke do clearly satisfy these conditions, under the usual type of regularity conditions. For the observed likelihood yoke we usually write $\hat{\Omega}$ and $\hat{\theta}$ instead of Ω' and θ', this yoke being given by

$$g(\theta; \hat{\theta}) = \ell(\theta; \hat{\theta}, a) - \ell(\hat{\theta}; \hat{\theta}, a).$$

The expected likelihood yoke

$$h(\theta; \theta') = E\{\ell(\theta) - \ell(\theta'); \theta'\}$$

equals $-I(\theta', \theta)$, i.e. minus the directed distance from $P_{\theta'}$ to P_θ as defined by (2.121).

Example 5.5 Yokes on a simplex
Let μ and μ' be points in the unit simplex ∇^{m-1} in $\mathbb{R}^m$, i.e.

$$\nabla^{m-1} = \{x \in \mathbb{R}^m : x_1 > 0, \ldots, x_m > 0, x_1 + \cdots + x_m = 1\}.$$

Let μ and μ' be elements of ∇^{m-1}. Writing θ for $(\mu_1, \ldots, \mu_{m-1})$ and θ' for $(\mu'_1, \ldots, \mu'_{m-1})$, we may define a yoke by

$$g(\theta; \theta') = -\frac{1}{2} \sum_{i=1}^{m} \frac{(\mu'_i - \mu_i)^2}{\mu'_i}. \tag{5.86}$$

Noting that

$$g(\theta; \theta') = -\frac{1}{2}\left(\sum \frac{\mu_i^2}{\mu'_i} - 1\right), \tag{5.87}$$

we find

$$g_r = -\frac{\mu_r}{\mu'_r} + \frac{\mu_m}{\mu'_m},$$

$$g_{;r} = \frac{1}{2}\left(\frac{\mu_r^2}{\mu_r'^2} - \frac{\mu_m^2}{\mu_m'^2}\right), \tag{5.88}$$

$$g_{rs} = -\delta_{rs}\frac{1}{\mu_r'} - \frac{1}{\mu_m'},$$

$$g_{r;s} = \delta_{rs}\frac{\mu_r}{\mu_r'^2} + \frac{\mu_m}{\mu_m'^2},$$

where δ_{rs} is the Kronecker delta.

In fact the yoke (5.86) is identical to the observed likelihood yoke for a parametric model on the unit simplex given by

$$p(y;\mu) = a(\lambda)\left(\prod_{i=1}^{m} y_i\right)^{-\frac{1}{2}} \exp\left\{-\frac{\lambda}{2}\sum_{i=1}^{m}\frac{(y_i-\mu_i)^2}{y_i}\right\},$$

where $y \in \nabla^{m-1}$, $\mu \in \nabla^{m-1}$, $0 < \lambda$, here assumed known, and $a(\lambda)$ is a norming constant which, remarkably, does not depend on μ. For a discussion of this and related simplex distributions, see Barndorff-Nielsen and Jørgensen (1991). □

Repeated differentiation of the equation $\not{g}_r = g_r(\theta, \theta) = 0$ yields the sequence of relations

$$\not{g}_{rs} + \not{g}_{r;s} = 0, \tag{5.89}$$

$$\not{g}_{rst} + \not{g}_{rs;t} + \not{g}_{rt;s} + \not{g}_{r;st} = 0, \tag{5.90}$$

and so on, the general form being

$$\not{g}_{R_m} + \sum_{R_m/2} \not{g}_{R_{m1};R_{m2}} = 0, \tag{5.91}$$

where we have used the convention for summing over ordered partitions, introduced in section 5.2. Formula (5.91) generalizes the expected and observed balance relations (5.8), (5.9) and (5.13).

Note that (5.89) and the condition that $[\not{g}_{rs}] < 0$ imply that $\not{g}_{r;s}$ is a symmetric positive definite matrix, even though the function g has not been assumed to be symmetric. This is a crucial fact that allows the development of a general theory of differential geometries derived from yokes, as indicated and discussed in Barndorff-Nielsen (1987b) and below. The so-called observed likelihood geometries and the expected likelihood geometries are particular cases of that general theory, corresponding to the observed likelihood yoke and the expected likelihood yoke, respectively. Note further that (5.89) or rather its equivalent form $\not{g}_{r;s} = -\not{g}_{rs}$ specializes to the identity between the two well-known forms for the expected information when g is taken to be the expected likelihood yoke. In fact, with $h(\theta;\theta')$ given by (5.68) we have

$$\begin{aligned} h_r &= E\{l_r(\theta);\theta'\}, \\ h_{rs} &= E\{l_{rs}(\theta);\theta'\}, \\ h_{r;s} &= E\{l_r(\theta)l_s(\theta');\theta'\} \end{aligned}$$

and hence

$$\not{h}_{r;s} = i_{rs} = -\not{h}_{rs},$$

where as usual i denotes the expected information matrix. Similarly, when g is the observed likelihood yoke formula (5.89) becomes equivalent to (5.14).

It is also of interest to observe that on introducing for an arbitrary yoke g the **normed yoke**

$$\bar{g}(\theta;\theta') = g(\theta;\theta') - g(\theta';\theta'), \tag{5.92}$$

which is again a yoke, and defining $h(\theta;\theta')$ by

$$h(\theta;\theta') = \bar{g}(\theta';\theta) \tag{5.93}$$

we have that h is also a yoke, constituting a kind of dual to g, and that

$$\not{h}_{r;s} = \not{g}_{r;s}. \tag{5.94}$$

Note that both the observed likelihood yoke and the expected likelihood yoke are normed. In many applications of these two likelihood yokes the norming is of no real importance because only mixed derivatives $g_{R_m;S_n}$ with $m \geqslant 1$ are considered.

Any yoke induces a collection of geometrical objects on Ω, including a Riemannian metric given by $\not{g}_{r;s}$ and a family of connections

$$\{\overset{\alpha}{\Gamma} : \alpha \in \mathbb{R}\}$$

given, in terms of the lower Christoffel symbols, by

$$\overset{\alpha}{\Gamma}_{rst} = \frac{1+\alpha}{2}\not{g}_{rs;t} + \frac{1-\alpha}{2}\not{g}_{t;rs}.$$

In particular,

$$\overset{1}{\Gamma}_{rst} = \not{g}_{rs;t}, \tag{5.95}$$

$$\overset{-1}{\Gamma}_{rst} = \not{g}_{t;rs}. \tag{5.96}$$

We shall be working with the generalizations of (5.95) and (5.96) given by $\not{g}_{t_1 \cdots t_n;r}$ and $\not{g}_{r;t_1 \cdots t_n}$ and with the derived quantities

$$\not{g}^{;r}_{t_1 \cdots t_n} = \not{g}^{rs}\not{g}_{t_1 \cdots t_n;s}, \tag{5.97}$$

$$\not{g}^{r}_{;t_1 \cdots t_n} = \not{g}^{rs}\not{g}_{s;t_1 \cdots t_n}, \tag{5.98}$$

where $\not{g}^{rs}$ denotes the inverse of the matrix $\not{g}_{r;s}$.

Example 5.6 Yokes on a simplex
For the yoke (5.86) we find

$$g_{rs;t} = \delta_{rst}\frac{1}{\mu_r'^2} - \frac{1}{\mu_m'^2},$$

$$g_{r;st} = -2\delta_{rst}\frac{\mu_r}{\mu_r'^3} + 2\frac{\mu_m}{\mu_m'^3},$$

where $\delta_{rst} = \delta_{rs}\delta_{rt}$, and hence

$$\overset{1}{\Gamma}_{rst} = \delta_{rst}\frac{1}{\mu_r^2} - \frac{1}{\mu_m^2},$$

$$\overset{-1}{\Gamma}_{rst} = -2\delta_{rst}\frac{1}{\mu_r^2} + 2\frac{1}{\mu_m^2}.$$

Furthermore, writing $\delta_{st_1 \cdots t_n} = \delta_{st_1} \cdots \delta_{st_n}$, we have

$$\not{g}_{t_1 \cdots t_n;s} = 0 \qquad (n > 2),$$

$$\not{g}_{s;t_1 \cdots t_n} = (-1)^{n-1}n!\left(\delta_{st_1 \cdots t_n}\frac{1}{\mu_r^n} - \frac{1}{\mu_m^n}\right) \qquad (n \geqslant 1).$$

□

One of the applications of the quantities (5.97)–(5.98) is to the construction of tensors. We proceed to discuss an important special case of this.

The higher-order derivatives of a real function f on Ω, which we denote by $f_{/R_n} = f_{/r_1 \cdots r_n}$, are symmetric in the indices, but they are not tensorial in nature. However, using the quantities (5.97) or (5.98) we are able to construct what we shall refer to as **tensorial derivatives** of f and which we shall denote by $f_{//R_n}$.

To define this type of derivative we let

$$\overset{1}{\Gamma}{}^{S_\nu}_{R_n} = \sum_{R_n/\nu} \phi^{;s_1}_{R_{n1}} \cdots \phi^{;s_\nu}_{R_{n\nu}}, \tag{5.99}$$

$$\overset{-1}{\Gamma}{}^{S_\nu}_{R_n} = \sum_{R_n/\nu} \phi^{s_1}_{;R_{n1}} \cdots \phi^{s_\nu}_{;R_{n\nu}}, \tag{5.100}$$

where again we are summing over ordered partitions. If θ, and hence Ω, is subjected to a smooth one-to-one transformation then both types of multi-array (5.99) and (5.100) transform in the same manner, and we now let

$$\Gamma^{S_\nu}_{R_n}$$

denote any collection of multi-arrays that transform in this way.

The tensorial derivatives of f relative to the collection of arrays $\Gamma^{S_\nu}_{R_n}$ are defined implicitly by the linear equations

$$f_{/R_n} = \sum_{\nu=1}^{n} f_{//S_\nu} \Gamma^{S_\nu}_{R_n} \tag{5.101}$$

for $n = 1, 2, \ldots$.

The higher-order tensorial derivatives are not obtainable by repeated covariant differentiation. They are, however, for any given value of θ, equal to the ordinary higher-order derivatives of f in a particular coordinate system around θ, namely that for which

$$\Gamma^{s}_{R_n}(\theta) = 0 \qquad (n > 1).$$

The existence of such a system follows from the transformation law for $\Gamma^{s}_{R_n}$ under change of coordinate system θ.

The first few of these tensorial derivatives are

$$\begin{aligned}
f_{//r} &= f_{/r}, \\
f_{//rs} &= f_{/rs} - \Gamma^{t}_{rs} f_{/t}, \\
f_{//rst} &= f_{/rst} - \Gamma^{u}_{rs} f_{/tu}[3] - (\Gamma^{u}_{rst} - \Gamma^{u}_{rv}\Gamma^{v}_{st}) f_{/u}, \\
f_{//rstu} &= f_{/rstu} - \Gamma^{v}_{rs} f_{//tuv}[6] - \Gamma^{v}_{rst} f_{//uv}[4] - \Gamma^{v}_{rs}\Gamma^{w}_{tu} f_{//vw}[3] - \Gamma^{v}_{rstu} f_{/v}.
\end{aligned} \tag{5.102}$$

Thus the first-order tensorial derivative equals the ordinary partial derivative. The second-order tensorial derivative is, in fact, equal to the covariant derivative, in the sense of differential geometry, with respect to the connection given by the Christoffel symbols Γ^t_{rs}. It is rather simple to check that these transform as tensors, but to handle the general case it is necessary to develop a general technique, which, however will not be discussed here.

The above construction of tensorial derivatives may be used with advantage in the study of various asymptotic statistical expansions. It leads, for instance, to the compact and invariant decompositions of Bartlett adjustments given by (5.30), (5.33)–(5.37) and (5.56)–(5.61).

The concepts of yokes and symmetric tensorial derivatives give rise to a definition of **invariant Taylor expansions**. Ordinary Taylor expansions are not invariant in that if a function f is approximated by a Taylor expansion up to and including polynomials of degree n, say, then the approximation will depend on which coordinate system on the domain of definition of f one employs.

Let

$$g^r = \not{g}^{rs} g_s, \tag{5.103}$$

$$g^{R_n} = g^{r_1} \cdots g^{r_n}. \tag{5.104}$$

Then f may be expanded around $\theta \in \Omega$ as

$$f(\theta') = f(\theta) + \sum_{\nu=1}^{\infty} \frac{1}{\nu!} f_{/\!/R_\nu}(\theta) g^{R_\nu}(\theta; \theta'). \tag{5.105}$$

where the tensorial derivatives $f_{/\!/R_\nu}$ are to be taken as those determined by $\overset{-1}{\Gamma}$ of (5.100), via (5.101). The quantity g^r behaves as a contravariant tensor in θ and as a scalar in θ', so that each of the terms

$$\frac{1}{\nu!} f_{/\!/R_\nu}(\theta) g^{R_\nu}(\theta; \theta') \tag{5.106}$$

is invariant under changes of coordinates. Moreover, for each ν, (5.106) is of the same order of magnitude in $\theta' - \theta$ as the corresponding term in the ordinary Taylor expansion of f in θ-coordinates.

Note that if for a fixed θ we transform θ' to a new set of coordinates γ given by (5.103), i.e.

$$\gamma^r = g^r(\theta; \theta') = \not{g}^{rs}(\theta) g_s(\theta; \theta'), \tag{5.107}$$

so that $\gamma = 0$ corresponds to $\theta' = \theta$, then in the new coordinate system the tensorial derivatives of f at 0 equal the ordinary derivatives at 0. This, in fact, establishes the validity of the equality in (5.105).

Dually, we may instead of $\overset{-1}{\Gamma}$ and g^r, of (5.100) and (5.103), take $\overset{1}{\Gamma}$

of (5.99) and

$$g^{;r} = \not{g}^{rs} g_{;s}. \tag{5.108}$$

Then (5.105) is again valid as an invariant Taylor expansion provided the tensorial derivatives $f_{//R_\nu}$ are now defined with $\Gamma = \overset{1}{\Gamma}$ in (5.101) and provided we replace g^{R_ν} by $g^{;R_\nu} = g^{;r_1} \cdots g^{;r_\nu}$.

5.7 Appendix: Differentiation of matrix quantities

We provide here a few known results on differentiation of matrix functions, both because of their repeated usefulness in the following and because the derivation to be presented illustrates the power of the index notation. Let $A = A(\theta)$ be an $m \times m$ matrix depending on a d-dimensional parameter θ with generic coordinates $\theta^r, \theta^s, \theta^t, \ldots$. The entries of A and its inverse A^{-1} will be denoted by a_{ij} and a^{ij}, respectively. We wish to show that

$$\{\log |A|\}_{/r} = a^{ji} a_{ij/r}, \tag{5.109}$$

$$\{\log |A|\}_{/rs} = -a^{jk} a^{li} a_{kl/r} a_{ij/s} + a^{ji} a_{kl/r} a_{ij/s} + a^{ji} a_{ij/rs}, \tag{5.110}$$

$$a^{ij}_{/r} = -a^{ik} a^{lj} a_{kl/r}. \tag{5.111}$$

To prove (5.109), let $A_{\bar{ij}}$ denote the $(m-1) \times (m-1)$ matrix obtained from A be deleting the ith row and the jth column. The cofactor of a_{ij} is defined as

$$C_{ij} = (-1)^{i+j} |A_{\bar{ij}}| \tag{5.112}$$

and we have for the determinant of A the formulae

$$A = \sum_{j=1}^{m} a_{ij} C_{ij} = \sum_{j=1}^{m} a_{ji} C_{ji}. \tag{5.113}$$

Furthermore, the inverse of A is given by

$$a^{ij} = |A|^{-1} C_{ji}. \tag{5.114}$$

Consequently,

$$\begin{aligned} \{\log |A|\}_{/r} &= |A|^{-1} |A|_{/r} \\ &= |A|^{-1} |A|_{/a_{ij}} a_{ij/r} \\ &= |A|^{-1} C_{ij} a_{ij/r} \\ &= a^{ji} a_{ij/r}, \end{aligned} \tag{5.115}$$

verifying (5.109).

By definition

$$a^{ik}a_{kj} = \delta^i_j \tag{5.116}$$

where δ^i_j is the Kronecker delta. Differentiation of (5.116) yields

$$a^{ik}_{/r}a_{kj} + a^{ik}a_{kj/r} = 0 \tag{5.117}$$

from which (5.111) follows. Finally, formula (5.110) is a simple consequence of (5.109) and (5.111).

Further results and exercises

5.1 Show that derivatives of joint cumulants of log-likelihood derivatives follow the same type of rules of calculation as do derivatives of joint moments of log-likelihood derivatives. Further, verify formula (5.9).

[Skovgaard, 1986; Section 5.2]

5.2 Derive the following expansions:

$$\begin{aligned}
\hat{j}_{rs} &= i_{rs} - (H_{rs} + \nu_{rst}\ell^t) - \tfrac{1}{2}\{(\nu_{rstu} + \nu_{rsv}\nu_{tuw}i^{vw})\ell^t\ell^u \\
&\qquad + (\nu_{rsu}H_{vt}i^{uv} + H_{rst})\ell^t\} + \cdots \\
\hat{j}^{rs} &= i^{rs} + i^{rt}i^{su}(\nu_{tuv}\ell^v + H_{tu}) \\
&\qquad \blacktriangledown + \tfrac{1}{2}(\nu_{tuvw} + \nu_{tuv}\nu_{wxy}i^{vw} + 2\nu_{sxv}\nu_{uyw}i^{vw})\ell^x\ell^y \\
&\qquad + (\nu_{tuv}H_{wx}i^{vw} + H_{sux} + H_{tv}\nu_{uxw}i^{vw}[2])\ell^x \\
&\qquad + H_{sv}H_{wu}i^{vw} \blacktriangledown + \cdots \\
\log(|j|/|\hat{j}|) &= \nu_{rst}i^{st}\ell^r + \tfrac{1}{2}(\nu_{rst}\nu_{uvw}i^{tu}i^{vw} \\
&\qquad + \nu_{rsu}\nu_{svw}i^{tv}i^{uw} + \nu_{rstu}i^{tu})\ell^r\ell^s \\
&\qquad \blacktriangledown + (H_{rs}\nu_{tuv}i^{st}i^{uv} + \nu_{rst}H_{uv}i^{su}i^{tv} + H_{rst}i^{st})\ell^r \blacktriangledown + \cdots.
\end{aligned}$$

[Sections 5.3 and 5.7]

5.3 Verify formulae (5.62) and (5.63).

[Section 5.4]

5.4 As a rule, differentiation of asymptotic relations preserves the asymptotic orders. However, care has to be exercised. Consider for instance the relation $\ell_\theta = O_p(n^{1/2})$, expressing the fact that, subject to standard conditions, the score is of order $n^{1/2}$, as follows from the asymptotic normality with mean 0 of ℓ_θ. Formal differentiation of that relation would give $\ell_{\theta\theta} = O_p(n^{1/2})$, whereas the appropriate

statement is $\ell_{\theta\theta} = O(n)$. Discuss this issue and formulate a resolution.
[Sections 5.3 and 5.4]

5.5 A linear combination of cumulants is said to be **stable** if to first asymptotic order it is unaffected by conditioning on an ancillary statistic. Show that the quantities $\lambda_{R_m;S_n}$ are stable in this sense, and that the same is therefore true of, for instance, $\lambda_{r,s,t,u} - \lambda_{rs,tu}[3]$ and $\lambda_{r,s,tu} + \lambda_{rs,tu}$.
[McCullagh (1987); Barndorff-Nielsen and Blæsild (1993a); Section 5.5]

5.6 Derive the expansion (5.55) for $\ell(\hat{\theta}) - \ell(\theta)$ directly by invariant Taylor expansion based on the observed likelihood yoke.
[Section 5.6]

Bibliographical notes

Mixed log model derivatives were introduced in Barndorff-Nielsen (1986b). In particular, the crucial relation (5.14) was noted there and was the starting point for the development of 'observed likelihood geometries', which have been further studied, *inter alia*, by Barndorff-Nielsen (1987a; 1988a) and Barndorff-Nielsen and Jupp (1987).

The technique and results for expected/observed likelihood expansions follow Lawley (1956) rather closely; see also McCullagh (1987). For the observed likelihood expansions the procedures were developed by Barndorff-Nielsen (1986b; 1986c; 1988a). The possibilities for translation between observed and expected expressions were studied systematically by Barndorff-Nielsen and Blæsild (1993a).

The concepts of yokes and of invariant Taylor expansions were introduced by Barndorff-Nielsen (1987b); see also Barndorff-Nielsen (1989).

McCullagh and Cox (1986) discussed the expected Bartlett adjustment R of (5.30) from the viewpoint of invariant derivation and decomposition. They considered, in particular, a decomposition of (5.30) into six invariant terms, i.e. scalars, rather than the decomposition into four such terms given by the formulae (5.33)–(5.37).

McCullagh and Cox's (1986) paper pointed the way towards the present systematic development of invariant Taylor expansions and towards a theory of certain types of geometric objects called **derivative strings, differential strings** and, more generally, **phyla**. There are also connections to stochastic analysis. For a brief general account, see Barndorff-Nielsen, Jupp and Kendall (1991).

Many of the results discussed in this chapter as well as other useful formulae for asymptotic statistical calculations are summarized by Barndorff-Nielsen, Blæsild, Pace and Salvan (1991).

For systematic discussion of tensor notation in statistics, see McCullagh (1984b; 1987).

Extensive discussions of the relations between statistical inference and differential geometry are given by Amari (1985; 1987a; 1987b); Barndorff-Nielsen, Cox and Reid (1986); Barndorff-Nielsen (1988a); and Murray and Rice (1993).

The application of computer algebra to handle expansions such as those discussed in this chapter is considered by Kendall (1993) and Andrews and Stafford (1993).

CHAPTER 6

Higher-order theory: Likelihood combinants

6.1 Introduction

We now discuss higher-order refinements of the classical theory of the limiting behaviour in distribution of the most important likelihood quantities. Some results of this type have, in fact, already been considered in sections 5.3–5.5.

Questions of distributional behaviour can be approached from first principles, by the techniques of Edgeworth expansions and using results of the type given in sections 5.3–5.5. Instead, however, we shall exploit the fact, which has gradually emerged, that to a large extent the essential results needed for asymptotic likelihood analysis are wrapped up in the so-called p^*-formula for the conditional distribution of the maximum likelihood estimator. The p^*-formula has, in particular, the remarkable property of giving excellent results for both exponential models and transformation models. Also, the error statements for p^* refer to relative error whereas in Edgeworth-type expansions the errors are absolute.

From the p^*-formula it is possible, rather directly, to obtain third-order distributional refinements for the score and profile score statistics, the log-likelihood ratio statistic, and the directed likelihood (or signed log-likelihood ratio). A special aspect of this is that the p^*-formula provides a means of calculating very accurate approximations to various types of interval probability.

All these results are expressed in terms of 'observed' quantities and are conditional on ancillaries.

Analogous results involving 'expected' calculations can in some cases be obtained simply from the 'observed' results by translation from 'observed' to 'expected' terms, using the 'dictionary' given in section 5.5. This further obviates the need for extensive, and more or less repetitive and long-winded, calculations.

Moreover, many of the results will be so phrased as to apply immediately in an unconditional setting.

A further advantage of the approach taken here is that it leads to better understanding of the error terms because in several cases it provides fairly explicit expressions for these terms.

We shall distinguish between different degrees of approximation by writing $\overset{\circ}{=}$, $\overset{\dagger}{=}$ or $\overset{*}{=}$ when, under ordinary repeated sampling, the relation in question is valid with relative error $O(n^{-1/2})$, $O(n^{-1})$ or $O(n^{-3/2})$, respectively. Further, we say that a statement holds with $^\circ$- or †- or *-accuracy, or equivalently, that it is correct to first, second or third order, if the statement has asymptotic relative error $O(n^{-1/2})$, $O(n^{-1})$ or $O(n^{-3/2})$ respectively. Likewise, if a combinant, i.e. a function of the data and the parameter, has a † or * as suffix this signifies that it yields results of † or *-accuracy. These conventions are adopted because, whenever possible, the approximate relations to be discussed will be given an 'invariant' form not referring to a particular sampling scheme. That is both mathematically natural and reflects that the relations provide good approximations in many circumstances where ordinary sample size is a poor indicator of the degree of accuracy. Generally speaking, it is the actual information content in the likelihood function rather than sample size that determines the accuracy.

Reparameterizations $(\psi, \chi) \to (\phi, \lambda)$ of the form

$$\phi = \phi(\psi), \quad \lambda = \lambda(\psi, \chi)$$

will be referred to as **interest-respecting reparameterizations** because they do not involve any essential change of the parameter of interest.

Further, the notational conventions introduced in Chapter 5 are used throughout.

6.2 The p^*-formula

We now consider a remarkably simple formula for the conditional distribution of a maximum likelihood estimator giving either a very close approximation to that distribution or, indeed, in some cases the exact distribution.

For a given ancillary statistic a, the log-likelihood can be considered as a function of θ and of the data, the latter via $(\hat{\theta}, a)$, so that we can write the log-likelihood ℓ as $\ell(\theta; \hat{\theta}, a)$ and the normed form $\bar{\ell}$ as

$$\bar{\ell}(\theta; \hat{\theta}, a) = \ell(\theta; \hat{\theta}, a) - \ell(\hat{\theta}; \hat{\theta}, a). \tag{6.1}$$

This way of considering the log-likelihood function has already been discussed earlier; see, in particular, sections 2.5, 5.2 and 5.4.

The ***p**-formula** is defined by

$$p^*(\hat{\theta}; \theta \mid a) = c\,|\hat{j}|^{1/2} e^{\bar{\ell}}. \tag{6.2}$$

The formula determines a probability density, with respect to Lebesgue measure, if the range space of $\hat{\theta}$ is an open subset of $\mathbb{R}^d$ and with respect to counting measure when the range space is discrete. In (6.2) $|\hat{j}|$ is the determinant of the observed information matrix for θ evaluated at $\theta = \hat{\theta}$ and considered as a function of $\hat{\theta}$ and a. Further, c is a normalizing constant, $c = c(\theta, a)$, chosen so that the total integral of (6.2) with respect to $\hat{\theta}$, keeping a fixed, is unity.

Before discussing the accuracy of $p^*(\hat{\theta}; \theta \mid a)$ as an approximation to the actual conditional density $p(\hat{\theta}; \theta \mid a)$ we note below three important properties of the p^*-formula. For properties 2 and 3 we assume for simplicity that, corresponding to each fixed value of a, the range of $\hat{\theta}$ is the full parameter space except possibly for its boundary points.

1 The function p^* is invariant under one-to-one transformation of the data y.
2 The expression p^* is parameterization invariant.
3 The norming constant c is parameterization invariant.

Property 1 is immediate from the definition (6.2) of p^*. Property 2 may be reformulated as:

3′ If θ and ω are alternative parameterizations then, in a natural extended notation,

$$p^*(\hat{\omega}; \omega \mid a) = |\partial\hat{\theta}/\partial\hat{\omega}|\, p^*(\hat{\theta}; \theta \mid a). \tag{6.3}$$

Because there is no ambiguity we have omitted the suffices from formulae such as (6.3).

The validity of (6.3) as well as the fact that the value of c does not depend on the parameterization, i.e. property 3, follow from

$$|j_\omega(\hat{\omega})| = |\partial\hat{\theta}/\partial\hat{\omega}|^2 |j_\theta(\hat{\theta})|. \tag{6.4}$$

Note that $\hat{\theta}$ and $\hat{\omega}$ are in one-to-one correspondence given the value of a.

As an approximation to $p(\hat{\theta};\theta|a)$ the p^*-formula (6.2) is, in broad generality, accurate to order $O(n^{-3/2})$, i.e.

$$p(\hat{\theta};\theta|a) = p^*(\hat{\theta};\theta|a)\{1 + O(n^{-3/2})\}$$

or written equivalently,

$$p(\hat{\theta};\theta|a) \stackrel{*}{=} p^*(\hat{\theta};\theta|a), \tag{6.5}$$

the error term being a relative error.

When θ is the mean value parameter of a (1, 1) exponential model we have that p^* equals the normalized saddlepoint approximation. This follows immediately from (4.42) and (4.51) of Barndorff-Nielsen and Cox (1989). More generally, for (m, m) exponential models p^* agrees with the normalized tilted approximation (Barndorff-Nielsen and Cox, 1989; section 6.5). In this special case the above statement on the $O(n^{-3/2})$-accuracy of the p^*-formula follows immediately from the properties of the normalized tilted approximation. A verification of (6.5) for curved exponential models will be given in section 7.4.

Furthermore, for a considerable range of models formula (6.5) is in fact exact, i.e. $\stackrel{*}{=}$ can be replaced by $=$. In particular, this is the case for virtually all transformation models. For such models, $c(\theta, a)$ is constant in θ. This follows from formula (2.95). In fact, under minor regularity assumptions, that formula specializes to

$$p(\hat{\theta};\theta|a) = c(a)e^{\ell-\hat{\ell}}$$

when the density is relative to invariant measure $\nu(d\hat{\theta})$. A direct calculation shows that this measure is of the form

$$\nu(d\hat{\theta}) = |\hat{j}|^{1/2}d\hat{\theta}. \tag{6.6}$$

Hence, with Lebesgue measure as the dominating measure, we have

$$\begin{aligned} p(\hat{\theta};\theta|a) &= c(a)|\hat{j}|^{1/2}e^{\ell-\hat{\ell}} \\ &= p^*(\hat{\theta};\theta|a), \end{aligned} \tag{6.7}$$

as claimed.

Instances of other interesting models for which p^* is exact, but which are not transformational, are given below, in Example 6.2 and in *Further results and exercises* 6.3–6.5. Under many but not all of those models the norming constant c equals $(2\pi)^{-d/2}$ exactly, i.e.

$\bar{c} = 1$ where $\bar{c}$ is defined by

$$\bar{c} = (2\pi)^{d/2} c. \tag{6.8}$$

Note, in relation to formula (6.5), that while an approximation to $p(\hat{\theta}; \theta | a)$ of $O(n^{-3/2})$-accuracy could also be obtained by Edgeworth expansion, not only would the error term be absolute rather than relative but also the exactness properties discussed above would be lost. Moreover, while Edgeworth approximations deteriorate fatally for large deviations from the centre of the approximated distribution, the p^*-formula generally has good large-deviation properties. Moreover, in a range of cases, even though p^* is not exact the approximation improves the more extreme the deviation.

As already mentioned, in some interesting instances we have $\bar{c} = 1$. In this connection note that comparing the right-hand side of (6.2) to the usual asymptotic normal distribution for $\hat{\theta}$ discussed in section 3.3, one sees that c must often be close to $(2\pi)^{-d/2}$ and that the error of this approximation may be expected to be at most $O(n^{-1/2})$ under repeated sampling. In fact, the approximation is ordinarily better, $O(n^{-1})$ for repeated sampling, i.e.

$$\bar{c} \stackrel{\dagger}{=} 1. \tag{6.9}$$

This will be verified in section 6.3.

It is convenient also to introduce the notation

$$p^\dagger(\hat{\theta}; \theta | a) = (2\pi)^{-d/2} |\hat{j}|^{1/2} e^{\bar{l}}. \tag{6.10}$$

We shall refer to this as the **$p^\dagger$-formula**. In view of the result (6.9), $p^\dagger(\hat{\theta}; \theta | a)$ yields in general an approximation to $p(\hat{\theta}; \theta | a)$ of asymptotic accuracy $O(n^{-1})$. Importantly, moreover, it has the property that this degree of approximation holds for essentially any ancillary a which is asymptotically distribution constant to order $O(n^{-1/2})$, i.e. $p^\dagger$ is stable in the sense to be defined in section 8.1. Note also that $p^\dagger$ is parameterization invariant, i.e. we have, in analogy with (6.3),

$$p^\dagger(\hat{\omega}; \omega | a) = |\partial\hat{\theta}/\partial\hat{\omega}| p^\dagger(\hat{\theta}; \theta | a). \tag{6.11}$$

As compared to $p^\dagger$, the p^*-formula gives at least one degree higher accuracy but at the cost of having to calculate the normalizing constant $c(\theta, a)$ and, more importantly, of having to specify an ancillary which, if not exact, has to be equivalent, to order $O(n^{-1})$, to an adjusted form of a directed likelihood statistic, distribution constant to order $O(n^{-1})$. In particular, under a (2, 1) curved exponential model, in order for p^* to be accurate to order $O(n^{-3/2})$ the ancillary has, to order $O(n^{-1})$, to agree with a biased adjusted version of the directed likelihood for testing the curved model against the ambient (2, 2) exponential model. (For a detailed discussion of this latter point see Barndorff-Nielsen, 1984.)

Example 6.1 Exponential distribution
Suppose that $Y_1, \dots, Y_n$ is a sample of size n from the exponential distribution of mean μ, i.e. from the density $\mu^{-1}e^{-y/\mu}$. Then $\ell(\mu; \hat{\mu}) = -n \log \mu - n\hat{\mu}/\mu$, where $\hat{\mu} = \bar{y}$, so that $\bar{\ell} = \ell - \hat{\ell} = n \log(\hat{\mu}/\mu) - n\hat{\mu}/\mu + n$, $\hat{j} = n/\hat{\mu}^2$ and the $p^\dagger$-formula (6.10) yields for the density of $\hat{\mu}$

$$p^\dagger(\hat{\mu}; \mu) = \bar{\Gamma}(n)^{-1}(\hat{\mu}/\mu)^{n-1}(1/\mu)e^{-n\hat{\mu}/\mu}, \tag{6.12}$$

where

$$\bar{\Gamma}(n) = (2\pi)^{1/2} n^{n-1/2} e^{-n}$$

is Stirling's aproximation to $\Gamma(n)$. It follows, in particular, that $p^*(\hat{\mu}; \mu)$, which is obtained from $p^\dagger(\hat{\mu}; \mu)$ by normalization, is exactly equal to $p(\hat{\mu}; \mu)$, in agreement with the fact that the model is of transformation type.

If, instead, the rate parameter $\rho = \mu^{-1}$ is used to specify the distribution, so that the density is $\rho e^{-\rho x}$, we have $\hat{\rho} = \bar{y}^{-1}$ and, with an obvious change of notation, $\ell(\rho; \hat{\rho}) = n \log \rho - n\rho/\hat{\rho}$, $\bar{\ell} = n \log(\rho/\hat{\rho}) - n\rho/\hat{\rho} + n$, $\hat{j} = n/\hat{\rho}^2$ and the $p^\dagger$-formula gives the density of $\bar{y}^{-1}$ in exact agreement with (6.12), illustrating in simple explicit form the invariance property (6.11). □

Example 6.2 Inverse Gaussian distribution
The density of the inverse Gaussian distribution with parameters $\lambda > 0$ and $\psi > 0$ is

$$\frac{\sqrt{\lambda}}{\sqrt{(2\pi)}} e^{\sqrt{(\lambda\psi)}} y^{-\frac{3}{2}} \exp\{-\tfrac{1}{2}(\lambda y^{-1} + \psi y)\}.$$

For samples $Y_1, \dots, Y_n$ of sizes $n > 1$ the p^*-formula for the joint distribution of $(\hat{\lambda}, \hat{\psi})$ is exact. It is also exact for the distribution of $\hat{\psi}$ when λ is known.

This is simple to verify from the facts that $\bar{Y}$ and $\sum(Y_i^{-1} - \bar{Y}^{-1})$ are independent and have the inverse Gaussian distribution with parameters $(n\lambda, n\psi)$ and the $\lambda^{-1}\chi^2_{n-1}$ distribution, respectively (Tweedie, 1957).

If ψ is considered as known one finds

$$p^*(\hat{\lambda}; \lambda, \psi) = c_0(\lambda, \psi)\hat{\lambda}^{-\frac{n}{2}-1}\{1 + \tfrac{1}{2}\sqrt{(\hat{\lambda}\psi)}\}^{\frac{1}{2}}$$
$$\exp\left[-\frac{n}{2}\sqrt{(\hat{\lambda}\psi)} - \frac{n}{2}\left(\frac{\lambda}{\hat{\lambda}}\right)\{1 + \sqrt{(\hat{\lambda}\psi)}\}\right], \tag{6.13}$$

where

$$\hat{\lambda} = 4\{\sqrt{(\psi + 4n^{-1}\textstyle\sum y_i^{-1})} - \sqrt{\psi}\}^{-2}$$

and $c_0(\lambda, \psi)$ is a norming constant which would typically be computed numerically for given λ, ψ.

In this case an explicit formula for $p(\hat{\lambda}; \lambda, \psi)$ is not available and p^* is not exact. The relative error in approximating $p(\hat{\lambda}; \lambda, \psi)$ by $p^*(\hat{\lambda}; \lambda, \psi)$ is $O(n^{-3/2})$. □

In the examples above, no ancillary statistic is involved, because in the prime exponential models the maximum likelihood estimates are minimal sufficient and determine the components of the p^*-formula.

Example 6.3 Location model
Suppose now that $Y_1, \ldots, Y_n$ form a sample from the location family $h(x - \mu)$; see Example 1.6 and section 2.5. The order statistics $Y_{(n1)}, \ldots, Y_{(nn)}$ are sufficient and usually minimal sufficient. The sample configuration

$$A = (Y_{(nn)} - Y_{(n1)}, \ldots, Y_{(n2)} - Y_{(n1)})$$

has a distribution not involving μ and hence A is ancillary. The log-likelihood is determined by the maximum likelihood estimate $\hat{\mu}$ and a, or equivalently by $Y_{(n1)}$ and a, which together determine the full set of order statistics. The observed information is a function only of a and hence the p^*-formula, which is exact, implies that the density of $\hat{\mu}$ is proportional to the normalized likelihood function. This result is proved directly in Example 2.18. □

Example 6.4 Location–scale model
For the location–scale model the p^*-formula takes the form

$$p^*(\hat{\mu}, \hat{\sigma}; \mu, \sigma | a) = c_0(a)\sigma^{-n}\hat{\sigma}^{n-2}\prod_{i=1}^{n} f\left(\frac{\hat{\mu} - \mu}{\sigma} + \frac{\hat{\sigma}}{\sigma}a_i\right) \tag{6.14}$$

where $a_i = (y_i - \hat{\mu})/\hat{\sigma}$, $a = (a_1, \ldots, a_n)$ is the configuration statistic, and

$c_0(a)$ is a norming constant. Since the location–scale model is a transformation model $p^*(\hat{\mu}, \hat{\sigma}; \mu, \sigma | a)$ is equal to the actual conditional density $p(\hat{\mu}, \hat{\sigma}; \mu, \sigma | a)$, in agreement also with Example 2.34. □

6.3 The score and the profile score

We now consider the application of the p^*-formula to the score vector. Given an ancillary a, the maximum likelihood estimator $\hat{\theta}$ and the score vector $u = \nabla \ell$ with coordinates ℓ_r will, in general, be in one-to-one correspondence for a region of values of $\hat{\theta}$ around the true parameter value θ, and this region will carry all the probability mass, except for an asymptotically negligible amount. The Jacobian of the transformation from $\hat{\theta}$ to the vector of derivatives $\ell_r = \ell_r(\theta; \hat{\theta}, a)$ is the matrix $\ell_{;}$ of mixed second-order log model derivatives $\ell_{r;s}$, as defined in section 5.2, in which the suffix r refers to derivatives with respect to θ and the suffix s to derivatives with respect to $\hat{\theta}$. From the results discussed in section 6.2 we therefore have that an approximation of high accuracy to the conditional distribution of the score vector given a is provided by

$$p(u; \theta | a) \doteq p^*(u; \theta | a), \tag{6.15}$$

where

$$p^*(u; \theta | a) = c(\theta, a) | \hat{j} |^{\frac{1}{2}} | \ell_{;} |^{-1} e^{\bar{\ell}}. \tag{6.16}$$

In line with definition (6.10), let

$$p^\dagger(u; \theta | a) = (2\pi)^{-d/2} | \hat{j} |^{1/2} | \ell_{;} |^{-1} e^{\ell - \hat{\ell}}, \tag{6.17}$$

so that

$$p^*(u; \theta | a) = \bar{c} p^\dagger(u; \theta | a), \tag{6.18}$$

$\bar{c}$ being the standardized norming constant of $p^*(\hat{\theta}; \theta | a)$; see (6.8).

We now derive asymptotic expansions of $p^\dagger(u; \theta | a)$ and $p^*(u; \theta | a)$ that are similar to Edgeworth expansions but in which the coefficients are determined by mixed derivatives of the log model function, rather than by cumulants.

Using the expansions in section 5.4 we find that $p^\dagger(u; \theta | a)$ may be expanded around the d-dimensional normal distribution with mean 0 and covariance matrix $\jmath$ as

$$p^\dagger(u; \theta | a) \doteq \phi(u; \jmath)(1 + S_1' + S_2'), \tag{6.19}$$

where the terms S_1' and S_2' are of order $O(n^{-1/2})$ and $O(n^{-1})$, respectively, and

$$S_1' = \tfrac{1}{6}\not{l}_{rst}h^{rst}(u;\not{j}), \tag{6.20}$$

$$\begin{aligned} S_2' = &-\tfrac{1}{24}\{\not{l}_{rstu} + \not{l}_{rs;tu}[6]\}h^{rstu}(u;\not{j}) - \tfrac{1}{8}\not{l}_{rstu}\not{j}^{rs}\not{j}^{tu} \\ &- \tfrac{1}{4}\not{l}_{rs;tu}\{\not{j}^{tu}h^{rs}(u;\not{j}) + 2\not{j}^{rs}\not{j}^{tu}\} \\ &+ \tfrac{1}{72}\not{l}_{rst}\not{l}_{uvw}\{h^{rstuvw}(u;\not{j}) - 9\not{j}^{rs}\not{j}^{tu}\not{j}^{vw}\} \\ &- \tfrac{1}{12}\not{l}_{rtv}\not{l}_{suw}\not{j}^{rs}\not{j}^{tu}\not{j}^{vw}. \end{aligned} \tag{6.21}$$

Here, as elsewhere, the covariance matrix $\not{j}$ has elements $\not{j}_{rs} = [-\partial^2\ell(\theta;\hat{\theta},a)/\partial\theta^r\partial\theta^s]_{\hat{\theta}=\theta}$, in which a is fixed and $\hat{\theta}$ is set equal to θ after differentiation. The tensors $\not{l}$ are given by the formulae (5.59)–(5.61) and the factors h are tensorial Hermite polynomials (Barndorff-Nielsen and Cox, 1989, p. 151), the first few of which are

$$\begin{aligned} h^{rs}(u;\not{j}) &= u^r u^s - \not{j}^{rs}, \\ h^{rst}(u;\not{j}) &= u^r u^s u^t - \not{j}^{rs}u^t[3], \\ h^{rstu}(u;\not{j}) &= u^r u^s u^t u^u - \not{j}^{rs}u^t u^u[6] + \not{j}^{rs}\not{j}^{tu}[3], \end{aligned}$$

with $u^r = \not{j}^{rs}u_s$, u_s being the sth coordinate of the score vector u.

Insertion of (6.19)–(6.21) in (6.18) and subsequent integration over u shows that

$$\bar{c} \doteq 1 + \bar{\not{C}} \tag{6.22}$$

where

$$\bar{\not{C}} = (\tfrac{1}{8}\not{l}_{rstu} + \tfrac{1}{2}\not{l}_{rs;tu})\not{j}^{rs}\not{j}^{tu} + (\tfrac{1}{8}\not{l}_{rst}\not{l}_{uvw} + \tfrac{1}{12}\not{l}_{rtv}\not{l}_{suw})\not{j}^{rs}\not{j}^{tu}\not{j}^{vw}. \tag{6.23}$$

Under repeated sampling $\bar{\not{C}}$ is typically $O(n^{-1})$. This, in particular, shows the validity of formula (6.9).

Combination of (6.19)–(6.21) and (6.22)–(6.23) now yields

$$p^*(u;\theta|a) \doteq \phi(u;\not{j})(1 + S_1 + S_2), \tag{6.24}$$

where

$$S_1 = \tfrac{1}{6}\not{l}_{rst}h^{rst}(u;\not{j}), \tag{6.25}$$

$$\begin{aligned} S_2 = &-\tfrac{1}{24}\{\not{l}_{rstu} + \not{l}_{rs;tu}[6]\}h^{rstu}(u;\not{j}) \\ &- \tfrac{1}{4}\not{l}_{rs;tu}\not{j}^{tu}h^{rs}(u;\not{j}) + \tfrac{1}{72}\not{l}_{rst}\not{l}_{uvw}h^{rstuvw}(u;\not{j}), \end{aligned} \tag{6.26}$$

S_1 and S_2 being of order $n^{-1/2}$ and n^{-1}, respectively, under ordinary repeated sampling. It follows from (6.25) and (6.26) that S_1 and S_2 are both parameterization invariant, the only variant part of the right-hand side of (6.24) being the multivariate normal density.

Indeed, the general structure of (6.25) and (6.26) could have been anticipated via an appeal to the required invariance followed by use of the standard procedure of producing invariants by contraction of tensors.

Example 6.5 The normal distribution

The simplest illustration of these formulae is obtained from the normal distribution of mean μ and variance σ^2. The score vector is

$$n\begin{pmatrix} (\bar{Y}-\mu)/\sigma^2 \\ -\frac{1}{2}\sigma^{-2}+\frac{1}{2}\sigma^{-4}\{(\bar{Y}-\mu)^2+\hat{\sigma}^2\} \end{pmatrix},$$

where $\hat{\sigma}^2 = \sum(Y_i - \bar{Y})^2/n$. Formula (6.24) specifies an asymptotic expansion of the distribution of the score vector, which could also be derived directly from the distribution of $(\bar{Y}, \hat{\sigma}^2)$, which are independently normal and proportional to chi-squared. □

Note that Edgeworth or saddlepoint approximations to the marginal distribution of u are in principle straightforward to establish, and that the repeated sampling properties of these approximations apply directly to u since under ordinary repeated sampling u is a sum of independent, identically distributed random variables. For the Edgeworth approximation with error $O(n^{-3/2})$ all we need to know are the cumulants of u up to order 4, while determination of the saddlepoint approximation requires knowledge of the cumulant transform of u.

Next we turn to the profile score for the interest parameter ψ, i.e.

$$\ell_{P\psi} = \ell_P(\psi)_{/\psi} = \ell_\psi(\psi, \hat{\chi}_\psi).$$

This is, under wide conditions, approximately normally distributed with mean 0 and covariance matrix

$$\operatorname{var}(\ell_{P\psi};\theta) \mathrel{\dot{=}} j_P(\hat{\psi}), \tag{6.27}$$

where $j_P(\psi)$ is the observed information calculated from the profile log likelihood. Our interest here is in obtaining second- and third-order refinements of this fact.

The mean of $\ell_{P\psi}$ is, unlike the mean of the full score vector, not in general exactly 0; see the more elementary discussion in section 4.3. By expanding $\ell_{P\psi}$ in the argument $\hat{\chi}_\psi$ around χ one finds, after some algebra, that the mean has an asymptotic expansion whose first term is generally of order $O(1)$ and is given by

$$E(\ell_{P\psi};\theta) = -\tfrac{1}{2}\mathrm{tr}\{i_{\chi\chi}^{-1}(\psi,\chi)(\tau_{\chi\chi,\psi} + \tau_{\chi,\chi,\psi})\}, \tag{6.28}$$

where $\tau_{\chi\chi,\psi}$ and $\tau_{\chi,\chi,\psi}$ behave as tensors under interest-respecting parameterizations and are defined by

$$\tau_{\chi\chi,\psi} = \nu_{\chi\chi,\psi} - \nu_{\chi\chi,\chi} i_{\chi\chi}^{-1} i_{\chi\psi}, \tag{6.29}$$

$$\tau_{\chi,\chi,\psi} = \nu_{\chi,\chi,\psi} - \nu_{\chi,\chi,\chi} i_{\chi\chi}^{-1} i_{\chi\psi}, \tag{6.30}$$

the quantities $\nu_{\chi\chi,\psi}$, $\nu_{\chi\chi,\chi}$, $\nu_{\chi,\chi,\psi}$ and $\nu_{\chi,\chi,\chi}$ being joint moments of log-likelihood derivatives.

Note in particular that if ψ and χ are orthogonal the final terms on the right-hand sides of (6.29) and (6.30) vanish. Further, it is easily shown on differentiating the equation defining orthogonality that when ψ is a scalar the bias is proportional to $\nu_{\psi\chi\chi}$ and hence vanishes when and only when $\nu_{\psi\chi\chi}$ is zero.

Example 6.6 The normal distribution

The profile score for the variance $\psi = \sigma^2$ is

$$-\tfrac{1}{2}n\psi^{-1}(1 - \hat{\psi}/\psi)$$

and because $E(\hat{\psi}) = (1 - n^{-1})\psi$, the (small) bias in the profile score is clear. As noted in section 4.3, the effect would be more striking in a linear model problem with a large number of linear parameters. □

To study the distribution of $\ell_{P\psi}$ in more detail we transform the p^*-formula for $(\hat{\psi}, \hat{\chi})$ into a probability density for $(\ell_{P\psi}, \hat{\chi}_\psi)$. Written in full, the two equations determining $\ell_{P\psi}$ and χ_ψ are

$$\ell_{P\psi} = \ell_\psi(\psi, \hat{\chi}_\psi; \hat{\psi}, \hat{\chi}, a), \quad \ell_\chi(\psi, \hat{\chi}_\psi; \hat{\psi}, \hat{\chi}, a) = 0. \tag{6.31}$$

As elsewhere, we use a tilde to indicate evaluation of functions of (ψ, χ) at the partial maximum likelihood point $(\psi, \hat{\chi}_\psi)$. Note that, with this convention, $\ell_{P/\psi\psi}$ is generally different from $\tilde{\ell}_{\psi\psi}$, because in $\ell_{P/\psi\psi}$ we have first inserted $(\psi, \hat{\chi}_\psi)$ in ℓ and then differentiated, whereas in $\tilde{\ell}_{\psi\psi}$ these two operations are taken in the opposite order; in formulae,

$$\ell_{P/\psi\psi} = \partial^2\ell(\psi, \hat{\chi}_\psi)/\partial\psi^2,$$

$$\tilde{\ell}_{\psi\psi} = [\partial^2\ell(\psi, \chi)/\partial\psi^2]_{\chi=\hat{\chi}_\psi}.$$

However, we do have $\ell_{P\psi} = \ell_{P/\psi} = \tilde{\ell}_\psi$ because $\ell_\chi(\psi, \hat{\chi}_\psi) = 0$.

Differentiation of the equations (6.31) with respect to $\hat{\psi}$ and $\hat{\chi}$ yields

$$\begin{aligned} \ell_{P/\psi\hat{\psi}} &= \tilde{\ell}_{\psi\chi}\hat{\chi}_{\psi/\hat{\psi}} + \tilde{\ell}_{\psi;\hat{\psi}}, \quad \tilde{\ell}_{\chi\chi}\hat{\chi}_{\psi/\hat{\psi}} + \tilde{\ell}_{\chi;\hat{\psi}} = 0, \\ \ell_{P/\psi\hat{\chi}} &= \tilde{\ell}_{\psi\chi}\hat{\chi}_{\psi/\hat{\chi}} + \tilde{\ell}_{\psi;\hat{\chi}}, \quad \tilde{\ell}_{\chi\chi}\hat{\chi}_{\psi/\hat{\chi}} + \tilde{\ell}_{\chi;\hat{\chi}} = 0, \end{aligned} \tag{6.32}$$

from which it follows that

$$(\tilde{\ell}_{\psi}, \hat{\chi}_{\psi})_{/(\hat{\psi},\hat{\chi})} = \begin{bmatrix} \tilde{\ell}_{\psi;\hat{\psi}} - \tilde{j}_{\psi\chi}\tilde{j}_{\chi\chi}^{-1}\tilde{\ell}_{\chi;\hat{\psi}} & \tilde{j}_{\chi\chi}^{-1}\tilde{\ell}_{\chi;\hat{\psi}} \\ \tilde{\ell}_{\psi;\hat{\chi}} - \tilde{j}_{\psi\chi}\tilde{j}_{\chi\chi}^{-1}\tilde{\ell}_{\chi;\hat{\chi}} & \tilde{j}_{\chi\chi}^{-1}\tilde{\ell}_{\chi;\hat{\chi}} \end{bmatrix}. \tag{6.33}$$

The above equations are true for vector parameters ψ and χ, as well as in the scalar case, by straightforward interpretation as vectors or matrices of the terms entering the equations. From (6.33) we have

$$|(\tilde{\ell}_{\psi}, \hat{\chi}_{\psi})_{/(\hat{\psi},\hat{\chi})}| = |\tilde{j}_{\chi\chi}|^{-1}|\tilde{\ell}_{;}| \tag{6.34}$$

and hence, transforming (6.2),

$$p^*(\tilde{\ell}_{\psi}, \hat{\chi}_{\psi}; \psi, \chi | a) = c|\hat{j}|^{\frac{1}{2}}|\tilde{j}_{\chi\chi}||\tilde{\ell}_{;}|^{-1}e^{\ell - \hat{\ell}} \tag{6.35}$$

There are now two key cases that may be considered, namely those where the minimal sufficient statistic for χ given ψ and a is either just $\hat{\chi}_{\psi}$ or when it is $(\tilde{\ell}_{\psi}, \hat{\chi}_{\psi})$. In the latter case an approximation to the marginal distribution of $\tilde{\ell}_{\psi}$ can be obtained by application of Laplace's method to (6.35), but we shall not develop that approach.

In the first case the joint conditional density of $(\tilde{\ell}_{\psi}, \hat{\chi}_{\psi})$ factorizes as

$$p(\tilde{\ell}_{\psi}, \hat{\chi}_{\psi}; \psi, \chi | a) = p(\hat{\chi}_{\psi}; \psi, \chi | a)p(\tilde{\ell}_{\psi}; \psi | \hat{\chi}_{\psi}, a), \tag{6.36}$$

and then it is pertinent to consider the conditional distribution of $\tilde{\ell}_{\psi}$ given $\hat{\chi}_{\psi}$ and a. An approximation to this distribution is obtained by substituting (6.35) for the left-hand side of (6.36) and by approximating the first factor on the right by means of the p^*-formula. The result is

$$p(\tilde{\ell}_{\psi}; \psi | \hat{\chi}_{\psi}, a) \doteq \{c(\psi, \chi | a)/c(\chi | \psi, a)\}\{|\hat{j}|^{\frac{1}{2}}|\tilde{j}_{\chi\chi}|^{\frac{1}{2}}/|\tilde{\ell}_{;}|\}e^{\tilde{\ell} - \hat{\ell}}, \tag{6.37}$$

where $c(\psi, \chi | a)$ and $c(\chi | \psi, a)$ are the normalizing constants from applying p^* to $(\hat{\psi}, \hat{\chi})$ and to $\hat{\chi}_{\psi}$ for given ψ, respectively.

A general class of models for which $\hat{\chi}_{\psi}$ is minimal sufficient has the form

$$p(y; \psi, \chi) = \exp\{\chi^{\mathrm{T}}t(y; \psi) - u(y; \psi) - h(\psi, \chi)\}. \tag{6.38}$$

Note that this model type is preserved under repeated sampling.

Provided y is in one-to-one correspondence with $(\hat{\psi}, \hat{\chi})$ no ancillary a is called for.

Example 6.7 Exponential model
As a simple case, suppose Y is two-dimensional, $Y = (U, V)$ say, and that Y follows a prime exponential model of order 2,

$$\exp\{\phi u + \chi v - k(\phi, \chi) - g(u, v)\}. \tag{6.39}$$

Let $\psi = \phi/\chi$ be the interest parameter. Then (6.37) becomes

$$p(\tilde{\ell}_\psi; \psi \mid \hat{\chi}_\psi) \doteq c_0 \hat{\chi} |\hat{\Sigma}|^{-\frac{1}{2}} e^{\tilde{\ell} - \hat{\ell}}, \tag{6.40}$$

where Σ is the covariance matrix of (U, V), and c_0 is given by

$$\begin{aligned} c_0 = &\{c(\psi, \chi \mid a)/c(\chi \mid \psi, a)\}(\tilde{k}_{/\phi\phi}\psi^2 + 2\tilde{k}_{/\phi\chi}\psi + \tilde{k}_{/\chi\chi})^{\frac{1}{2}}\hat{\chi}_\psi^{-1} \\ &\cdot \exp\{\hat{\chi}_\psi(\psi u + v) - k(\psi\hat{\chi}_\psi, \hat{\chi}_\psi)\}. \end{aligned} \tag{6.41}$$

The likelihood equation for $\hat{\chi}_\psi$ is

$$\psi u + v = \psi k_{/\phi}(\psi\hat{\chi}_\psi, \hat{\chi}_\psi) + k_{/\chi}(\psi\hat{\chi}_\psi, \hat{\chi}_\psi),$$

showing that $\psi u + v$ depends on ψ and $\hat{\chi}_\psi$ only. Hence, the factor c_0 depends on ψ, χ, and $\hat{\chi}_\psi$ only, and has thus the role of an approximate norming constant. □

6.4 The maximum likelihood estimator revisited

In analogy with formula (6.24) we may derive a further approximation to the conditional distribution of the maximum likelihood estimator from $p^*(\hat{\theta}; \theta \mid a)$, directly via Taylor expansion in $\hat{\theta}$ around θ. Alternatively we may draw on the expansion already obtained for $p^*(u; \theta \mid a)$, i.e. (6.24), by noting that

$$\begin{aligned} p^*(\hat{\theta}; \theta \mid a) &= p^*(u; \theta \mid a) |\ell_{;}| \\ &= \{\phi(u; \jmath)/\phi(\hat{\theta} - \theta; \jmath^{-1})\} |\ell_{;}| (1 + S_1 + S_2) \phi(\hat{\theta} - \theta; \jmath^{-1}) \end{aligned} \tag{6.42}$$

and then expanding each of the three factors in front of $\phi(\hat{\theta} - \theta; \jmath^{-1})$. The resulting expression is

$$p^*(\hat{\theta}; \theta \mid a) \doteq \phi(\hat{\theta} - \theta; \jmath^{-1})(1 + M_1 + M_2), \tag{6.43}$$

where M_1 and M_2 are given by

$$2M_1 = \delta^{rst}(\ell_{rs;t} + \tfrac{2}{3}\ell_{rst}) - \delta^t \jmath^{rs}(\ell_{rs;t} + \ell_{rst}) \tag{6.44}$$

$$
\begin{aligned}
24M_2 = {} & 3(\delta^{rstuvw} - \not{\jmath}^{rs}\not{\jmath}^{tu}\not{\jmath}^{vw}[15])(\not{\ell}_{rs;t} + \tfrac{2}{3}\not{\ell}_{rst})(\not{\ell}_{uv;w} + \tfrac{2}{3}\not{\ell}_{uvw}) \\
& + (\delta^{rstu} - \not{\jmath}^{rs}\not{\jmath}^{tu}[3])\{(6\not{\ell}_{rs;tu} + 8\not{\ell}_{rst;u} + 3\not{\ell}_{rstu}) \\
& - 6\not{\jmath}^{vw}(\not{\ell}_{vw;u} + \not{\ell}_{uvw})(\not{\ell}_{rs;t} + \tfrac{2}{3}\not{\ell}_{rst})\} \\
& - 3(\delta^{rs} - \not{\jmath}^{rs})\{2\not{\jmath}^{tu}(\not{\ell}_{tu;rs} + \not{\ell}_{rtu;s} + \not{\ell}_{stu;r} + \not{\ell}_{rstu}) \\
& + (2\not{\jmath}^{tv}\not{\jmath}^{uv} - \not{\jmath}^{tu}\not{\jmath}^{vw})(\not{\ell}_{tu;r} + \not{\ell}_{rtu})(\not{\ell}_{vw;s} + \not{\ell}_{rvw})\}
\end{aligned}
\tag{6.45}
$$

with $\delta^r = \hat{\omega}^r - \omega^r$ and $\delta^{rs\cdots} = \delta^r\delta^s\cdots$.

From (6.43) the leading term of an asymptotic expansion for the bias of $\hat{\theta} - \theta$ is readily obtained by multiplying on both sides of that relation with $\hat{\theta}^r - \theta^r$ and integrating over $\hat{\theta}$. The result is

$$E(\hat{\theta}^r; \theta) = \theta^r + \not{b}^r + \cdots, \tag{6.46}$$

where

$$\not{b}^r = -\tfrac{1}{2}\not{\jmath}^{rs}\not{\jmath}^{tu}\not{\ell}_{s;tu}. \tag{6.47}$$

The bias correction $\not{b}^r$ is asymptotically equivalent to

$$b^r = -\tfrac{1}{2}i^{rs}i^{tu}(\lambda_{s,tu} + \lambda_{s,t,u}), \tag{6.48}$$

as follows from the dictionary in section 5.5 for translation between observed and expected quantities. (The formulae (6.47) and (6.48) agree with the formulae (5.54) and (5.26), respectively.)

A more direct derivation of the bias correction follows by expanding the maximum likelihood estimating equation to second order in the form

$$\ell_r + (\hat{\theta}^u - \theta^u)\ell_{ru} + \tfrac{1}{2}(\hat{\theta}^u - \theta^u)(\hat{\theta}^t - \theta^t)\ell_{rut} \doteq 0.$$

Take expectations term by term to obtain

$$
\begin{aligned}
& b^u E(\ell_{ru}) + \operatorname{cov}(\hat{\theta}^u - \theta^u, \ell_{ru}) + \tfrac{1}{2}i^{ut}E(\ell_{rut}) \\
& \quad + \tfrac{1}{2}\operatorname{cov}\{(\hat{\theta}^u - \theta^u)(\hat{\theta}^t - \theta^t), \ell_{rut}\} \doteq 0.
\end{aligned}
$$

In the first term $E(\ell_{ru}) = -i_{ru}$. For the second term we can use the first-order expansion $\hat{\theta}^u - \theta^u = i^{ut}\ell_t$ to obtain

$$\operatorname{cov}(\hat{\theta}^u - \theta^u, \ell_{ru}) = i^{tu}E(\ell_t\ell_{ru}) = i^{tu}\lambda_{t,ru}.$$

The third term is $\tfrac{1}{2}i^{tu}\lambda_{rut}$, and the last term is negligible. Thus

$$b^u i_{ru} = i^{tu}\lambda_{t,ru} + \tfrac{1}{2}i^{tu}\lambda_{rut},$$

a set of linear equations for the bias with solution

$$b^r = i^{rs}i^{tu}(\lambda_{t,su} + \tfrac{1}{2}\lambda_{sut}). \tag{6.49}$$

Differentiation of the identity $\lambda_{r,s} + \lambda_{rs} = 0$ shows that (6.49) and (6.48) are equivalent; see also (5.9).

Writing

$$\theta^* = \hat{\theta} - \not{b} \tag{6.50}$$

for the bias-corrected maximum likelihood estimator, we find, from the above and after some algebra,

$$p^*(\theta^*; \theta | a) \stackrel{*}{=} \phi(\theta^*; \jmath^{-1})(1 + S_1 + S_2)(1 + D_1 + D_2), \tag{6.51}$$

where S_1 and S_2 are the invariant terms given by (6.25) and (6.26) and where

$$D_1 = -\tfrac{1}{2}\not{\ell}_{t;rs} h^{rst}(\nabla \ell; \jmath^{-1}), \tag{6.52}$$

$$D_2 = -\tfrac{1}{2}\not{\ell}_{r;stu} H^{rstu}(\nabla \ell) + \tfrac{1}{8}\not{\ell}_{t;rs}\not{\ell}_{u;vw} H^{rstuvw}(\nabla \ell). \tag{6.53}$$

The quantities $H^{rstu}(\nabla \ell)$ and $H^{rstuvw}(\nabla \ell)$ in (6.53) are polynomials in the coordinates of the score vector $\nabla \ell = [\ell_r]$. They transform as contravariant tensors and they are given explicitly by

$$H^{rstu}(\nabla \ell) = \ell^{rstu} - \jmath^{rs}\ell^{tu}[3], \tag{6.54}$$

where $\ell^r = \jmath^{rs}\ell_s$, and by

$$\begin{aligned} H^{rstuvw}(\nabla \ell) = {} & \ell^{rstuvw} + \jmath^{tu}\ell^{rsvw} - 4\jmath^{uw}\ell^{rstv} - \jmath^{vw}\ell^{rstu} \\ & + 4\jmath^{rs}\jmath^{uw}\ell^{tv} + 4\jmath^{ru}\ell^{stvw} - 4\jmath^{ru}\jmath^{tv}\ell^{sw} \\ & + \jmath^{rs}\jmath^{vw}\ell^{rs} - 2\jmath^{rs}\jmath^{tu}\ell^{vw} + 4\jmath^{rt}\jmath^{uw}\ell^{sv} - 4\jmath^{ru}\jmath^{st}\ell^{vw} \\ & + \jmath^{rs}\jmath^{tu}\jmath^{vw}. \end{aligned} \tag{6.55}$$

The terms D_1 and D_2, as well as the bias correction $\not{b}$ to $\hat{\theta} - \theta$, can be made to disappear by choosing a coordinate system in which $\not{\ell}_{r;st} = \not{\ell}_{r;stu} = 0$.

Example 6.8 Normal mean and standard deviation

A simple illustration of these formulae is provided by taking $Y_1, \ldots, Y_n$ to be independently normally distributed with mean μ and variance σ^2 and defining the components of θ to be $\theta^1 = \mu$, $\theta^2 = \sigma$. We concentrate on the bias corrections (6.49).

The first three derivatives of the log-likelihood are easily evaluated in explicit form and their expectations, and those of products such as $(\partial \ell / \partial \mu)\,(\partial^2 \ell / \partial \mu \partial \sigma)$ evaluated to give

$$\begin{aligned} i_{11} &= n/\sigma^2, & i_{12} &= 0, & i_{22} &= 2n/\sigma^2; \\ i^{11} &= \sigma^2/n, & i^{12} &= 0, & i^{22} &= \sigma^2/(2n); \end{aligned}$$

$$\begin{array}{llll} \lambda_{111}=0, & \lambda_{112}=2n/\sigma^3, & \lambda_{122}=0, & \lambda_{222}=10n/\sigma^3; \\ \lambda_{1,11}=0, & \lambda_{1,12}=-2n/\sigma^3, & \lambda_{1,22}=0; & \\ \lambda_{2,11}=0, & \lambda_{2,12}=0, & \lambda_{2,22}=-6n/\sigma^3. & \end{array}$$

For example $\lambda_{2,22}=E\{(\partial\ell/\partial\sigma)(\partial^2\ell/\partial\sigma^2)\}$; this is the most complex of the expressions for evaluation.

Thus the bias terms are respectively

$$\begin{aligned} b^1 &= i^{1s}i^{tu}(\lambda_{t,su}+\tfrac{1}{2}\lambda_{sut}) \\ &= (\sigma^2/n)^2(\lambda_{1,11}+\tfrac{1}{2}\lambda_{111}+\tfrac{1}{2}\lambda_{2,12}+\tfrac{1}{4}\lambda_{122}) \\ &= 0, \end{aligned}$$

as is to be expected because $\hat{\mu}$ is exactly unbiased, and

$$\begin{aligned} b^2 &= i^{2s}i^{tu}(\lambda_{t,su}+\tfrac{1}{2}\lambda_{sut}) \\ &= \tfrac{1}{2}(\sigma^2/n)^2(\lambda_{1,21}+\tfrac{1}{2}\lambda_{211}+\tfrac{1}{2}\lambda_{2,22}+\tfrac{1}{4}\lambda_{222}) \\ &= -3\sigma/(4n). \end{aligned}$$

A direct calculation, using the proportionality of $\hat{\sigma}$ to a chi-variable with $n-1$ degrees of freedom, yields

$$E(\hat{\sigma})=\sqrt{\frac{2}{n}}\frac{\Gamma(\frac{1}{2}n-\frac{1}{2})}{\Gamma(\frac{1}{2}n)}\sigma$$

and asymptotic expansion via Stirling's formula confirms that

$$E(\hat{\sigma})=\sigma\left\{1-\frac{3}{4n}+O(n^{-2})\right\}.$$

More generally, note that if V is the standard unbiased estimate of σ^2, then when $Y_1,\ldots,Y_n$ are independently and identically distributed with variance σ^2 and fourth cumulant ratio ρ_4 we have

$$E(V)=\sigma^2,\quad \operatorname{var}(V)=2\sigma^4(1+\tfrac{1}{2}\rho_4)/n+O(n^{-2}).$$

It follows after a Taylor expansion of V around σ^2 that

$$\begin{aligned} E(\surd V) &= \sigma+\tfrac{1}{2}[d^2(\surd V)/dV^2]_{V=\sigma^2}\operatorname{var}(V)+O(n^{-2}) \\ &= \sigma\left\{1-\frac{1+\frac{1}{2}\rho_4}{4n}+O(n^{-2})\right\}, \end{aligned}$$

so that, because $\hat{\sigma}=\surd\{V(n-1)/n\}$,

$$E(\hat{\sigma})=\sigma\left\{1-\frac{3+\frac{1}{2}\rho_4}{4n}+O(n^{-2})\right\}.$$

□

In more complicated examples it may be difficult to calculate the expectations involved in (6.49). Consistent estimates, yielding a bias correction to the same order of accuracy, can be obtained by replacing λ_{rst}, $\lambda_{r,st}$ respectively by

$$n^{-1}\sum\partial^3\ell_i/\partial\theta^r\partial\theta^s\partial\theta^t, \quad n^{-1}\sum(\partial\ell_i/\partial\theta^r)(\partial^2\ell_i/\partial\theta^s\partial\theta^t),$$

where ℓ_i is the log-likelihood from the ith component observation.

6.5 The likelihood ratio

As a further application of the p^*-formula and some of the results of Chapter 5 we consider the distribution of the likelihood ratio statistic and Bartlett adjustments of this. We again suppose that the likelihood function involves the data via the maximum likelihood estimate and an ancillary statistic and can thus be written $\ell(\theta;\hat{\theta},a)$, occasionally abbreviated to $\ell(\theta)$. We write, as in Chapter 5,

$$\ell_r = \partial\ell(\theta;\hat{\theta},a)/\partial\theta^r, \quad \ell_{;r} = \partial\ell(\theta;\hat{\theta},a)/\partial\hat{\theta}^r, \tag{6.56}$$

with r replaced by 1 when the parameter is one-dimensional.

6.5.1 *Point hypotheses*

When the hypothesis to be tested is a point hypothesis, i.e. when there are no nuisance parameters, the likelihood ratio statistic is given by

$$w = 2\{\ell(\hat{\theta};\hat{\theta},a) - \ell(\theta;\hat{\theta},a)\}. \tag{6.57}$$

We may write the p^*-formula for the distribution of $\hat{\theta}$ in the form

$$p^*(\hat{\theta};\theta\,|\,a) = c\,|\hat{j}|^{1/2}e^{-w/2}, \tag{6.58}$$

where $\hat{j}_{rs} = -\ell_{rs}(\hat{\theta};\hat{\theta},a)$, and hence the conditional density of w, to order $O(n^{-3/2})$, is given by

$$p^*(w;\theta\,|\,a) = ce^{-w/2}\int_{\hat{\theta}|w} |\hat{j}|^{1/2}d\hat{\theta}, \tag{6.59}$$

the integral being over all $\hat{\theta}$ yielding the fixed value w of the likelihood ratio statistic.

To evaluate the integral we must transform from $\hat{\theta}$ to a new set of coordinates of which w is one. This is simplest when θ is one-dimensional and we therefore consider that case first.

Example 6.9 Exponential mean

Let $Y_1, \ldots, Y_n$ be independently exponentially distributed with mean μ. Then $\hat{\mu} = \sum Y_j/n = \bar{Y}$ and we can write

$$\ell(\mu; \hat{\mu}) = -n \log \mu - n\hat{\mu}/\mu,$$
$$w = 2n\{(\hat{\mu}/\mu - 1) - \log(\hat{\mu}/\mu)\}. \tag{6.60}$$

We have that

$$\hat{j} = -\ell_{11;}(\hat{\mu}; \hat{\mu}) = n/\hat{\mu}^2 \tag{6.61}$$

and in this case the p^*-formula is exact because the density of $\hat{\mu}$, being of the gamma form with index $n-1$, is a transformation model. To derive the distribution of w we have thus to carry out the two-to-one transformation (6.60) from $\hat{\mu}$ to w. We return to this below. □

In the general one-parameter case we have from (6.58) and $\ell_1(\hat{\theta}; \hat{\theta}, a) = 0$ that

$$\begin{aligned} \partial w/\partial\hat{\theta} &= 2\{\ell_{;1}(\hat{\theta}; \hat{\theta}, a) - \ell_{;1}(\theta; \hat{\theta}, a)\} \\ &= -2(\theta - \hat{\theta})\ell_{1;1}(\hat{\theta}; \hat{\theta}, a) \\ &\quad - (\theta - \hat{\theta})^2 \ell_{2;1}(\hat{\theta}; \hat{\theta}, a) + \cdots. \end{aligned} \tag{6.62}$$

But on differentiating the maximum likelihood equation we have

$$\ell_{11}(\hat{\theta}; \hat{\theta}, a) + \ell_{1;1}(\hat{\theta}; \hat{\theta}, a) = 0 \tag{6.63}$$

so that $\ell_{1;1}(\hat{\theta}; \hat{\theta}, a) = \hat{j}$, a special case of the formulae (5.13) and (5.14). Thus

$$\begin{aligned} \partial w/\partial\hat{\theta} &= -2(\theta - \hat{\theta})\hat{j} - (\theta - \hat{\theta})^2 \ell_{11;1}(\hat{\theta}; \hat{\theta}, a) + \cdots \\ &= -2(\theta - \hat{\theta})\hat{j}\{1 + c_1'(\hat{\theta}, a)(\theta - \hat{\theta}) + c_2'(\hat{\theta}, a)(\theta - \hat{\theta})^2 + \cdots\}, \end{aligned} \tag{6.64}$$

say. Thus, by (6.59), the derived density of w is, with error $O(n^{-3/2})$

$$\frac{c}{2} e^{-\frac{1}{2}w} \sum \frac{1}{\hat{j}^{1/2}|\theta - \hat{\theta}|\{1 + \hat{c}_1'(\theta - \hat{\theta}) + \hat{c}_2'(\theta - \hat{\theta})^2 + \cdots\}}, \tag{6.65}$$

where the sum is over those values of $\hat{\theta}$, typically two, generating a given w.

In the special case of Example 6.9 we have that

$$\begin{aligned}\partial w/\partial\hat{\mu} &= 2n(\hat{\mu}-\mu)/(\mu\hat{\mu})\\ &= -2(\mu-\hat{\mu})\hat{j}(\hat{\mu}/\mu),\end{aligned}$$

which is of the form (6.64).

To complete the above calculation we have to express the summand in (6.65) in terms of w via the expansion of (6.57) around $\hat{\theta}$

$$w = \hat{j}(\hat{\theta}-\theta)^2 + \tfrac{1}{3}\hat{\ell}_{111;}(\hat{\theta}-\theta)^3 + \cdots.$$

From this expansion we find

$$\sqrt{w} = \hat{j}^{\frac{1}{2}}(\hat{\theta}-\theta)\{1 + \tfrac{1}{3}(\ell_{111;}/\hat{j})(\hat{\theta}-\theta) + \cdots\}^{1/2},$$

where $\hat{\ell}_{111;} = \ell_{111;}(\hat{\theta};\hat{\theta},a)$, so that on inversion

$$\hat{\theta}-\theta = \sqrt{w}/\hat{j}^{\frac{1}{2}} + \tfrac{1}{6}\hat{\ell}_{111;}w/\hat{j} + \cdots, \tag{6.66}$$

where the first term takes both signs. Therefore from (6.65) the density of w is

$$\frac{c}{2}w^{-\frac{1}{2}}e^{-\frac{1}{2}w}\sum(1 + \hat{c}''_{\frac{1}{2}}\sqrt{w} + \hat{c}''_1 w + \cdots),$$

where the sum is over the two signs of $\sqrt{w}$, and $\hat{c}''_{\frac{1}{2}}$ and $\hat{c}''_1$ are functions of $\hat{\theta}$. Expansion of $\hat{c}''_{\frac{1}{2}}$ and $\hat{c}''_1$ around θ yields the density in the form

$$p^*(w;\theta|a) = \frac{c}{2}w^{-\frac{1}{2}}e^{-\frac{1}{2}w}\{1 + c_1 w + O(n^{-3/2})\}, \tag{6.67}$$

where c_1, which is $O(n^{-1})$, is a function of θ and a and incorporates a term from the expansion of $c'_{\frac{1}{2}}$.

The linear dependence on w in (6.67) is exactly of the form required to ensure that a suitable multiple of w has the chi-squared distribution with 1 degree of freedom with error $O(n^{-3/2})$. For this we define $b = E(w)$, and take w/b as a modified statistic which has therefore not only the correct mean for the chi-squared distribution with 1 degree of freedom but also, remarkably, the correct density with error $O(n^{-3/2})$.

The argument when θ is multi-dimensional is similar but more complex because we have to transform from θ to w and $d-1$ further variables, then integrating out the latter. Thus the p^*-based approximation to the distribution of w is

$$p^*(w;\theta|a) = \int_{\hat\theta|w} p^*(\hat\theta;\theta|a)d\hat\theta$$

$$= ce^{-\frac{1}{2}w}\int_{\hat\theta|w} |\hat{j}(\hat\theta;\theta,a)|^{1/2}d\hat\theta, \qquad (6.68)$$

as already indicated in (6.59). Here $\hat\theta|w$ signifies integration over the set, generally a $d-1$ dimensional manifold, of $\hat\theta$-values which give rise to the fixed value of w. It is not possible, in general, to find an explicit expression for the integral in (6.68), but an approximation of appropriate accuracy can be obtained by expansion of the integrand followed by termwise integration.

Since the integral is parameterization invariant we may, without loss of generality, assume that the parameterization is such that the set of $\hat\theta$ which gives the fixed value of w is the ellipsoid $\{\hat\theta: \jmath_{rs}(\hat\theta-\theta)^r(\hat\theta-\theta)^s = w\}$. Next, introducing new variables $\xi_r = \hbar_{rs}(\hat\theta-\theta)^s$, where $\hbar$ is a square root of the observed information matrix $\jmath$, and transforming these further, using the angles $\phi_1,\ldots,\phi_{d-1}$ of spherical polar coordinates, we find that under p^* the joint density of w and the vector $\phi = (\phi_1,\ldots,\phi_{d-1})$ is

$$p^*(w,\phi;\theta|a) = \bar{c}q_d(w)H, \qquad (6.69)$$

where as usual $\bar{c} = (2\pi)^{d/2}c$ is the standardized normalizing constant of the p^*-formula, q_d denotes the density of the χ^2 distribution on d degrees of freedom, and where

$$H = (|\hat{j}|/|\jmath|)^{1/2}|S^{d-1}|^{-1}J(\phi) \qquad (6.70)$$

where $J(\phi)$ depends on ϕ only and where $|S^{d-1}| = 2\pi^{d/2}/\Gamma(d/2)$ is the surface area of the unit sphere in $\mathbb{R}^d$. Specifically, $J(\phi)$ equals the Jacobian determinant of the transformation from $(\xi_1,\ldots,\xi_d)$ to $(\sqrt{w},\phi_1,\ldots,\phi_{d-1})$ evaluated at $w=1$. From the likelihood expansion (5.63) we find, on writing γ_r for $w^{-1/2}\xi_r$, that $\{|\hat{j}|/|\jmath|\}^{1/2}$ has an expansion of the form

$$\{|\hat{j}|/|\jmath|\}^{1/2} = 1 + c^r(\theta)\gamma_r\sqrt{w} + c^{rs}(\theta)\gamma_r\gamma_s w + \cdots, \qquad (6.71)$$

where the coefficients $c^r(\theta)$ and $c^{rs}(\theta)$ are functions of θ and a only. Inserting (6.70) and (6.71) in (6.69) and integrating termwise with respect to ϕ we find

$$p^*(w;\theta|a) \doteq \bar{c}q_d(w)\{1 + c_1(\theta)w\}, \tag{6.72}$$

where we have neglected all correction terms beyond the first two groups. The order of the coefficient $c_1(\theta)$ is typically given by $|j|^{-1/d}$. Hence, under ordinary repeated sampling we have in wide generality that, to order $O(n^{-3/2})$,

$$p^*(w;\theta|a) \stackrel{*}{=} \bar{c}q_d(w)\{1 + c_0(\theta)w/n\}, \tag{6.73}$$

where $c_0(\theta)$ is independent of n but depends in general on a.

We may now draw several important conclusions. First, integration of (6.73) yields

$$1 \stackrel{*}{=} \bar{c}\{1 + dc_0(\theta)/n\}. \tag{6.74}$$

Second, defining b by

$$b = d^{-1}E(w;\theta) \tag{6.75}$$

it follows from (6.73), on multiplying both sides by w and then integrating, that

$$db \stackrel{*}{=} \bar{c}d\{1 + (2 + d)c_0(\theta)/n\}, \tag{6.76}$$

and comparing (6.74) and (6.76) we find

$$b \stackrel{*}{=} 1 + 2c_0(\theta)/n. \tag{6.77}$$

Formulae (6.74) and (6.77) together yield the following important relation between the quantity b and the standardized normalizing constant $\bar{c}$

$$\bar{c} \stackrel{*}{=} b^{-d/2}. \tag{6.78}$$

Next, let w^* denote the Bartlett adjusted version of w defined by

$$w^* \stackrel{*}{=} b^{-1}w. \tag{6.79}$$

The adjustment obviously implies that w^* has exactly the mean value d of the limiting χ^2_d distribution. However, a much stronger statement is true: the factor b^{-1} adjusts not only the mean but the whole distribution, for it follows from (6.73) and (6.77) that

$$p(w^*;\theta|a) \stackrel{*}{=} q_d(w), \tag{6.80}$$

to order $O(n^{-3/2})$ under ordinary repeated sampling. This remarkable result is peculiar to the log-likelihood ratio statistic; no similar general result of equal simplicity is valid for the other main test statistics, such as w_u and w_p.

Example 6.10 Fisher's gamma hyperbola
For a sample of n from Fisher's gamma hyperbola model

$$p(x, y; \psi) = \exp(-\psi x - \psi^{-1} y), \tag{6.81}$$

we have that $(\hat{\psi}, a)$ is minimal sufficient, where $\hat{\psi} = \sqrt{(\bar{Y}/\bar{X})}$ is the maximum likelihood estimate of ψ and where $a = n\sqrt{(\bar{X}\bar{Y})}$ is ancillary, $\bar{X}$ and $\bar{Y}$ denoting the sample means. Setting $\zeta = \log \psi$ we have that the conditional distribution of $\hat{\zeta}$ given a has probability density function

$$p(\hat{\zeta}; \zeta \mid a) = \{2K_0(2a)\}^{-1} \exp\{-2a \cosh(\hat{\zeta} - \zeta)\}, \tag{6.82}$$

where K_0 is a Bessel function.

The normalizing constant of the p^*-approximation to $p(\hat{\zeta}; \zeta \mid a)$ is

$$c = \{\sqrt{(2a)} e^{2a} 2K_0(a)\}^{-1}$$

from which, using a well-known asymptotic expansion for K_0, we find that

$$\bar{c} = 1 + \frac{1}{16a} + O(a^{-2}). \tag{6.83}$$

On the other hand, the log-likelihood ratio statistic for the hypothesis $\zeta = 0$, say, is $w = 4a(\cosh \hat{\zeta} - 1)$ and

$$\begin{aligned} E(w) &= 4a\left\{\frac{K_1(2a)}{K_0(2a)} - 1\right\} \\ &= 1 - \frac{1}{8a} + O(a^{-2}), \end{aligned} \tag{6.84}$$

in agreement with (6.77). □

In many of the standard problems of normal theory $b \geqslant 0$, so that the Bartlett adjustment reduces the size of the test statistic. Example 6.10 shows that negative values of b may arise.

In general, when w is given in reasonably simple explicit form in terms of statistics with asymptotically known properties direct asymptotic expansion of $E(w)$ will usually be feasible for determining b, to the appropriate order.

It has been shown in section 6.3 that the standardized normalizing constant $\bar{c}$ is of the form $\bar{c} = 1 + \not{C}$ where $\not{C}$ is given by (6.23) and is of order $O(n^{-1})$ under ordinary repeated sampling. Hence it follows from (6.78) that the Bartlett factor b is of the form

$$b = 1 + \not{B}, \tag{6.85}$$

where

$$\begin{aligned}\not{B} &= -\frac{2}{d}\not{C} \\ &= -d^{-1}\{(\tfrac{1}{4}\not{t}_{rstu} + \not{t}_{rs;tu})\not{j}^{rs}\not{j}^{tu} \\ &\quad + (\tfrac{1}{4}\not{t}_{rst}\not{t}_{uvw} + \tfrac{1}{6}\not{t}_{rtv}\not{t}_{suw})\not{j}^{rs}\not{j}^{tu}\not{j}^{vw}\}.\end{aligned} \tag{6.86}$$

From the rules for translation between observed and expected expressions, discussed in section 5.5, we have that, the error being $O(n^{-3/2})$,

$$\not{B} \stackrel{+}{=} B, \tag{6.87}$$

where

$$B = -d^{-1}\{(\tfrac{1}{4}t_{rstu} + t_{rs;tu})i^{rs}i^{tu} + (\tfrac{1}{4}t_{rst}t_{uvw} + \tfrac{1}{6}t_{rtv}t_{suw})i^{rs}i^{tu}i^{vw}\}. \tag{6.88}$$

The expression for $\not{B}$ is in agreement with the formulae (5.56)–(5.58), and that for B with (5.30), (5.33–37).

6.5.2 Composite hypotheses

We proceed to consider the general case that concerns the testing of composite hypotheses. Here, too, we approach the question from the viewpoint of conditionality, and the p^*-formula is again a basic tool.

Suppose the parameter θ, of dimension d, is partitioned as (ψ, χ) with χ of dimension d_0, and that the hypothesis determining the likelihood ratio statistic

$$w = 2\{\ell(\hat{\theta}) - \ell(\tilde{\theta})\}$$

is given by fixing the value of ψ, so that $\tilde{\theta} = (\psi, \hat{\chi}_\psi)$. Let $\mathcal{M}$ and $\mathcal{M}_\psi$ denote the full model and the submodel under test, respectively.

A main point of the discussion below is the derivation of the generalization of formula (6.78) to

$$\bar{c}/\bar{c}_0 \stackrel{\bullet}{=} b^{-d/2}. \tag{6.89}$$

Here $\bar{c}$ and $\bar{c}_0$ are the standardized normalizing constants of the

p^*-formula for $\mathscr{M}$ and $\mathscr{M}_\psi$, respectively, and b is defined by

$$E(w) = (d - d_0)b.$$

The development to be given will, moreover, show that the adjusted likelihood ratio statistic

$$w^* = (\bar{c}/\bar{c}_0)^{2/(d-d_0)}w \tag{6.90}$$

is asymptotically chi-squared with $d - d_0$ degrees of freedom.

In order to apply the p^*-formula to the model $\mathscr{M}_\psi$ it is, in general, necessary to specify an auxiliary statistic a_ψ, say, which together with the ancillary a for $\mathscr{M}$ constitutes an ancillary for $\mathscr{M}_\psi$. This leads us to distinguish between several basically different situations, which we now outline before showing the above results.

The first possibility is that no such further ancillary statistic is required, and that we may assume that $(\hat{\chi}_\psi, \hat{\psi})$ is a one-to-one transformation of $(\hat{\chi}, \hat{\psi})$.

Example 6.11 Normal mean

Suppose that $Y_1, \ldots, Y_n$ are independent and normally distributed with mean ψ and variance χ, and take as null hypothesis $\psi = 0$. Then, in the usual notation,

$$\hat{\psi} = \bar{Y}, \quad \hat{\chi} = s'^2 = n^{-1}\sum(Y_i - \bar{Y})^2, \quad \hat{\chi}_\psi = s_0^2 = n^{-1}\sum Y_i^2$$

and no ancillary statistic has to be introduced to apply p^* to $\mathscr{M}$ and $\mathscr{M}_\psi$. □

Assuming that a is indeed ancillary for $\mathscr{M}_\psi$ and applying p^* twice, we obtain as approximation to the conditional density of $\hat{\psi}$ given $\hat{\chi}_\psi$

$$p^*(\hat{\psi}; \psi, \chi | \hat{\chi}_\psi, a) = (c/c_0)\{|\hat{j}|/|\tilde{j}_{\chi\chi}|\}^{1/2}e^{-w/2}. \tag{6.91}$$

The second possibility is that for the fixed value of ψ there is, exactly or to an adequate approximation, an ancillary statistic for χ, a_ψ say, such that $(\hat{\chi}_\psi, a_\psi)$ is a one-to-one transformation of $(\hat{\chi}, \hat{\psi})$. In particular, it may be that $a_\psi = \hat{\psi}$. In this case p^* can be adapted to give, approximately, the joint density of $(\hat{\chi}_\psi, a_\psi)$, and the conditional density of $\hat{\chi}_\psi$ given a_ψ, and then the ratio of the two will give the marginal density of a_ψ. The latter, being independent of χ, is then appropriate for testing the given value of ψ, and the approximation has the form

$$p^*(a_\psi; \psi | a) = (c/c_0)\{|\hat{j}|/|\tilde{j}_{\chi\chi}|\}^{1/2}|(\hat{\psi}, \hat{\chi})_{/(\hat{\chi}_\psi, a_\psi)}|e^{-w/2}. \tag{6.92}$$

Example 6.12 Two exponential distributions
Let $X_1,\ldots,X_n$ and $Y_1,\ldots,Y_n$ be independently exponentially distributed with parameters respectively ψ/χ and $\psi\chi$. Then, with $\bar{X}$ and $\bar{Y}$ denoting the sample means, we have $\hat{\psi}=(\bar{X}\bar{Y})^{-1/2}$ and $\hat{\chi}=\hat{\chi}_\psi=(\bar{X}/\bar{Y})^{1/2}$. Now $\hat{\psi}$ has a distribution independent of χ, i.e. $\hat{\psi}$ is ancillary for χ and so the discussion above applies with $a_\psi=\hat{\psi}$. In this example

$$\begin{aligned} w &= -4n-2n\log(\bar{X}\bar{Y})+4n(\bar{X}\bar{Y})^{1/2} \\ &= -4n+4n\log\hat{\psi}+4n\hat{\psi}^{-1}. \end{aligned}$$

Furthermore, (6.92) is exactly, and not just approximately, equal to the probability density function of $a_\psi=\hat{\psi}$ because both the full model and that for a fixed ψ are transformation models. □

Example 6.13 Normal mean
We may also tackle Example 6.11 by this second route, noting that when $\psi=0$, $a_\psi=\hat{\psi}/\sqrt{\hat{\chi}_\psi}$ has a distribution independent of χ. In the present case

$$w=-n\log(1-a_\psi^2).$$ □

The final possibility is a combination of the two discussed above and will be described only briefly. This is that the ancillary statistic a_ψ for χ, after minimal sufficient reduction, is of dimension less than that of ψ. One may then proceed by introducing a statistic b_ψ so that $(\hat{\chi}_\psi, a_\psi, b_\psi)$ is a one-to-one transformation of $(\hat{\chi},\hat{\psi})$; an approximate expression for

$$\frac{p(\hat{\chi}_\psi,a_\psi,b_\psi;\chi,\psi)}{p(\hat{\chi}_\psi;\chi,\psi\,|\,a_\psi)}=p(a_\psi;\chi,\psi)p(b_\psi;\psi\,|\,\hat{\chi}_\psi,a_\psi)$$

can now be obtained by means of p^* and this will be of a form similar to (6.91) and (6.92). Example 6.14 illustrates this.

Example 6.14 A special trivariate distribution
Let X, Y, Z have the joint density

$$\psi_1^2\exp\{-\psi_1(\chi^{-1}x+\chi y)\}\psi_2\sqrt{(x/y)}\exp\{-\psi_2\sqrt{(x/y)}z\} \quad (6.93)$$

taking as null hypothesis $\psi_1=\psi_2=1$. Then $\hat{\psi}_1=(XY)^{-1/2}$, $\hat{\chi}=\hat{\chi}_\psi=(X/Y)^{1/2}$, $\hat{\psi}_2=\{\sqrt{(X/Y)}Z\}^{-1}$ and under the null hypothesis $\hat{\psi}_1=a_\psi$ is exactly ancillary. Thus direct application of p^* gives the joint density of $\hat{\chi}, \hat{\psi}_1, \hat{\psi}_2$ and the conditional density of $\hat{\chi} = \hat{\chi}_\psi$ given $\hat{\psi}_1$. The further statistic b_ψ could then be taken as $\hat{\psi}_2$. □

Having thus indicated the various situations that can arise, we proceed to discuss the distribution of w and its Bartlett adjusted versions, for simplicity restricting ourselves to the two first situations.

To obtain the density of w from (6.91) or (6.92) we have in effect to integrate over all $\hat{\psi}$ or, respectively, over all a_ψ that produce the required value of w. Let $\phi = (\phi_1, \phi_2)$ denote the transformation on the parameter space mapping $(\hat{\chi}, \hat{\psi})$ onto $(\hat{\chi}_\psi, \hat{\psi})$, with a corresponding transformation of parameters. For instance, in Example 6.11 ϕ maps (χ, ψ) onto $(\chi + \psi^2, \psi)$. The observed information in terms of the parameter ϕ will be written $j^{[\phi]}$, so that

$$|\hat{j}|^{1/2} = |(\hat{\chi}_\psi, \hat{\psi})_{/(\hat{\chi}, \hat{\psi})}|\,|j^{[\phi]}(\hat{\chi}_\psi, \hat{\psi})|^{1/2}$$

$$|j_{\chi\chi}(\hat{\chi}_\psi, \hat{\psi})|^{1/2} = |\hat{\chi}_{\psi/\hat{\chi}}|\,|j^{[\phi_1]}_{\hat{\psi}}(\hat{\chi}_\psi)|^{1/2},$$

where in the information for the partial parameter ϕ_1, the second component ϕ_2 is fixed at the value indicated by the suffix.

Thus the expression (6.91) can be written

$$(c/c_0)|\bar{j}|^{1/2} e^{-w/2}, \tag{6.94}$$

where

$$|\bar{j}| = \frac{|j^{[\phi_1]}_{\hat{\psi}}(\hat{\chi}_\psi)|}{|j^{[\phi_1]}_{\psi}(\hat{\chi}_\psi)|\,|\hat{j}^{\psi\psi}|}$$

and we have used formula (5.16). Here $|\bar{j}|$ plays the role of $|\hat{j}|$ in the discussion of the first case above. Expression (6.94) has now to be integrated over $\hat{\psi}$ for given $\hat{\chi}_\psi$ and w.

The analogue of (6.68) is

$$p^*(w; \theta | a) = (c/c_0) e^{-w/2} \int_{\hat{\psi}|w, \hat{\chi}_\psi} |\bar{j}|^{1/2} d\hat{\psi}, \tag{6.95}$$

and the previous type of argument can be applied to give

$$p^*(w; \theta | a) = (\bar{c}/\bar{c}_0) q_{d-d_0}(w)\{1 + hw/n + O(n^{-3/2})\} \tag{6.96}$$

and to obtain formula (6.89).

Example 6.15 Normal mean

In Example 6.11 $\hat{\psi} = \bar{Y}$, $\hat{\chi} = \sum(Y_i - \bar{Y})^2/n$, $\hat{\chi}_\psi = \sum(Y_i - \psi)^2/n$, and it is easily shown that the normalizing constants c, c_0 in the approximations to the distributions of $(\hat{\chi}, \hat{\psi})$ and $\hat{\chi}_\psi$ are

$$c = \frac{2^{-(n-1)/2} n^{n/2-1} e^{-n/2}}{\sqrt{\pi}\Gamma(n-1)},$$

$$c_0 = \frac{2^{(n-1)/2} n^{n/2-1} e^{-n/2}}{\Gamma(n/2)},$$

so that by (6.89) the Bartlett factor is

$$b \doteq (\bar{c}_0/\bar{c})^2 = 1 + 3/(2n) + O(n^{-2})$$

and this can be checked directly by evaluating $E(w)$ from

$$w = n \log\{1 + t^2/(n-1)\},$$

where t is the Student t statistic. In this example (6.91) gives the conditional distribution of $\bar{y}$ given $\hat{\chi}_\psi$ and this can be transformed into the density of t, yielding exactly the correct form for the conditional distribution given $\hat{\chi}_\psi$; the exactness arises from p^* being exact for arbitrary transformation models. □

Example 6.16 Comparison of k variances

Suppose we have $s_1^2, \ldots, s_k^2$ independent with densities

$$\frac{(\frac{1}{2}f_i)^{\frac{1}{2}f_i}}{\Gamma(\frac{1}{2}f_i)(\sigma_i^2)^{\frac{1}{2}f_i}}(s_i^2)^{\frac{1}{2}f_i - 1}\exp\left[-\tfrac{1}{2}f_i\frac{s_i^2}{\sigma_i^2}\right].$$

The maximum likelihood estimate of σ_i^2 is s_i^2 and the corresponding normalization constant in p^* turns out to be

$$c_i = \frac{(\frac{1}{2}f_i)^{\frac{1}{2}(f_i-1)} e^{-f_i/2}}{\Gamma(\frac{1}{2}f_i)}, \tag{6.97}$$

for $i = 1, \ldots, k$. For $s_1^2, \ldots, s_k^2$ combined, the normalization constant is $c = c_1 \cdots c_k$. Under the null hypothesis $\sigma_1^2 = \cdots = \sigma_k^2 = \sigma^2$, say, we have $\hat{\sigma}^2 = \sum f_i s_i^2/f_+$, where $f_+ = f_1 + \cdots + f_k$, and the corresponding normalization constant c_0 in p^* is as (6.97) but with f_i replaced by f_+. Note that under the null hypothesis p^* gives the unconditional density of $\hat{\sigma}^2$. Hence, denoting the first-order Stirling approximation to Γ by Γ_0, we have

$$\bar{c}/\bar{c}_0 = \frac{\Pi\Gamma_0(\frac{1}{2}f_i)/\Gamma(\frac{1}{2}f_i)}{\Gamma_0(\frac{1}{2}f_+)/\Gamma(\frac{1}{2}f_+)}.$$

Since

$$\Gamma(\lambda) = \Gamma_0(\lambda)\left[1 + \frac{1}{12\lambda} + \cdots\right]$$

we find

$$\bar{c}/\bar{c}_0 = 1 - \tfrac{1}{6}\left(\sum f_1^{-1} - f_+^{-1}\right) + \cdots$$

and hence

$$b = \{(2\pi)^{(k-1)/2}(c/c_0)\}^{-(k-1)/2}$$
$$= 1 + \frac{1}{3(k-1)}\left(\sum f_i^{-1} - f_+^{-1}\right) + \cdots,$$

which agrees with Bartlett's (1937) adjustment factor, which in this instance can also be evaluated fairly directly by taking the expectation of the explicit form of w. □

Next, we consider the derivation of the density of w from (6.92), that is, from the marginal density of the ancillary statistic a_ψ for χ under the null hypothesis ψ. If we introduce a transformation ϕ mapping $(\hat{\chi}, \hat{\psi})$ onto $(\hat{\chi}_\psi, a_\psi)$, with a corresponding transformation of parameters from (χ, ψ) to (χ_ψ, α), we have that (6.92) may be rewritten as

$$p^*(a_\psi; \psi \mid a) = (c/c_0)\{|\hat{j}^{[\phi]}||\tilde{j}_{\chi\chi}|\}^{1/2} e^{-w/2}$$

because at ψ the density of a_ψ does not depend on χ_ψ. If we write this as

$$p^*(a_\psi; \psi \mid a) = (c/c_0)|\overset{\times}{j}|^{1/2} e^{-w/2},$$

say, we can proceed as before. Note that, to the order considered, the quantities involved depend only on a_ψ and w and not on $\hat{\chi}_\psi$.

Example 6.17 The gamma hyperbola generalized
Let $(X_1, Y_1), \ldots, (X_n, Y_n)$ be independently distributed with the density

$$p(x, y; \psi, \chi) = \psi^2 \exp\{-\psi(\chi^{-1}x + \chi y)\}.$$

Then the log-likelihood is

$$\ell(\psi, \chi) = n(2 \log \psi - \psi\chi^{-1}\bar{x} - \psi\chi\bar{y})$$

from which $\hat{\psi} = (\bar{X}\bar{Y})^{-1/2}$, $\hat{\chi} = (\bar{X}/\bar{Y})^{1/2} = \hat{\chi}_\psi$, for all ψ. For any fixed ψ, $\hat{\psi}$ is ancillary for χ; the likelihood ratio statistic for testing $\psi = 1$

is $w = 4n(\hat{\psi}^{-1} - 1 + \log \hat{\psi})$. Direct calculation gives

$$E(w) = 1 + \tfrac{11}{24}n^{-1} + O(n^{-3/2}). \tag{6.98}$$

The argument from (6.92) yields the density of $\hat{\psi}$ as

$$(\bar{c}/\bar{c}_0)\psi^{-1/2}\hat{\psi}^{-1/2}e^{-w/2} \tag{6.99}$$

and the density of w is obtained by transforming from $\hat{\psi}$ to w and integrating (summing) over the two values of $\hat{\psi}$ leading to a preassigned w. Direct calculation converts (6.99) into the density for w

$$(\bar{c}/\bar{c}_0)q_1(w)\{1 + \tfrac{11}{48}wn^{-1} + O(n^{-3/2})\};$$

on multiplying by w and integrating we have that

$$E(w) = (\bar{c}/\bar{c}_0)\{1 + \tfrac{11}{16}n^{-1} + O(n^{-3/2})\}.$$

On the other hand, direct evaluation of the normalizing constants gives, to order $O(n^{-3/2})$,

$$\bar{c} = 1 - \tfrac{1}{6}n^{-1},$$

$$\bar{c}_0 = 1 + \tfrac{1}{16}n^{-1},$$

whence

$$\bar{c}/\bar{c}_0 = 1 - \tfrac{11}{48}n^{-1} + O(n^{-3/2})$$

in agreement with (6.89) and (6.98). □

Note that since the approximate χ^2 distribution of w^* does not depend on the ancillary a, the adjusted statistic w^* is independent of a to the order of approximation in question, in particular, to order $O(n^{-3/2})$ under repeated sampling.

The constants entering the derivation of (6.89), including the constants $\bar{c}, \bar{c}_0$, and b, are in general functions of the parameter θ and of the ancillary a which is held constant throughout the argument. In many, though not all, problems this ancillary can be taken in a form differing by $O_p(n^{-1/2})$ from a constant α, say, the functions being differentiable at α. We can then, to the order in question, replace a by α in those constants.

Example 6.18 Exponential mean

The correction terms in (6.83) and (6.84) may be expressed as functions of n and an ancillary a_n with a limiting standard normal distribution by letting

$$a_n = \{a - E(a)\}/\sqrt{\text{var}(a)}. \tag{6.100}$$

Thus, in the notation of the discussion above $\alpha = 0$. Furthermore, since $E(a) = \{\Gamma(n+\frac{1}{2})/\Gamma(n)\}^2$ and $\text{var}(a) = n^2 - \{E(a)\}^2$,

$$a = n\left\{1 + \frac{a_n}{\sqrt{(2n)}} + O(n^{-1})\right\}$$

and hence (6.84) may be re-expressed as

$$E(w) = 1 - \frac{1}{8n} + O(n^{-3/2}). \qquad \square$$

It is further important that the above asymptotic results generally hold to the same asymptotic order if in the constants in question one replaces θ by $\hat{\theta}$.

Actually, somewhat stronger assertions than those derived here are valid; see *Further results and exercises* 6.10.

6.6 The directed likelihood

A variety of adjustments to the directed likelihood statistic r improve the asymptotic standard normal approximation to the distribution of this statistic and/or eliminate the effects of nuisance parameters. In the following we discuss a particular of these adjusted versions of the directed likelihood, the **modified directed likelihood** r^*. Other adjusted versions will be considered and compared to r^* in section 8.4.

All of these adjusted versions as well as the directed likelihood itself may be used (i) as test statistics; (ii) to calculate approximate interval probabilities; (iii) as components of ancillary statistics; and (iv) in residual analysis.

Let the parameter θ be partitioned as (ψ, χ) where ψ, the parameter of interest, is of dimension 1 while the nuisance parameter χ is of dimension $d-1$. The directed likelihood $r = r_\psi$ for inference on ψ is defined by

$$r = \text{sgn}(\hat{\psi} - \psi)[2\{\ell(\hat{\psi}, \hat{\chi}) - \ell(\psi, \hat{\chi}_\psi)\}]^{1/2}. \qquad (6.101)$$

For brevity we shall occasionally write $\hat{\theta}_\psi$ for $(\psi, \hat{\chi}_\psi)$ and, as usual, a tilde or a circumflex above a combinant indicates evaluation at $\hat{\theta}_\psi$ or $\hat{\theta}$, respectively.

We now define the modified directed likelihood r^* by

$$r^* = r + r^{-1}\log(u/r), \qquad (6.102)$$

where u is given below by formulae (6.103)–(6.105) or, equivalently, by formula (6.108). It will be verified below that under ordinary repeated sampling r^* follows the standard normal distribution to order $O(n^{-3/2})$ conditionally on an appropriate ancillary, and hence also unconditionally. In comparison, the directed likelihood r itself is standard normal to order $O(n^{-1/2})$ only. The difference $r^* - r$ is $O_p(n^{-1/2})$.

The quantity u in (6.102) is defined by

$$u = C\check{u}, \tag{6.103}$$

where

$$\check{u} = j_P(\hat{\psi})^{-1/2}\bar{\ell}_{P/\hat{\psi}}(\psi), \tag{6.104}$$

$$C = C(\psi) = |\ell_{\chi;\hat{\chi}}(\hat{\theta}_\psi)|/\{|j_{\chi\chi}(\hat{\theta}_\psi)||j_{\chi\chi}(\hat{\theta})|\}^{1/2}. \tag{6.105}$$

In (6.104), $j_P(\hat{\psi})$ is the observed profile information on ψ, evaluated at $\psi = \hat{\psi}$. Furthermore,

$$\bar{\ell}_P = \ell_P(\psi) - \ell_P(\hat{\psi}) = \ell(\hat{\theta}_\psi) - \ell(\hat{\theta})$$

is the normed profile log-likelihood and $\bar{\ell}_{P/\hat{\psi}}$ stands for the derivative of $\bar{\ell}_P$ with respect to $\hat{\psi}$ when $\bar{\ell}_P$ is considered as a function of $\psi, \hat{\psi}, \hat{\chi}_\psi$ and a. Thus

$$\bar{\ell}_{P/\hat{\psi}} = \bar{\ell}_{;\hat{\psi}} - \hat{\ell}_{;\hat{\psi}} + \hat{\chi}_{/\hat{\psi}}(\bar{\ell}_{;\hat{\chi}} - \hat{\ell}_{;\hat{\chi}}). \tag{6.106}$$

The quantity $\hat{\chi}_{/\hat{\psi}}$ is expressible in terms of mixed derivatives, as follows on differentiation of the likelihood equation for $\hat{\chi}_\psi$, the result being

$$\hat{\chi}_{/\hat{\psi}} = -\tilde{\ell}_{\chi;\hat{\psi}}(\tilde{\ell}_{\chi;\hat{\chi}})^{-1}. \tag{6.107}$$

The factors C and $\check{u}$ that determine u, cf. formula (6.103), are both invariant under interest-respecting reparameterizations. They may in fact be interpreted as relating respectively to two separate adjustments of r, one correcting for the influence of the nuisance parameters and the other adjusting for the lack of standard normality of r. This will be discussed in detail at the end of the present section.

Note in this connection that C is likely to be close to 1 when the data are informative about χ, because $\ell_{\chi;\hat{\chi}}(\hat{\theta}) = j_{\chi\chi}(\hat{\theta})$; see formula (5.14).

The formula for u may alternatively be expressed as

$$u = |\ell_{;\theta}(\hat{\theta}) - \ell_{;\theta}(\hat{\theta}_\psi) \quad \ell_{\chi;\theta}(\hat{\theta}_\psi)|/\{|j_{\chi\chi}(\hat{\theta}_\psi)||j(\hat{\theta})|\}^{3/2}. \tag{6.108}$$

In fact, the form (6.103) is obtainable from (6.108) by application of the formula for the determinant of a block matrix (see, for instance, Rao, 1973, p. 28) and by the identities (5.15) and (3.31).

In the case $d = 1$, i.e. when there are no nuisance parameters present, the expression (6.108) for u reduces to

$$u = \hat{j}^{-\frac{1}{2}}\{\ell_{;\hat{\psi}}(\hat{\psi}) - \ell_{;\hat{\psi}}(\psi)\}. \tag{6.109}$$

For a one-dimensional exponential family

$$\exp\{\theta y - k(\theta) - h(y)\} \tag{6.110}$$

the expression simplifies further to

$$u = k^{1/2}(\hat{\theta})(\hat{\theta} - \theta). \tag{6.111}$$

It may be noted that in the special case where $\hat{\chi}_\psi = \hat{\chi}$ we have from the partial likelihood equation $\ell_\chi(\psi, \hat{\chi}_\psi; \hat{\psi}, \hat{\chi}, a) = 0$ that

$$\ell_{\chi\psi}(\psi, \hat{\chi}; \hat{\psi}, \hat{\chi}, a) = 0 = \ell_{\chi;\hat{\psi}}(\psi, \hat{\chi}; \hat{\psi}, \hat{\chi}, a), \tag{6.112}$$

$$\ell_{\chi\chi}(\psi, \hat{\chi}; \hat{\psi}, \hat{\chi}, a) + \ell_{\chi;\hat{\chi}}(\psi, \hat{\chi}; \hat{\psi}, \hat{\chi}, a) = 0. \tag{6.113}$$

Relation (6.112) shows, in particular, that ψ and χ are orthogonal with respect to the observed information, and (6.113) may be rewritten as

$$j_{\chi\chi}(\hat{\theta}_\psi) = \ell_{\chi;\hat{\chi}}(\psi, \hat{\chi}; \hat{\psi}, \hat{\chi}, a) = \ell_{\chi;\hat{\chi}}(\hat{\theta}_\psi).$$

It follows that when $\hat{\chi}_\psi = \hat{\chi}$ the formula (6.108) for u simplifies to

$$u = j_{\hat{\psi}\hat{\psi}}^{-\frac{1}{2}}(\hat{\theta})\{\ell_{;\hat{\psi}}(\hat{\psi}, \hat{\chi}) - \ell_{;\hat{\psi}}(\psi, \hat{\chi})\}. \tag{6.114}$$

Thus u has the character of a 'dual' to a score test statistic.

Example 6.19 (m, m) *exponential models*

Consider a (m, m) exponential model with log-likelihood function $\ell = s^{\mathrm{T}}\phi - k(\phi)$, and let (η_1, ϕ_2) be a mixed parametrization of that model. With η_1 and ϕ_2 playing the roles of χ and ψ, respectively, we find

$$\ell_{;\hat{\eta}_1} = \phi_1 + \phi_2 s_{2/s_1}, \quad \ell_{;\hat{\phi}_2} = \phi_2 s_{2/\hat{\phi}_2}, \quad \ell_{\eta_1;\hat{\eta}_1} = \phi_{1/\eta_1},$$
$$\ell_{\eta_1;\hat{\phi}_2} = 0, \quad \ell_{\eta_1\eta_1} = -\phi_{1/\eta_1} + (s_1 - \eta_1)\phi_{1/\eta_1\eta_1}, \tag{6.115}$$

where

$$s_{2/s_1} = [\eta_2(\eta_1, \phi_2)_{/\eta_1}]_{(\eta_1, \phi_2) = (\hat{\eta}_1, \hat{\phi}_2)}$$

and similarly for $s_{2/\hat{\phi}_2}$; recall that $\hat{\eta} = s$. Furthermore, writing Σ for the variance matrix of the canonical statistic s, and Δ for the inverse matrix of Σ, we have $j(\hat{\eta}_1, \hat{\phi}_2) = i(\hat{\eta}_1, \hat{\phi}_2)$ and

$$\begin{aligned} i(\eta_1, \phi_2) &= \begin{bmatrix} \Sigma_{11}^{-1} & 0 \\ 0 & \Delta_{22}^{-1} \end{bmatrix} \\ &= \begin{bmatrix} \phi_{1/\eta_1} & 0 \\ 0 & \eta_{2/\phi_2} \end{bmatrix} \end{aligned} \tag{6.116}$$

cf. Barndorff-Nielsen (1988a, Lemma 2.3). Combining the above results and using the fact that $\hat{\eta}_{1\phi_2} = \hat{\eta}_1 = s_1$ we obtain

$$C(\psi) = \{|\Sigma_{11}(s_1, \hat{\phi}_2)|/|\Sigma_{11}(s_1, \phi_2)|\}^{1/2} \tag{6.117}$$

and

$$\check{u} = \Delta_{22}(s_1, \hat{\phi}_2)^{-1/2}(\hat{\phi}_2 - \phi_2). \tag{6.118}$$

Thus, in this special case, $\check{u}$ is simply the Wald statistic for testing ϕ_2.

Note, incidentally, that

$$\ell_{\eta_1;\hat{\eta}_1}(s_1, \phi_2) = j_{\eta_1\eta_1}(s_1, \phi_2). \tag{6.119}$$

□

Example 6.20 Location–scale model

Let $Y_1, \ldots, Y_n$ be a sample from a location–scale model

$$\sigma^{-n} \prod_{i=1}^{n} f\left(\frac{y_i - \mu}{\sigma}\right),$$

and suppose the interest parameter ψ is of the form

$$\psi = \mu + b\sigma + \tfrac{1}{2}c\sigma^2 \tag{6.120}$$

for certain known constants b and c.

Note that if $c = 0$ the constant b is the $F(b)$ quantile of the distribution function F determined by the known probability density function f. Also, if $b = 0$ and $c = 1$ and if f is the density of the standard normal distribution then the parameter ψ is the logarithm of the mean of the associated log-normal distribution.

Letting $g = -\log f$ and $a = (a_1, \ldots, a_n) = \hat{\sigma}^{-1}(y_1 - \hat{\mu}, \ldots, y_n - \hat{\mu})$, the sample configuration, we have that the log-likelihood function is of the form

$$\ell(\psi, \sigma) = -n \log \sigma - \sum g(\bar{a}_i),$$

where

$$\begin{aligned}\bar{a}_i &= \sigma^{-1}(y_i - \mu)\\ &= q(a_i + p + s + t)\end{aligned}$$

and

$$\begin{aligned}p &= (\hat{\psi} - \psi)/\hat{\sigma},\\ q &= \hat{\sigma}/\sigma,\\ s &= b(\sigma - \hat{\sigma})/\hat{\sigma},\\ t &= \tfrac{1}{2}c(\sigma^2 - \hat{\sigma}^2)/\hat{\sigma}.\end{aligned}$$

The only somewhat unfamiliar object in the definition of r^* is the first determinant in (6.108), and the quantities necessary for calculating that, in the present example, are

$$\begin{aligned}\ell_{;\hat{\psi}} &= -\sigma^{-1}\sum g'(\bar{a}_i),\\ \ell_{\sigma;\hat{\psi}} &= \sigma^{-2}\sum\{g'(\bar{a}_i) + (\bar{a}_i - b - c\sigma)g''(\bar{a}_i)\},\\ \bar{\ell}_{\sigma;\hat{\sigma}} &= \sigma^{-2}\sum(a_i - b - c\hat{\sigma})\{g'(\bar{a}_i) + (\bar{a}_i - b - c\sigma)g''(\bar{a}_i)\}.\end{aligned}$$

Suppose next that the parameter of interest is σ. The quantities needed to calculate the terms C and $\check{u}$ are then, writing for brevity $w_i = a_i\hat{\sigma} + \hat{\mu} - \mu$,

$$\begin{aligned}\ell_{\sigma;\hat{\sigma}} &= \sigma^{-2}\{\sum a_i\{g'(\tilde{w}_i/\sigma) + (\tilde{w}_i/\sigma)g''(\tilde{w}_i/\sigma)\},\\ \bar{j}_{\sigma\sigma} &= \sigma^{-2}\{\sum(\tilde{w}_i/\sigma)^2 g''(\tilde{w}_i/\sigma) + 2\sum(\tilde{w}_i/\sigma)g'(\tilde{w}_i/\sigma) - n\}\\ \hat{j}_{\sigma\sigma} &= \hat{\sigma}^{-2}\{\sum a_i^2 g''(a_i) + n\},\\ j_P(\sigma)^{-1} &= \sigma^2\sum g''(\tilde{w}_i/\sigma)/D,\end{aligned}$$

where

$$\begin{aligned}D = &\{\sum g''(\tilde{w}_i/\sigma)\}\{\sum(\tilde{w}_i/\sigma)^2 g''(\tilde{w}_i/\sigma) + 2\sum(\tilde{w}_i/\sigma)g'(\tilde{w}_i/\sigma) - 1 - \sum g''(\tilde{w}_i/\sigma)\}\\ &- \{\sum(\tilde{w}_i/\sigma)g'(\tilde{w}_i/\sigma)\}^2,\end{aligned}$$

$$\bar{\ell}_{P/\hat{\sigma}} = \sigma^{-1}\{n - \sum a_i g'(\tilde{w}_i/\sigma)\}.$$

For a study of higher-order inference on $\psi = \mu + \frac{1}{2}\sigma^2$ under the normal law, see Jensen (1988). □

Example 6.21 Gamma distribution

Suppose that both the parameters λ and α of the gamma distribution are unknown and that the mean of the distribution $\mu = \lambda/\alpha$ is the parameter of interest. Taking the shape parameter λ as the nuisance

Table 6.1 *Comparison of critical levels of the similar test and of the directed likelihood and modified directed likelihood tests, for the hypothesis that the mean of the gamma distribution equals 1 and for sample size 2. Levels: similar test* ε; *directed likelihood* ε^0; *modified directed likelihood* ε^*

y	$\log(\bar{y}/\bar{y}')$	r	r^*	ε	ε^0	ε^*
5.755	2.000	1.63	1.64	0.950	0.948	0.950
3.034	0.1000	2.27	1.73	0.950	0.988	0.958
0.194	0.0135	−2.96	−1.86	0.050	0.0015	0.031
1.935	0.0135	2.51	1.76	0.950	0.994	0.961

parameter we find by formula (6.108) that, for a sample of size n,

$$\log(u/r) = \{\Psi'(\hat{\lambda}_\mu)/\Psi'(\hat{\lambda})\}^{-1/2}(n\hat{\lambda})^{1/2}(1 - \bar{y}/\mu)/r, \qquad (6.121)$$

where $\Psi'(\lambda) = d\log\Gamma(\lambda)/d\lambda$ is the digamma function.

There exists an exact similar test of any specified value of μ. For samples of size $n = 2$ and 3 some critical values of this test, determined by numerical integration, are provided by Jensen (1986b). For larger sample sizes it becomes intractable to determine the critical values and one has to turn to approximations such as that provided by r^* or those derived by Jensen (1986b). Suppose the value to be tested is $\mu = 1$. In Table 6.1 the nominal critical levels obtained by referring r and r^* to the standard normal distribution are compared to the actual critical level of the similar test, for $n = 2$ and a few combinations of $\bar{y}$ and $\bar{y}'$, where $\bar{y}'$ denotes the geometric mean of the observations. □

Further numerical illustrations may be found in section 6.7.

6.6.1 *Derivation of* r^* *and some properties*

We assume that there is a fixed (i.e. independent of sample size) interval around the true value of ψ such that in that interval $\hat{\psi}$ is in one-to-one correspondence with r given ψ, $\hat{\chi}_\psi$ and the ancillary a. This a very mild restriction, and in many instances of interest the interval can in fact be taken to be the whole range of possible values of $\hat{\psi}$. For simplicity of presentation we shall act as if the latter were the case, it being a relatively straightforward, but quite technical, matter to complete the arguments in the general situation. Similar remarks apply to the adjusted versions of r to be considered in section 8.4.

The first step in deriving the modified directed likelihood r^* is to transform the p^*-formula (6.2) for the distribution of the maximum likelihood estimator $\hat{\theta} = (\hat{\psi}, \hat{\chi})$ into a formula for the joint density of r and some complementary variable, by the ordinary rule for transformation of probablity densities. The most convenient and natural choice for a complementary variable is the partial maximum likelihood estimator $\hat{\chi}_\psi$ for χ. To determine the Jacobian matrix $(r, \hat{\chi}_\psi)_{/(\hat{\psi}, \hat{\chi})}$ for the transformation from $(\hat{\psi}, \hat{\chi})$ to $(r, \hat{\chi}_\psi)$ we consider the equations that determine r and $\hat{\chi}_\psi$, namely

$$\tfrac{1}{2}r^2 = \ell(\hat{\psi}, \hat{\chi}; \hat{\psi}, \hat{\chi}, a) - \ell(\psi, \hat{\chi}_\psi; \hat{\psi}, \hat{\chi}, a), \tag{6.122}$$

$$\ell_\chi(\psi, \hat{\chi}_\psi; \hat{\psi}, \hat{\chi}, a) = 0. \tag{6.123}$$

These, by differentiation, yield the equations

$$\hat{\theta}_{/r}\{\ell_{;\hat{\theta}}(\hat{\theta}) - \ell_{;\hat{\theta}}(\hat{\theta}_\psi)\} = r, \tag{6.124}$$

$$\hat{\theta}_{/\hat{\chi}_\psi}\{\ell_{;\hat{\theta}}(\hat{\theta}) - \ell_{;\hat{\theta}}(\hat{\theta}_\psi)\} = 0, \tag{6.125}$$

$$\hat{\theta}_{/r}\ell_{\chi;\hat{\theta}}(\hat{\theta}_\psi) = 0, \tag{6.126}$$

$$\ell_{\chi\chi}(\hat{\theta}_\psi) + \hat{\theta}_{/\hat{\chi}_\psi}\ell_{\chi;\hat{\theta}}(\hat{\theta}_\psi) = 0. \tag{6.127}$$

Here, as elsewhere, $\hat{\theta}_\psi = (\psi, \hat{\chi}_\psi)$ and $\hat{\theta}_{/r}$ means the partial derivative of $\hat{\theta}$ with respect to the directed likelihood r keeping $\hat{\chi}_\psi$ and a fixed, etc. From (6.124)–(6.127) we find

$$(\hat{\psi}, \hat{\chi})_{/(r, \hat{\chi}_\psi)}[\ell_{;\hat{\theta}}(\hat{\theta}) - \ell_{;\hat{\theta}}(\hat{\theta}_\psi) \quad \ell_{\chi;\hat{\theta}}(\hat{\theta}_\psi)] = \begin{bmatrix} r & 0 \\ 0 & j_{\chi\chi}(\hat{\theta}) \end{bmatrix}. \tag{6.128}$$

The Jacobian determinant is therefore

$$|(\hat{\psi}, \hat{\chi})_{/(r, \hat{\chi}_\psi)}| = r|j_{\chi\chi}(\hat{\theta}_\psi)|\,|\ell_{;\hat{\theta}}(\hat{\theta}) - \ell_{;\hat{\theta}}(\hat{\theta}_\psi) \quad \ell_{\chi;\hat{\theta}}(\hat{\theta}_\psi)|^{-1}. \tag{6.129}$$

Using the formula for the determinant of a block matrix, this may be re-expressed as

$$|(r, \hat{\chi}_\psi)_{/(\hat{\psi}, \hat{\chi})}| = \{|\ell_{\chi;\hat{\chi}}(\hat{\theta}_\psi)|/|j_{\chi\chi}(\hat{\theta}_\psi)|\}r^{-1}\bar{\ell}_{\mathrm{P}/\hat{\psi}}, \tag{6.130}$$

where $\bar{\ell}_{\mathrm{P}/\hat{\psi}}$ has been defined in connection with formula (6.106).

The p^*-formula for the density of $(r, \hat{\chi}_\psi)$ can consequently be cast in the form

$$p^*(r, \hat{\chi}_\psi; \psi, \chi)|a) = c(\psi, \chi, a)(r/u)e^{\bar{\ell}_{\mathrm{P}}}|j_{\chi\chi}(\hat{\theta}_\psi)|^{1/2}e^{\ell - \bar{\ell}}, \tag{6.131}$$

where u is defined by formula (6.103) or, equivalently, by (6.108).

Transformation of (6.131) to the joint density of r^* and $\hat{\chi}_\psi$ and some rewriting, including introduction of the p^*-formula for the conditional distribution of $\hat{\chi}_\psi$ given a and r^*, yields

$$p^*(r^*, \hat{\chi}_\psi; \theta | a) = \phi(r^*) p^*(\hat{\chi}_\psi; \theta | r^*, a) H, \tag{6.132}$$

where ϕ is the density of the standard normal distribution and the factor H is given by

$$H = (\bar{c}/\bar{c}_\psi) \exp\{\tfrac{1}{2}(r^* - r)^2\} |\partial r^*(r, \hat{\chi}_\psi, a)/\partial r|^{-1} \tag{6.133}$$

with $\bar{c} = (2\pi)^{d/2} c(\theta, a)$ and $\bar{c}_\psi = (2\pi)^{(d-1)/2} c_\psi(\theta, r^*, a)$, $c_\psi = c_\psi(\theta, r^*, a)$ being the norming constant of the p^*-formula for the conditional distribution of $\hat{\chi}_\psi$ given r^*, a and ψ. By expansion of the various terms in H it may be shown that

$$H = 1 + O(n^{-3/2}). \tag{6.134}$$

The asymptotic standard normality of r^* to order $O(n^{-3/2})$ as well as the validity of $p^*(\hat{\chi}_\psi; \theta | r^*, a)$ as an approximation to $p(\hat{\chi}_\psi; \theta | r^*, a)$ to the same asymptotic order are immediate consequences of (6.132) and (6.134).

The rather explicit form of the factor H given by (6.134) enables one, in certain cases, to evaluate the error in considering r^* as standard normal much more accurately than the general $O(n^{-3/2})$ statement. In particular, the relative error can sometimes be shown to be bounded, or even to go to 0, in the extreme tails of the distribution.

Example 6.22 Location model

For a location model $\Pi f(y_i - \mu)$ we have, with $g = -\log f$, $a_i = y_i - \hat{\mu}$ and $a = (a_1, \ldots, a_n)$, that

$$r^2 = 2\{\textstyle\sum g(\hat{\mu} - \mu + a_i) - \sum g(a_i)\}, \tag{6.135}$$

$$u = \{\textstyle\sum g''(a_i)\}^{-1/2} \sum g'(\hat{\mu} - \mu + a_i). \tag{6.136}$$

Suppose, for instance, that $g(y) = \lambda\sqrt{(1 + y^2)}$, i.e. the density f is that of a symmetric hyperbolic distribution with precision parameter λ, here assumed known. A detailed analysis shows that the relative error of the approximations to the tail probabilities of the conditional distribution of $\hat{\mu}$ given a, obtained by treating r^* as standard normal, tends to zero as one goes farther and farther out in the tails. □

Example 6.23 Inverse Gaussian distribution

Under the inverse Gaussian distribution IG(λ, ψ) with the precision parameter λ considered as known, we have

$$r = \lambda^{1/4}\hat{\psi}^{-1}(\hat{\psi}^{1/2} - \psi^{1/2}),$$
$$u = \tfrac{1}{2}\lambda^{1/4}\hat{\psi}^{-3/4}(\hat{\psi} - \psi).$$

Using the facts that the p^*-formula is exact for the inverse Gaussian model and that in the present case the norming constant c of p^* equals $(2\pi)^{-1/2}$ one finds by using (6.133) that the relative error of using $\Phi(r^*)$ as an approximation to the tail probability of the inverse Gaussian tends to 0 at both ends of the distribution, whatever the sample size. □

6.6.2 *Large-deviation properties*

Jensen (1992) has studied the large-deviation properties of r^* in detail when the model is an exponential family and has proved more rigorously that in wide generality r^* is asymptotically standard normal to order $O(n^{-1})$ in a large-deviation region where the directed likelihood r itself is of order $O(n^{1/2})$. In regions of the form $|r| < c$, for an arbitrary positive constant c, the accuracy of the standard normal approximation is $O(n^{-3/2})$, as indicated earlier. The method of proof is based on a standard saddlepoint approximation, on Laplace's method and on Temme's (1982) technique for uniform expansion of integrals; these techniques are outlined in Barndorff-Nielsen and Cox (1989, sections 6.5, 6.2 and 3.9).

It follows that under a curved exponential family and when nuisance parameters are present there exists an approximate similar test of accuracy $O(n^{-1})$ in the large-deviation region. This applies in particular to the Behrens–Fisher problem, and a small numerical study by Jensen (1992) indicates that the distributional accuracy obtained by using r^* is overall better and more reliable than that of the Welch–Aspin solution (Welch, 1949) with error $O(n^{-3})$.

6.6.3 *Discrete distributions*

The above discussion is for continuous distributions. For non-continuous distributions closely analogous results are not available in any great generality. However, for distributions whose support has a lattice structure some such results do exist. In particular, suppose that the support of the one-dimensional exponential family (6.110) is an uninterrupted sequence of integers and let r^* again be defined as (6.108) but with u given by

$$\{k''(\hat{\theta})\}^{1/2}\{1-\exp(\hat{\theta}-\theta)\} \tag{6.137}$$

rather than by (6.108). Then we again have that r^* is asymptotically $N(0,1)$ distributed with error $O(n^{-1})$ for $\sqrt{n}|\hat{\theta}-\theta|<c\sqrt{n}$ and with error $O(n^{-3/2})$ for $\sqrt{n}|\hat{\theta}-\theta|<c$. Applications of this type of result may be found in Pierce and Peters (1992).

6.6.4 *Decomposition of the correction term*

Pierce and Peters (1992) have considered the modified directed likelihood in the case where the interest parameter is a one-dimensional component of the canonical parameter in an (m,m) exponential model and have proposed a decomposition of the adjustment term into a sum of two terms, interpreted as correcting, respectively, for the possible effect of nuisance parameters and for the deviation from standard normality of the directed likelihood itself. In the following we discuss a generalization of the Pierce–Peters decomposition to the general setting of modified directed likelihood and modified profile likelihood, the latter as defined in section 8.2.

We may write r^* as

$$r^* = r + z_{\mathrm{NP}} + z_{\mathrm{INF}}, \tag{6.138}$$

where

$$z_{\mathrm{NP}} = r^{-1}\log C, \tag{6.139}$$

$$z_{\mathrm{INF}} = r^{-1}\log(\check{u}/z) \tag{6.140}$$

where $\check{u}$ and C are defined by formulae (6.104) and (6.105).

The two terms of the decomposition may be considered as expressing respectively an adjustment for the presence of the nuisance parameters (if any) and an adjustment to obtain a closer approximation to the standard normal distribution.

Example 6.24 (m,m) exponential models
Consider again an (m,m) exponential model with canonical parameter ϕ, canonical statistic s, mean value parameter $\eta = E(S;\theta)$ and cumulant function $k(\phi)$. The log-likelihood function is then

$$\ell = \ell(\psi,\chi) = s^{\mathrm{T}}\phi - k(\phi),$$

where $\phi = \phi(\psi, \chi)$. Under standard regularity conditions we have $s = \hat{\eta}$. The quantities C and $\check{u}$ that determine the two adjustment terms z_{NP} and z_{INF} are

$$C = |\hat{\phi}_{/\chi}^{\text{T}} \hat{\Sigma} \tilde{\phi}_{/\chi}| / \{ |\hat{\phi}_{/\chi}^{\text{T}} \hat{\Sigma} \hat{\phi}_{/\chi}| \, |\tilde{\phi}_{/\chi}^{\text{T}} \tilde{\Sigma} \tilde{\phi}_{/\chi} - (s - \tilde{\eta})^{\text{T}} \tilde{\phi}_{/\chi\chi}| \}^{1/2}, \tag{6.141}$$

$$\check{u} = \{ \tilde{j}^{\psi\psi} \}^{-1/2} (\tilde{\theta} - \hat{\theta})^{\text{T}} \hat{\Sigma} [\hat{\theta}_{/\psi} - \hat{\theta}_{/\chi} \{ \hat{\theta}_{/\chi}^{\text{T}} \hat{\Sigma} \hat{\theta}_{/\chi} \}^{-1} \{ \hat{\theta}_{/\chi}^{\text{T}} \hat{\Sigma} \hat{\theta}_{/\psi} \}]. \tag{6.142}$$

When ψ is one of the coordinates of the canonical parameter ϕ these expressions reduce to those given in Example 6.19. □

Numerical investigations (Pierce and Peters, 1992) indicate that whereas the term z_{NP} is often appreciable and yields a very substantial improvement, the term z_{INF} has fairly typically a minor effect only, provided the ψ-specific information $\hat{j}_{\psi\psi \cdot \chi} = \{ \hat{j}^{\psi\psi} \}^{-1}$ is not too small relative to the number of nuisance parameters. This is discussed further in section 8.5.

6.7 Approximate interval probabilities

As indicated at the beginning of section 6.6, one use of the modified directed likelihood is to the approximate calculation of tail and, more generally, interval probabilities. A numerical illustration of this has, in fact, already been given in Example 6.21. Here we discuss and illustrate this type of application in some detail. We also consider a related type of approximation which constitutes a generalization of the Lugannani–Rice formula, a saddlepoint approximation applying to tail areas of one-dimensional distribution functions.

It is to be noted that while the discussion here, and in section 6.6, is in terms of a parametric statistical model, the tail-area approximations to be discussed are applicable much more generally, by the technique of exponential tilting.

The term $r^{-1} \log(u/r)$ in r^*, formula (6.102), is of order $O_{\text{p}}(n^{-1/2})$. Furthermore, Taylor expansion in ψ around $\hat{\psi}$ shows that

$$u/r = 1 + O_{\text{p}}(|\hat{\psi} - \psi|) \tag{6.143}$$

and hence, in the neighbourhood of $\psi = \hat{\psi}$, we have

$$\Phi(r^*) \doteq \Phi^*, \tag{6.144}$$

where

$$\Phi^* = \Phi(r) + (r^{-1} - u^{-1}) \phi(r). \tag{6.145}$$

and Φ is the standard normal distribution function. Thus, the quantity Φ^* offers an alternative to $\Phi(r^*)$ for the approximate calculation of tail probabilities.

For brevity, in the rest of this section we restrict the discussion to the case where no nuisance parameters are present, so that $\psi = \theta$.

If, moreover, the model is an exponential family of order 1 then Φ^* equals the Lugannani–Rice tail area approximation; see Barndorff-Nielsen and Cox (1989). Looked at from a slightly different angle, the Lugannani–Rice formula for approximating a one-dimensional distribution function F emerges by embedding F in the corresponding exponentially tilted family of one-dimensional distribution functions and then calculating the quantities r and u, necessary for the determination of r^*, and finally applying (6.145).

Further, if for $\hat{\theta} \to \pm\infty$ we have $|r| \to \infty$ and

$$r^{-1}\log(u/r) \to 0, \tag{6.146}$$

$$u/r^3 \to 0, \tag{6.147}$$

as is often true, then, at both ends, the ratio of the tail approximations provided by $\Phi(r^*)$ and $\Phi(r) + (r^{-1} - u^{-1})\phi(r)$ tends to 1, i.e. the two approximations give ultimately the same approximation. Note that this statement refers to $\hat{\theta} \to \pm\infty$, not a sample size tending to infinity. When the conditions (6.146) and (6.147) are not satisfied $\Phi(r^*)$ may give better approximations than Φ^* and vice versa, cf. Examples 6.26 and 6.30 below.

Example 6.25 Gamma distribution
Consider a single observation from the gamma distribution

$$\frac{\alpha^\lambda}{\Gamma(\lambda)} y^{\lambda-1} e^{-\alpha y}.$$

Suppose that the shape parameter λ is known and that the interest parameter ψ equals $\log\alpha$, which is a location parameter. Then

$$r^2 = 2\lambda\{e^{\hat{\psi}-\psi} - (\hat{\psi} - \psi) - 1\}, \tag{6.148}$$

$$u = \lambda^{-\frac{1}{2}}(e^{\hat{\psi}-\psi} - 1) \tag{6.149}$$

and hence the conditions (6.146) and (6.147) are satisfied. □

Example 6.26 Inverse Gaussian distribution
As noted earlier, in the present case the relative error of the

approximations to the tail probabilities based on r^* tends to 0 both for $\hat{\psi} \to 0$ and for $\hat{\psi} \to \infty$. On the other hand, in the case where $4\lambda\psi = 1$ the relative error of the approximation based on Φ^* is unbounded as $\hat{\psi} \to 0$. □

The ratio r/u can be expanded in the neighbourhood of $\hat{\theta} = \theta$, where $r = u = 0$, as

$$r/u = 1 - \tfrac{1}{6}\not{t}_3 r + \cdots, \qquad (6.150)$$

where

$$\not{t}_3 = -(\not{\ell}_3 + 3\not{\ell}_{2;1})\hat{j}^{-\frac{3}{2}} \qquad (6.151)$$

is the standardized observed skewness, which is of order $n^{-\frac{1}{2}}$ under repeated sampling (cf. Barndorff-Nielsen, 1986a, formula (3.16)).Thus for $\hat{\theta} = \theta$ we have

$$\Phi^* = \Phi(r) + (r^{-1} - u^{-1})\phi(r) = \tfrac{1}{2} + \frac{1}{6\sqrt{(2\pi)}}\not{t}_3, \qquad (6.152)$$

$$\Phi(r^*) = \Phi(\tfrac{1}{6}\not{t}_3). \qquad (6.153)$$

Hence Φ^* and $\Phi(r^*)$ agree exactly at $\hat{\theta} = \theta$ only when the observed skewness equals 0.

Next we present some numerical illustrations and comparisons of $\Phi(r^*)$ and Φ^* and the corresponding exact tail probabilities for a variety of distributions. The distributions are the negative exponential distribution, the inverse Gaussian distribution with known precision λ, the von Mises distribution with known precision λ, and the Cauchy distribution. For the first two we denote the sample size by n and the sum of the observations by y_+. In the von Mises case, λ plays the role of the sample size and we consider a single observed angle y. The tables show the values of the upper tail probabilities based on the approximate normality of the directed likelihood r, on its modified form r^* and on the approximation Φ^*. We let $\bar{\Phi}(r) = 1 - \Phi(r)$, $\bar{\Phi}(r^*) = 1 - \Phi(r^*)$, and $\bar{\Phi}^* = 1 - \Phi^*$.

Example 6.27 Exponential distribution
Table 6.2 shows exact and approximate tail probabilities of the negative exponential distribution $\lambda e^{-\lambda y}$ for $\lambda = 1$ and sample sizes $n = 1$ and 5. □

Table 6.2 *Exact and approximate upper tail probabilities. Exponential distribution with mean 1*

	y_+	Exact	$\bar{\Phi}(r)$	$\bar{\Phi}(r^*)$	$\bar{\Phi}^*$
$n=1$	0.5	0.6065	0.7329	0.6070	0.6043
	1.0	0.3679	0.5000	0.3695	0.3670
	3.0	0.0498	0.0897	0.0505	0.0500
	5.0	0.0^2674	0.0144	0.0^2689	0.0^2681
	7.0	0.0^3912	0.0^2220	0.0^3938	0.0^3926
	9.0	0.0^3123	0.0^3329	0.0^3127	0.0^3126
$n=5$	1	0.9^2634	0.9^2778	0.9^2634	0.9^2633
	3	0.8153	0.8537	0.8153	0.8152
	5	0.4405	0.5000	0.4407	0.4405
	10	0.0293	0.0399	0.0293	0.0293
	15	0.0^3857	0.0^2134	0.0^3859	0.0^3858
	20	0.0^4169	0.0^4295	0.0^4170	0.0^4170
	25	0.0^6267	0.0^6506	0.0^6268	0.0^6268

Example 6.28 Inverse Gaussian distribution
We consider again the inverse Gaussian distribution of Example 6.2, now with $\lambda = \psi = 1$. Exact and approximate tail probabilities are given in Table 6.3. □

Table 6.3 *Exact and approximate upper tail probabilities. Inverse Gaussian distribution with $\lambda = \psi = 1$*

	y_+	Exact	$\bar{\Phi}(r)$	$\bar{\Phi}(r^*)$	$\bar{\Phi}^*$
$n=3$	1	0.9645	0.9772	0.9639	0.9638
	2	0.6782	0.7602	0.6734	0.6724
	3	0.3927	0.5000	0.3864	0.3848
	5	0.1156	0.1856	0.1120	0.1108
	10	0.0^2548	0.01343	0.0^2519	0.0^2505
	20	0.0^4174	0.0^4718	0.0^4163	0.0^4155
$n=5$	1	0.9^4466	0.9^4683	0.9^4461	0.9^4460
	3	0.8334	0.8759	0.8318	0.8315
	5	0.4147	0.5000	0.4115	0.4108
	10	0.0378	0.0569	0.0331	0.0328
	20	0.0^3148	0.0^3398	0.0^3143	0.0^3141
	25	0.0^5994	0.0^4317	0.0^5957	0.0^5937

Table 6.4 *Exact and approximate upper tail probabilities. for von Mises distribution with $\psi = 0$ (λ = known)*

	y	Exact	$\bar{\Phi}(r)$	$\bar{\Phi}(r^*)$	$\bar{\Phi}^*$
$\lambda = 1$	$\pi/6$	0.3288	0.3023	0.3261	0.3261
	$\pi/4$	0.2562	0.2220	0.2540	0.2540
	$\pi/2$	0.1098	0.0787	0.1212	0.1217
	$3\pi/4$	0.0403	0.0323	0.0921	0.0955
	$5\pi/6$	0.0254	0.0267	0.1089	0.1182
$\lambda = 2$	$\pi/6$	0.2524	0.2321	0.2468	0.2468
	$\pi/4$	0.1631	0.1395	0.1564	0.1564
	$\pi/2$	0.0375	0.0228	0.0339	0.0340
	$3\pi/4$	0.0^2920	0.0^2449	0.0124	0.0126
	$5\pi/6$	0.0^2543	0.0^2315	0.0126	0.0132

Example 6.29 von Mises distribution
The probability density function of the von Mises distribution is given by

$$\{2\pi I_0(\lambda)\}^{-1}\exp\{\lambda\cos(y-\psi)\}.$$

Table 6.4 shows exact and approximate tail probabilities for $\psi = 0$ and known λ. □

Example 6.30 Cauchy model
As an extreme test of r^* we consider a single observation y from the Cauchy location model

$$\pi^{-1}\{1+(y-\psi)^2\}^{-1}. \tag{6.154}$$

In this case, and with $\psi = 0$, we have

$$\begin{aligned} r &= \operatorname{sgn}\hat{\psi}\{2\log(1+\hat{\psi}^2)\}^{\frac{1}{2}}, \\ u &= \sqrt{2}\hat{\psi}(1+\hat{\psi}^2)^{-1}. \end{aligned} \tag{6.155}$$

For $\psi = 0$ the lower tail probability of the distribution (6.154) is

$$F(y) = \pi^{-1}\arctan y + \tfrac{1}{2} \tag{6.156}$$

and Table 6.5 illustrates the accuracy of the approximations to $F(y)$ provided by $\Phi(r)$, $\Phi(r^*)$ and Φ^*. □

A more detailed study of the precise relation between the two types of approximation $\Phi(r^*)$ and Φ^* has been undertaken by Jensen

Table 6.5 *Exact and approximate upper tail probabilities lower tail probabilities (per cent)*

	$y=-100$	$y=-30$	$y=-5$	$y=-1$
F	0.32	1.06	6.28	25.00
$\Phi(r)$	0.001	0.01	0.53	11.95
$\Phi(r^*)$	0.15	0.61	4.69	22.84
Φ^*	0.28	0.94	5.58	23.22

(1992). In particular, he has shown that in the setting of exponential models the approximation (6.144) holds with errors of orders $O(n^{-3/2})$ and $O(n^{-1})$ in regions of the form $|\hat{\theta}-\theta|<c/\sqrt{n}$ and $|\hat{\theta}-\theta|<c$ respectively. Together with the large-deviation properties of r^* discussed in section 6.6.2 this establishes that the large-deviation behaviour of Φ^* is similar to that of $\Phi(r^*)$.

We conclude by indicating, still for the case of no nuisance parameters, an approach to the approximation Φ^* which is somewhat more direct than that presented earlier, via $\Phi(r^*)$, and which, for a range of cases, yields a useful expression for the error term. This derivation rests on Temme's method, discussed in Barndorff-Nielsen and Cox (1989, section 3.9).

Our starting points is the $p^\dagger$-approximation (6.10) which in integrated form gives the approximation

$$P\{\hat{\theta}\leqslant\bar{\theta}\,|\,a;\theta\}\stackrel{\dagger}{=}I \tag{6.157}$$

where $\bar{\theta}$ denotes an arbitrary parameter value and

$$I=P^\dagger\{\hat{\theta}\leqslant\bar{\theta}\,|\,a;\theta\}=(2\pi)^{-1/2}\int_{-\infty}^{\bar{\theta}}\hat{\jmath}^{1/2}e^{\ell-\hat{\ell}}\,d\hat{\theta}. \tag{6.158}$$

Transformation of the variable of integration from $\hat{\theta}$ to the directed likelihood r, assumed to be a one-to-one increasing function of $\hat{\theta}$ for given θ and a, gives

$$I=\int_{-\infty}^{\bar{r}}(r/u)\phi(r)dr,$$

the upper limit of integration $\bar{r}$ being the value that r takes when $\hat{\theta}$ equals $\bar{\theta}$. Letting $s=\hat{\jmath}^{-1/2}r$ and transforming to integration over s, we obtain

$$I=\int_{-\infty}^{\bar{s}}q(s)\phi(s;\hat{\jmath}^{-1})ds, \tag{6.159}$$

where $\phi(\cdot;\tau)$ denotes the normal density with mean 0 and variance τ and where $q(s) = \jmath^{1/2}r/u$, the right-hand side of this latter equation being considered as a function of s for given θ and a. Defining a function h by

$$h(s) = \{q(s) - q(0)\}/s$$

we find, by Temme's method, that

$$I = \bar{\Phi}^* + R, \tag{6.160}$$

with

$$\bar{\Phi}^* = \Phi(\bar{r}) + (\bar{r}^{-1} - \bar{u}^{-1})\phi(\bar{r})$$

and where the error term R is given by

$$R = \jmath^{-1}\int_{-\infty}^{\bar{s}} h'(s)\phi(s;\jmath^{-1})ds. \tag{6.161}$$

Furthermore, under mild regularity conditions ensuring that the behaviour of $h'(s)$ is not too drastic, the method shows that $\bar{\Phi}^* \doteq P^{\dagger}\{\hat{\theta} \leqslant \bar{\theta} | a; \theta\}$ with relative error $O(\jmath^{-1})$ uniformly over the whole range of values of $\hat{\theta}$.

Further results and exercises

6.1 Repeat the calculations of Example 6.8 taking $\log\sigma$ as the parameter of interest. How are the results modified if $Y_1, \ldots, Y_n$ follow a normal theory linear model?

[Section 6.4]

6.2 Show, at least heuristically, how the p^*-formula can be derived by applying Laplace's method (as described, for instance, by Barndorff-Nielsen and Cox, 1989, section 6.2), not directly but 'inversely'. (Laplace's method gives an approximate formula for an integral. However, if the value of the integral happens to be known one can 'inversely' draw conclusions about the integrand from the Laplace formula.)

[Barndorff-Nielsen, 1990b; Section 6.2]

6.3 The hyperboloid model of exponential order 3 may be defined as the joint distribution of two random variables U and V having probability density function

$$p(u, v; \chi, \phi, \lambda) = (2\pi)^{-1}\lambda e^{\lambda}\sinh u \exp\{-\lambda(\cosh u \cosh \chi - \sinh u \sinh \chi \cos(v - \phi))\},$$

where $0 \leqslant u$, $0 \leqslant v < 2\pi$ and where the parameters satisfy $0 \leqslant \chi$, $0 \leqslant \phi < 2\pi$, and $0 < \lambda$. Show that for a sample of size $n > 1$ the p^*-formula for the distribution of the maximum likelihood estimator $(\hat{\chi}, \hat{\phi}, \hat{\lambda})$ is exact.

[Barndorff-Nielsen, 1983; Section 6.2]

6.4 Suppose $U \sim \mathrm{IG}(\chi, \psi)$ and that $V | U = u \sim \mathrm{IG}(\kappa u^2, \lambda)$. Show that for samples of size $n > 1$ the p^*-formula is exact for the joint distribution of $(\hat{\chi}, \hat{\psi}, \hat{\kappa}, \hat{\lambda})$. Discuss the distributional properties of the estimates.

[Barndorff-Nielsen, 1983; Section 6.2]

6.5 Generalize the result of the previous exercise to more general combinations of inverse Gaussian distributions. It is also to some extent possible to combine with gamma distributions and Gaussian distributions.

[Barndorff-Nielsen and Blæsild, 1988; Section 6.2]

6.6 In the setting of section 6.3, study in detail the special case where $Y = (U, V)$, U and V being independent and gamma distributed with common shape parameter n, corresponding to repeated sampling from negative exponential populations, and with $\theta = (\psi, \chi)$ where $E(U) = (\psi\chi)^{-1}$ and $E(V) = \chi^{-1}$. In particular, calculate the correction terms S_1 and S_2, given by formulae (6.25) and (6.26), and the right-hand side of formula (6.40).

[Section 6.3]

6.7 With $Y = (U, V)$ as defined in the previous exercise, calculate the bias corrections (6.47) and (6.48) and the correction terms D_1 and D_2 given by formulae (6.52) and (6.53).

[Section 6.4]

6.8 The two bias corrections (6.47) and (6.48) can be expressed as respectively

$$\not{b}^r = -\tfrac{1}{2}\not{j}^{tu}\not{\Gamma}_{tu}^{-1}{}^{r},$$

$$b^r = -\tfrac{1}{2}i^{tu}\Gamma_{tu}^{-1}{}^{r},$$

where $\bar{\Gamma}_{tu}^{-1r}$ and Γ_{tu}^{-1r} are the Christoffel symbols of the observed and expected -1-connections of statistical differential geometry. It follows from this that the bias corrections can be made to vanish by suitable choice of the parameterization.

[Barndorff-Nielsen, 1986b; Section 6.4]

6.9 Show that the statement that $w^* \sim q_d(w)$ to order $O(n^{-3/2})$ is equivalent to

$$w \sim (1 - \tfrac{1}{2}db_1/n)q_d(w) + \tfrac{1}{2}db_1 q_{d+2}(w)/n + O(n^{-3/2}),$$

where the coefficient b_1 is determined by

$$E(w) = d\{1 + b_1/n + O(n^{-3/2})\}.$$

Give the corresponding condition in terms of Laplace transforms of the density.

[Barndorff-Nielsen and Cox, 1984a; Section 6.5]

6.10 The discussion given in section 6.5 shows that the Bartlett adjusted log-likelihood ratio statistic $w^* = w/b$ is asymptotically distributed as χ^2 to order $O(n^{-3/2})$. A closer analysis reveals that the error term is in fact of order $O(n^{-2})$, even when, as is generally the case, the adjustment term b involves parameter estimates determined by maximum likelihood. An essential element in the proof is the odd–even property of cumulant expansions (cf. Further results and exercises 4.6 in Barndorff-Nielsen and Cox, 1989).

[Barndorff-Nielsen and Hall, 1988; Section 6.5]

6.11 Show that the Bartlett adjustment for independent and identically distributed components with a scalar parameter θ can be found by the following argument exploiting invariance under reparameterization. Let $\tilde{R}_1$, $\tilde{R}_2$, $\tilde{R}_3$ denote the score standardized to unit variance, the deviation of the observed information from its linear regression on $\tilde{R}_1$ and the deviation of the third log-likelihood derivative from its linear regression on $\tilde{R}_1$, $\tilde{R}_2$, but standardized to zero mean and unit variance. Let ρ_3, ρ_4 denote the standardized third and fourth cumulants of $\tilde{R}_1$. Then we have the following results:

(a) Examination of the structure of the Bartlett factor and its variance shows that it has the form

$$k_0 + k_1\rho_3^2 + k_2\rho_4 + k_1' \operatorname{var}(\tilde{R}_2) + k_2' \operatorname{cov}(\tilde{R}_1^2, \tilde{R}_2),$$

where the ks are constants not depending on the particular distribution.

(b) For a prime exponential family $\tilde{R}_2 = 0$ and examination of normal, inverse Gaussian and exponential cases show that

$$k_0 = 0, \quad k_1 = \tfrac{5}{12}, \quad k_2 = -\tfrac{1}{4}.$$

(c) Direct evaluation via the expansion of $\mathrm{var}(\tilde{R}_2)$, $\mathrm{cov}(\tilde{R}_1^2, \tilde{R}_2)$ in terms of cumulants shows that $k_1' = 1/4$, $k_2' = -1/2$.

In the multiparameter case there is a corresponding expansion in terms of six invariant scalar functions of log-likelihood derivatives.

[McCullagh and Cox, 1986; Section 6.5]

6.12 Use Laplace's method to derive the asymptotic expression for the Bessel function K_0 used for formula (6.57).

[Section 6.5]

6.13 Suppose the model is a (m, m) exponential family and that it is desired to test an affine hypothesis about this, i.e. in a suitable formulation it is desired to test that some of the coordinates of the canonical parameter θ are zero. Similar tests can be constructed based on the conditional distribution given the minimal sufficient statistic under the null hypothesis. Discuss in detail how the Bartlett adjusted likelihood ratio statistic w^* and the modified directed likelihood r^*, calculated unconditionally, agree to order $O(n^{-3/2})$ with the similar tests.

[Jensen, 1986a; Sections 6.5 and 6.6]

6.14 In the single-parameter case, let

$$P_\theta^*\{\hat{\theta} \leqslant \bar{\theta} \mid a\} = \int_{-\infty}^{\bar{\theta}} p^*(\hat{\theta}; \theta \mid a) d\hat{\theta}.$$

Show that the relative error

$$\Phi(r^*)/P_\theta^*\{\hat{\theta} \leqslant \bar{\theta} \mid a\}$$

tends, for $\hat{\theta} \to -\infty$, towards

$$H^* = \bar{c}\exp\{\tfrac{1}{2}r^{-2}\log^2(r/u)\}\left|1 + r^{-2}\left\{\log(r/u) - 1 + (r/u)\frac{du}{dr}\right\}\right|.$$

Here

$$\frac{du}{dr}=\frac{r}{u}\{1+\tfrac{1}{2}\hat{j}^{-3/2}(\hat{\ell}_{3;}+\hat{\ell}_{2;1})u+\hat{j}^{-1}(\hat{\ell}_{;2}-\ell_{;2})\}.$$

[Barndorff-Nielsen and Chamberlin, 1991; Section 6.6]

6.15 In the situation of Example 6.20 a parameter χ which is orthogonal to $\psi=\mu+b\sigma+c\sigma^2/2$ in the sense of observed information can, by solving the appropriate equation, i.e. $\chi_{/\psi}\jmath_{\chi\chi}+\jmath_{\chi\psi}=0$ (cf. section 2.7), be given explicitly as

$$\chi=\mu+\mathscr{E}(a_i)\sigma-c^{-1}\{\mathscr{V}(a_i)+n/\sum g''(a_i)\}\log\{b+c\sigma-\mathscr{E}(a_i)\}$$

where

$$\mathscr{E}(a_i)=\sum a_i g''(a_i)/\sum g''(a_i)$$

$$\mathscr{V}(a_i)=\sum\{a_i-\mathscr{E}(a_i)\}^2 g''(a_i)/\sum g''(a_i).$$

This is provided $c\neq 0$. If $c=0$ we may define χ by

$$\chi=\mu-[\{\mathscr{V}(a_i)+n/\sum g''(a_i)\}/\{b+\mathscr{E}(a_i)\}+\mathscr{E}(a_i)]\sigma.$$

In the special case, of normal distributions and $b=0$ and $c=1$, the previous formula for χ reduces to

$$\chi=\mu-2c^{-1}\log\sigma.$$

[Section 6.6]

6.16 Show that the function $h'(s)$, which together with $\jmath$ and by (6.161) determines the accuracy of the approximation $\Phi^*\doteq P_\theta^\dagger\{\hat{\theta}\leqslant\bar{\theta}|a\}$, may be expressed as

$$h'(s)=\jmath r^{-2}[1-(r/u)^3\{1+\hat{j}^{-1}(\hat{\ell}_{;2}-\ell_{;2})+\tfrac{1}{2}\hat{j}^{-3/2}u(\hat{\ell}_{3;}+\ell_{3;})\}].$$

Use this to find an expression for the limit value of the relative error of the approximation as $\hat{\theta}\to-\infty$, imposing slight regularity conditions to make the limit value exist.

Write out the form $h'(s)$ takes in the case of a (1, 1) exponential model and note in particular that $h'(\cdot)$ in this case does not depend on sample size.

[Barndorff-Nielsen, 1990c; Section 6.7]

6.17 In some contexts the monotonicity of asymptotic expansions as a function of their argument is a matter of concern and then the following argument is sometimes helpful. Consider the expansion

$$f_n(z) = z/\sqrt{n} + a_2 z^2/n + a_3 z^3/n^{3/2} + O(n^{-2})$$

and suppose $f_n(z)$ is known to be monotonic in z. Now

$$\begin{aligned} f_n'(z) &= 1/\sqrt{n} + 2a_2 z/n + 3a_3 z^2/n^{3/2} + O(n^{-2}) \\ &= \frac{1}{\sqrt{n}} \exp\left\{ -(2a_2 \frac{z}{\sqrt{n}} + (3a_3 - 2a_2^2)\frac{z^2}{n} \right\} + O(n^{-2}) \end{aligned}$$

and the exponential is, of course, positive. This suggests replacing $f_n(z)$ by

$$g_n(z) = \frac{1}{\sqrt{n}} \int_0^z \exp\left\{ 2a_2 \frac{u}{\sqrt{n}} + (3a_3 - 2a_2^2)\frac{u^2}{n} \right\} du.$$

There are three cases according as $3a_3 - 2a_2^2$ is negative, zero or positive, leading respectively to representations in terms of the standardized normal integral, the exponential function and Dawson's integral.

6.18 Let $A_1, \ldots, A_m$ be independently and identically normally distributed with mean μ_A and variance σ_A^2. Suppose that, given $A_1, \ldots, A_m$, the log-likelihood has the form $\sum \ell_j(\theta, A_j)$. Show that the Laplace expansion for the log-likelihood in the 'random effects' model is of the form

$$\begin{aligned} &\sum \{\ell_j(\theta, \mu_A) - \tfrac{1}{2}\log(1 - \sigma_A^2 \ell_j^{(2)})\} \\ &\quad + \tfrac{1}{2}\sum \sigma_A^2 \{\ell_j^{(1)}\}^2/(1 - \sigma_A^2 \ell_j^{(2)}) + \tfrac{1}{2}\sum(\ell_j^{(3)}\ell_j^{(1)} + \tfrac{1}{4}\ell_j^{(4)})\sigma_A^4 + o(\sigma_A^4), \end{aligned}$$

where $\ell_j^{(r)}$ is $[\partial^r \ell_j(\theta_j, a)/\partial a_r]_{a=\mu_A}$, and such that it is exact for the normal theory random-effects model and always corresponds to a positive likelihood. This can be applied to, for example, matched pair binary response data.

[Solomon and Cox, 1992]

Bibliographic notes

The introduction and systematic study of the p^*-formula (Barndorff-Nielsen, 1980; 1983; 1984; 1988a; Barndorff-Nielsen and Cox, 1984a; 1984b) builds on work by Fisher (1934), Efron and Hinkley (1978), Barndorff-Nielsen and Cox (1979), Daniels (1954; 1956; 1958; 1980), Cox (1980), Durbin (1980) and Hinkley (1980).

Although formula (6.2) for p^* is the same irrespective of the type of model, its roots in the theory of transformation models are quite different from its roots in the theory of (m, d) exponential models. Various general derivations of p^* have however been given recently. See Fraser and Reid (1988), Skovgaard (1990b), and Barndorff-Nielsen (1989; 1990b).

For further discussion of p^*, see McCullagh (1987), Fraser (1988), Reid (1988) and Barndorff-Nielsen (1988a; 1988b; 1989).

In general, application of (6.2) requires the determination of an ancillary a. Saddlepoint approximations to the marginal distribution of $\hat{\theta}$, as well as to other estimators defined via estimating equations, have been discussed by Skovgaard (1985a; 1985b), Field (1982), Daniels (1983), and Hougaard (1985); see also Field and Ronchetti (1990) and the geometrically motivated work of Pázman (1984; 1986; 1987; 1989). These techniques do not involve ancillaries and are, in general, at odds with the viewpoint of conditionality.

The derivation of the invariant expansion (6.24)–(6.26) of the p^*-formula for the conditional distribution of the score vector is due to Mora (1992), and section 6.4 is based on Barndorff-Nielsen (1986b; 1988a) and Mora (1992). Edgeworth expansions for the distribution of the maximum likelihood estimator in curved exponential families are discussed by Amari and Kumon (1983); cf. also Amari (1985) and Mora (1992).

Integration of p^* so as to obtain an expression for the distribution of, for instance, a component of $\hat{\omega}$ or some combinant can be achieved, at least approximately, by means of these expansions or by use of Laplace's method. For a discussion of the latter approach, see Barndorff-Nielsen (1990b).

The material on Bartlett adjustments of the likelihood ratio in section 6.5 is from Barndorff-Nielsen and Cox (1984a). For a more direct approach, following Lawley (1956), see Hayakawa (1977) and Cordeiro (1983; 1987), and, for a special class of problems, Box (1949). Section 6.6 follows mainly Barndorff-Nielsen (1986a; 1991; 1993a) and Barndorff-Nielsen, Peters and Pierce (1993).

When no nuisance parameters are present and if the model is an exponential family of order 1 formula (6.145) equals the Lugannani and Rice (1980) tail-area approximation, which has been further discussed by Daniels (1987). The extension of the Lugannani–Rice approximation to more general one-parameter models was briefly introduced by Barndorff-Nielsen (1988b) and more fully treated by

Barndorff-Nielsen (1990a; 1990c) and Fraser (1990). See also Skovgaard (1987) which derives Lugannani–Rice type approximations for conditional distributions. The correction (6.137) for lattice distributions is due to Daniels (1987).

The first systematic discussions of properties of directed likelihood are due to Lawley (1956) and Sprott (1973; 1975). The use of directed likelihoods as approximate ancillaries was considered in Barndorff-Nielsen (1980; 1984), Pedersen (1981) and Jensen (1992). Pierce and Schafer (1986) discussed the use of directed likelihood in residual analysis. Adjustment of directed likelihoods to improve the degree of asymptotic normal approximation was first considered by Lawley (1956), who derived bias and standard deviation adjustments; cf. also McCullagh (1982; 1984a; 1984b; 1987), Barndorff-Nielsen (1984; 1986a) and Jensen (1986a; 1987a). The modified directed likelihood r^* was introduced by Barndorff-Nielsen (1986a) and has been further studied by Barndorff-Nielsen (1991; 1992; 1993), Jensen (1992) and Pierce and Peters (1992). Note, however, that the modification of directed likelihood considered by Jensen (1992) differs somewhat from that of this chapter except for prime exponential models. Closely related issues have been studied from somewhat different viewpoints by Fraser and by DiCiccio and coworkers. Their work is best approached via Fraser (1991), DiCiccio and Martin (1993) and DiCiccio, Field and Fraser (1990).

CHAPTER 7

Higher-order theory: some further results and tools

7.1 Introduction

The concepts of ancillarity and sufficiency have played a prominent part in the previous chapters of this book. It is in the spirit of higher-order asymptotics to study approximate versions of these concepts.

The conjunction of an exact or approximate ancillary statistic and the maximum likelihood estimator constitutes what is termed a **conditionality structure**. Jacobians of transformations to such structures occur as important elements in some of the developments discussed in this and the next chapter.

By a **conditionality resolution** we mean a conditionality structure $(\hat{\theta}, a)$ and an approximate expression for $p(\hat{\theta}; \theta | a)$, the associated model function for inference on θ. The particular approximation $p^*(\hat{\theta}; \theta | a)$ has been discussed and applied extensively in the previous chapter, and some consideration was also given the associated less accurate approximation $p^\dagger(\hat{\theta}; \theta | a)$. A sketch of the derivations of p^* and $p^\dagger$ for curved exponential models will be given here.

Several other aspects of higher-order asymptotics that relate in a direct manner to material in Chapter 6 are also considered in the present chapter, in particular the effect of sampling rules and a type of second-order asymptotic independence.

7.2 Approximate sufficiency and ancillarity

The question of how to construct ancillary statistics as parts of conditionality structures $(\hat{\theta}, a)$ has, except for some rather special cases, an exact solution only for transformation models and, in a degenerate sense, for (m, m) exponential models. Under the former type of model the maximal invariant is an obvious ancillary, while for (m, m) exponential models no ancillary is required for inference on θ since the maximum likelihood estimator $\hat{\theta}$ is minimal sufficient. In the present section we discuss the question of construction of approximate ancillaries. First, however, we consider the topic of approximate sufficiency.

The normed log-likelihood function $\bar{\ell}(\cdot) = \ell(\cdot) - \ell(\hat{\theta})$ is a minimal sufficient statistic, and the same is true of any version of the log-likelihood function of the form $\ell(\cdot) - \ell(\theta_0)$ for some fixed value θ_0 of the parameter. (For a mathematically rigorous formulation, see Barndorff-Nielsen, 1978, section 4.2.) This implies that also the score $u = \nabla \ell$ can be considered as a minimal sufficient statistic.

Since $\bar{\ell}$ can be approximated to any desired degree of accuracy by Taylor expansion around $\hat{\theta}$ it is obvious that the maximum likelihood estimator $\hat{\theta}$ together with the log-likelihood derivative arrays $\{\hat{\ell}_{r_1 r_2}\}, \ldots, \{\hat{\ell}_{r_1 \cdots r_m}\}$, up to some finite order m and evaluated at $\hat{\theta}$, must constitute an approximately sufficient statistic. More specifically, the approximate sufficiency of

$$s = (\hat{\theta}, \{\hat{\ell}_{r_1 r_2}\}, \ldots, \{\hat{\ell}_{r_1 \cdots r_m}\})$$

can in essence be expressed by saying that to order $O(n^{-m/2})$ the distribution of the data Y can be factorized as

$$p(y; \theta) \doteq p(s; \theta) q(y \mid s)$$

for a certain function $q(y \mid s)$ which does not depend on θ. For a rigorous mathematical formulation, see Michel (1978). In particular, the maximum likelihood estimator itself is asymptotically sufficient to order $O(n^{-1/2})$.

From the approximately sufficient statistic s an approximate conditionality structure can, in wide generality, be constructed by making a linear transformation of the elements of $\{\hat{l}_{r_1 r_2}\}, \ldots, \{\hat{l}_{r_1 \cdots r_m}\}$ into a set of random variables $a_1, \ldots, a_q$ that approximately follow a standard q-dimensional normal distribution. The transformation will depend on $\hat{\theta}$, and the approximate normality is a consequence of the joint asymptotic normality of the various derivatives ℓ_{r_1}, $\ell_{r_1 r_2}, \ldots, \ell_{r_1 \cdots r_m}$.

We now outline this construction in the simplest case of a one-dimensional parameter and with $m = 2$. Thus as the approximately sufficient statistic we take $(\hat{\theta}, \hat{j})$, as usual denoting the observed and expected information by j and i, respectively. We aim to find a function of $(\hat{\theta}, \hat{j})$ with a fixed distribution, in fact a standard normal distribution. The observed information $\hat{j}$ is approximately equal to $\hat{\imath}$, a function of $\hat{\theta}$, and to determine approximate properties of $\hat{j}-\hat{\imath}$ we consider the Taylor expansion of $\hat{j}-\hat{\imath}$ around θ which may be written as

$$\hat{j}-\hat{\imath}=j-i-(\hat{\theta}-\theta)\{\ell_3(\theta)+2\nu_{2,1}(\theta)+\nu_{1,1,1}(\theta)\}+\cdots.$$

Here and below we use the notation $\nu_{1,2}=E(\ell_\theta\ell_{\theta\theta};\theta)$, $\nu_{1,1,1}=E(\ell_\theta^3;\theta)$, etc. By formulae (5.19) and (5.6) we then find that, with error $O(n^{-\frac{1}{2}})$, under ordinary repeated sampling,

$$\begin{aligned}\hat{j}-\hat{\imath}&\doteq j-i-i^{-1}\ell_1(\theta)\{\ell_3(\theta)+2\nu_{2,1}(\theta)+\nu_{1,1,1}(\theta)\}\\&=j-i+i^{-1}\nu_{2,1}\ell_1(\theta)-i^{-1}\ell_1(\theta)\{\ell_3(\theta)-\nu_3(\theta)\}.\end{aligned}$$

The first-order approximation to $\hat{j}-\hat{\imath}$ is $j-i+i^{-1}\nu_{2,1}(\theta)\ell_1(\theta)$ and hence we have, writing $\nu_{2,2}$ for $\nu_{2,2}(\theta)$, etc.,

$$\begin{aligned}E(\hat{j}-\hat{\imath})&\doteq 0,\\ \operatorname{var}(\hat{j}-\hat{\imath})&\doteq \nu_{2,2}-i^2-i^{-1}\nu_{2,1}^2,\end{aligned}$$

with errors $O(1)$ and $O(n^{\frac{1}{2}})$, respectively. Letting

$$\gamma=i^{-1}(\nu_{2,2}-\nu_2^2-i^{-1}\nu_{2,1}^2)^{1/2}$$

we find, all in all, that the statistic a defined by

$$a=(\hat{\imath}\hat{\gamma})^{-1}(\hat{j}-\hat{\imath}) \tag{7.1}$$

is asymptotically standard normal to order $O(n^{-\frac{1}{2}})$, a being, as required, a function of $(\hat{\theta},\hat{j})$.

Further, the first-order term in the expansion of the covariance of $\hat{\theta}$ and a is zero, i.e. $\hat{\theta}$ and a are independent under their asymptotic joint normal distribution. However, for small samples it can make a major difference whether one conditions on a or not.

The statistic (7.1) is commonly referred to as the **Efron–Hinkley ancillary**. Efron and Hinkley (1978) introduced and studied the properties of a and gave some striking numerical examples of the effects of conditioning on that ancillary.

Skovgaard (1985a) has discussed the multivariate parameter case with respect to second-order ancillarity, using a multivariate analogue of the Efron–Hinkley ancillary, of the form $a=\hat{\Gamma}^{-\frac{1}{2}}(\hat{j}-\hat{\imath})$ where $\Gamma^{\frac{1}{2}}$ is any square-root of the variance matrix of the information matrix difference $\hat{j}-\hat{\imath}$. The dimension of a equals the rank of Γ. A simple direct calculation shows that $\hat{\theta}$ and a are asymptotically uncorrelated.

Skovgaard determines the Edgeworth approximation with error $O(n^{-1})$ to the joint distribution of $\hat{\theta}$ and a and thence to the conditional distribution of $\hat{\theta}$ given a, the result for the conditional density of $Z = \sqrt{n}(\hat{\theta} - \theta)$ being expressible as

$$p(z|a) \doteq \phi_d\{z;(i + \Gamma^{\frac{1}{2}}a)^{-1}\}\{1 + \lambda_z iz + \tfrac{1}{6}\lambda_{z,z,z}i^3z^3 - \tfrac{1}{2}i\lambda_{z,z,z}iz\} + \cdots, \tag{7.2}$$

where λ_z and $\lambda_{z,z,z}$ are the mean vector and third-order cumulant array of Z. Note that the correction terms in the braces do not involve the ancillary a. Furthermore, the formula for the Edgeworth approximation of the density of a shows that the term of order $n^{-\frac{1}{2}}$ depends on θ so that in this sense a is only first-order asymptotically ancillary. However, in any moderate deviation region such as $\{\theta: |\theta - \theta_0| < c/\sqrt{n}\}$, where θ_0 is an arbitrary fixed value of the parameter, the distribution of a is constant except for terms of order $O(n^{-1})$. In this sense a is **locally second-order ancillary**.

The Efron–Hinkley ancillary (7.1) and multivariate analogues of this, in particular the score ancillaries discussed below, are constructed by purely local considerations, i.e. from properties of moderate deviations around $\hat{\theta}$. This alone makes it plausible that more effective and appropriate conditioning can in many cases be achieved by employing an alternative ancillary that, in some measure, also captures the large-deviation structure of the problem.

The general character of the directed likelihood derived from an ambient parametric model in which the model under study is exactly or approximately embedded and the specific properties of the directed likelihood and its modified version, discussed in section 6.6, indicate that it will generally be advantageous to use these as ancillaries, when they can be introduced in a natural way.

A particular aspect of this is that typically a directed likelihood ancillary will be variation independent of the maximum likelihood estimator $\hat{\theta}$, whereas this will not be the case for the Efron–Hinkley ancillary and its multivariate analogues. This type of variation independence is desirable as it agrees with the more general conception that an ancillary should not contain any information about the value of the maximum likelihood estimator, only about its 'precision'.

On the other hand, because the Efron–Hinkley ancillary and its generalizations are determined solely by aspects of the log-likelihood at $\hat{\theta}$, they are often considerably simpler to calculate.

As an illustration we consider an example of a (2, 1) exponential model. This example was devised so as to make the calculation analytically feasible as far as possible while still exhibiting a difference between the Efron–Hinkley ancillary a and the directed likelihood r for testing the model against the ambient (2, 2) exponential model. In particular, it would not do to take a (2, 1) exponential transformation model because both a and r would be equivalent to the maximal invariant statistic and thus would lead to the same conditional model.

Example 7.1 Comparison of the Efron–Hinkley ancillary and the directed likelihood ancillary
Let U and $V-1$ be independent and exponentially distributed with parameters χ and ψ, i.e. with probability density

$$p(u, v; \chi, \psi) = \chi\psi e^{-\chi u - \psi v + \psi}. \tag{7.3}$$

We shall be working with the curved exponential submodel of (7.3) determined by letting χ be the function of ψ given by

$$\chi(\psi)\psi e^{\psi} = 1. \tag{7.4}$$

The model thus defined is a (2, 1) exponential model with model function

$$p(u, v; \psi) = \exp\{-\chi(\psi)u - \psi v\}. \tag{7.5}$$

The mean parameter of the full model (7.3) is $\eta = (\chi^{-1}, \psi^{-1})$ with domain of variation $H = (0, \infty)^2$, and the model (7.5) is determined by the curve H_0 in H given by

$$\eta(\psi) = (\psi e^{\psi}, 1 + \psi^{-1}). \tag{7.6}$$

The maximum likelihood estimator in the submodel is the unique solution of

$$\chi'(\psi) = -v/u. \tag{7.7}$$

Thus the observation pairs (u, v) giving rise to a fixed value of $\hat{\psi}$ are half-lines with end points at (0, 0) and slopes

$$-\chi'(\hat{\psi}) = (1 + \hat{\psi}^{-1})\hat{\psi}^{-1}e^{-\hat{\psi}}.$$

By introducing the notation

$$\Psi_{\pm} = \{1 + (1 + \hat{\psi})^{\pm 2}\}^{-1/2}$$

the affine ancillary $a = (\hat{j}/\hat{\imath} - 1)/\hat{\gamma}$ (cf. (7.1)) and the directed likelihood

r for testing the model (7.5) against the model (7.3) may be expressed as

$$a = \{v - 1 - \hat{\psi}^{-1}\}\hat{\psi}\hat{\Psi}_{-}^{-1}$$
$$\tfrac{1}{2}r^2 = \{\hat{\Psi}_{+} + \hat{\Psi}_{-}\}a - \log\{1 + \hat{\Psi}_{-}a\} - \log\{1 + \hat{\Psi}_{+}a\},$$

the sign of r being the same as that of a.

An important difference between a and r is that r and $\hat{\psi}$ are variation independent whereas a and $\hat{\psi}$ are not. This exemplifies a general point made earlier.

The difference between the two ancillaries is, moreover, illustrated

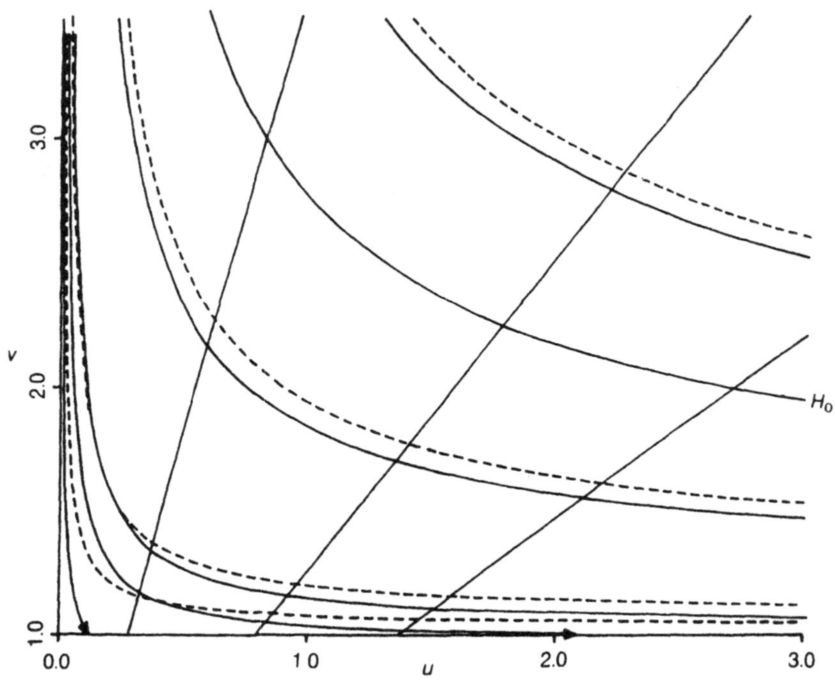

Figure 7.1. *Level curves for $\hat{\psi}$, a and r. The three straight lines represent, from left to right, $\hat{\psi} = 0.5$, 0.8, 1.0. The mean curve H_0 corresponds to $a = r = 0$. Further level curves correspond to $a = 0.5$, -0.5, -1.0, -1.1, -1.25 (solid curves), and $r = 0.5$, -0.5, -1.5, -2.0 (broken curves). The arrowheads indicate the points where the level curves for a smaller than -1 hit the u-axis, giving $\psi = \psi_a$, where ψ_a is the upper bound of the possible values of $\hat{\psi}$ for the particular a, that is, $\psi_a = (a^2 - 1)^{-1/2} - 1$.*

in Fig. 7.1 which shows their level curves (constant-value curves) in the (u, v)-plane.

The level curves for a and r are nearly identical for numerically small values of the two ancillaries. Moreover, the level curves are similar in shape for all positive values. In contrast, the level curves deviate considerably for a slightly smaller than -1. The effect of this on the conditional inference given a or r is that for observations yielding positive or negative but numerically small values of the two ancillaries, similar conclusions will be drawn, whereas observations below H_0, yielding negative values of a and r, may result in quite different conclusions. For instance, for $(u, v) = (0.049, 1.199)$, giving $a = -1.25$ and $r = -2.48$, the conditional distribution of $\hat{\psi}$ given a is concentrated on the interval $(0, \psi_a) = (0, 0.33)$ while if $\psi = 1.0$ most of the mass of the conditional distribution of $\hat{\psi}$ given r is concentrated beyond 0.33.

The marginal densities for a and r are shown in Fig. 7.2 in two rather extreme cases, $\psi = 0.1$ and $\psi = 2.5$, which correspond to the mean vector $\eta(\psi)$ of (U, V) being $(0.1, 11.1)$ and $(30.5, 1.4)$, respectively. It is seen that the densities for r are almost identical and normal shaped, though not standard normal, as the mode varies from -0.6 to -0.52. On the other hand, the densities for a vary considerably and are far from having a normal shape.

From the viewpoint of likelihood-based inference, to evaluate the performance of an approximate ancillary it is essential to compare the conditional likelihood functions given the ancillary to the corresponding unconditional likelihoods. Ideally these should be the same, as happens if and only if the density of the conditioning statistic does not depend on the parameter.

Figure 7.3 provides such a comparison, for a single observation, $(u, v) = (1.38, 1.02)$. The conditional log-likelihood given r agrees well with the unconditional, while the conditional log-likelihood given a deviates rather drastically from the unconditional. The close agreement shown by r is, in fact, typical of all practically interesting values of r.

The above discussion is for a single observation (u, v) from (7.5). Under repeated sampling from (7.5) with sample size n the difference, as n increases, between conditioning on a and on r disappears rather quickly in the normal deviation region but persists for larger deviations, as shown in further numerical results due to Pedersen (1981).

□

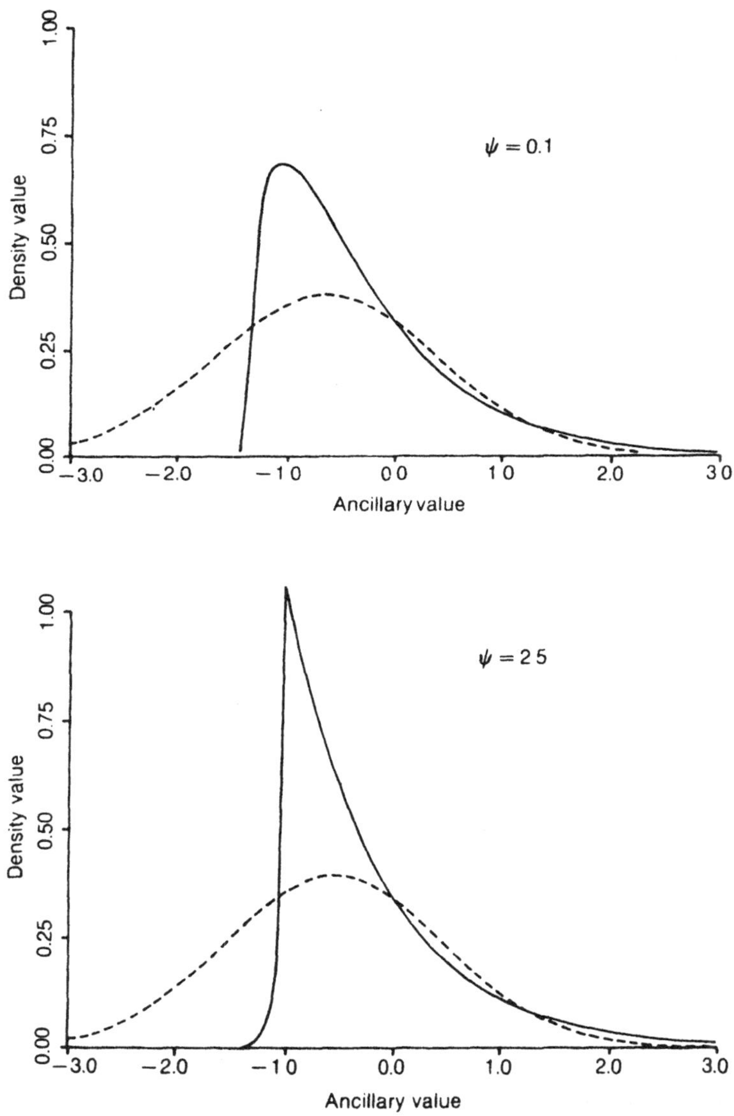

Figure 7.2 *The marginal probability densities of a and r for $\psi = 0.1$ and 2.5. a: solid curves; r: broken curves.*

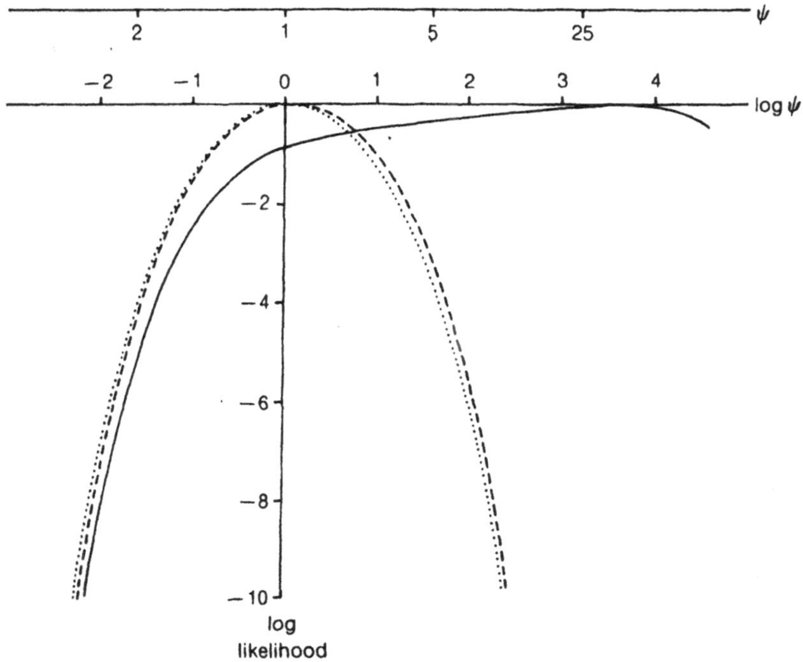

Figure 7.3 *The conditional log-likelihood functions for* $\log\psi$ *given* a *(solid curve) and* r *(broken curve), together with the unconditional log-likelihood function (dotted curve). The observation* $(u, v) = (1.38, 1.02)$ *corresponds to* $\hat{\psi} = 1.0$, $a = -1.1$ *and* $r = -2.58$.

We go on to construct, in a fairly general setting, an approximate ancillary based on components of the score vector. For (m, d) exponential models the ancillary is of dimension $m - d$ and the construction may be viewed as a generalization of the Efron–Hinkley ancillary. Later on we shall, in the same setting, discuss a related construction of approximate ancillaries based on directed likelihoods.

Suppose $\mathscr{M}$ is a submodel of a parametric model $\mathscr{M}^0$. Let ζ be the parameter of the model $\mathscr{M}^0$ and let $m > d$ denote the dimension of ζ. We shall indicate maximum likelihood values under $\mathscr{M}^0$ by a prime prefix, using a circumflex for maximum likelihood under $\mathscr{M}$. Suppose, moreover, that a conditionality structure $('\zeta, a^0)$ has been established for $\mathscr{M}^0$. How can this structure be modified into a conditionality structure $(\hat{\theta}, a)$ for $\mathscr{M}$?

Generic coordinates of ζ and $'\zeta$ will be denoted respectively by

ζ^i, ζ^j, etc. and $'\zeta^i, '\zeta^j$, etc. It will also be convenient to consider ζ as a one-to-one function of a parameterization of $\mathcal{M}^0$ of the form (θ, ψ), with ψ being of dimension $m-d$, and such that $\mathcal{M}$ is the submodel of $\mathcal{M}^0$ corresponding to a particular value, ψ_0 say, of ψ. We shall use ℓ to denote the log-likelihood function under $\mathcal{M}^0$, though ℓ^0 would be a more consistent notation. Accordingly, $\ell(\zeta(\theta, \psi_0))$ is the likelihood function under $\mathcal{M}$.

What we seek is a statistic b of dimension $m-d$ which together with a^0 can serve as ancillary $a = (b, a^0)$ in a conditionality structure $(\hat{\theta}, a)$.

In broad generality the score vector $\ell_i = \ell_i(\zeta)$ will, conditionally on a^0 as well as unconditionally, be approximately normally distributed with mean value 0 and precision $\jmath(\zeta)$, where $\jmath(\zeta)$ is the observed information metric for ζ, i.e.

$$\jmath(\zeta) = -[\ell_{ij}(\zeta; \hat{\zeta}, a^0)]_{\hat{\zeta}=\zeta}.$$

Let Z_0 be the submanifold of the domain of variation Z for ζ corresponding to $\mathcal{M}$ and let $B = B(\theta; a^0)$ be an $(m-d) \times m$ matrix whose $m-d$ row vectors constitute an orthonormal basis for the orthogonal complement in $\mathbb{R}^m$ to the tangent space $(TZ_0)_\theta$ of Z_0 at $\zeta(\theta, \psi_0)$, orthonormality and orthogonality being with respect to the inner product determined by $\jmath$. Such a matrix can be constructed from an arbitrary $(m-d) \times m$ matrix $A = A(\theta, a^0)$, whose row vectors span the orthogonal complement to $(TZ_0)_\theta$, specifically by the formula

$$B = (A\jmath^{-1}A^{\mathrm{T}})^{-\frac{1}{2}}A\jmath^{-1}. \tag{7.8}$$

Let B^i_κ denote the (κ, i)th element of B, where the index κ runs from 1 to $m-d$. Assuming approximate normality of the score vector ℓ_i, as indicated above, we have that the vector $B^i_\kappa \ell_i$ is approximately distributed as $N_{m-d}(0, I)$ and the same is true of the statistic

$$b = [b_\kappa] = [\hat{B}^i_\kappa \hat{\ell}_i], \tag{7.9}$$

where $\hat{\zeta} = \zeta(\hat{\theta}, \psi_0)$ and $\hat{B} = B(\hat{\theta}, a^0)$. The statistic $a = (b, a^0)$, with b given by (7.9), will thus, in broad circumstances, be applicable as the ancillary component of a conditionality structure $(\hat{\theta}, a)$. We shall refer to (7.9) as the **score ancillary**.

When $d = m-1$ the complementary ancillary b is, up to its sign, uniquely determined by the method of construction given above.

However, for $m-d>1$ the expression (7.9) for b depends on the choice of B, and two different choices will in general lead to different conditional models.

An exception to the latter statement occurs for models which, conditionally on a^0, are transformation models. Under such a model all the various versions of b will, in general, be one-to-one functions of the maximal invariant statistic and will therefore lead to the same conditional model for inference on θ.

Another exception is that where under $\mathcal{M}^0$ there is a statistic v and a corresponding partition (θ, ψ) of ζ with independently varying components θ and ψ such that

$$p(y;\zeta|a^0) = p(v;\psi|a^0)p(y;\theta|a^0,v), \tag{7.10}$$

i.e. v is a cut in the conditional model given a^0, and in particular, v is S-ancillary with respect to θ. If, as above, the model $\mathcal{M}$ is determined by fixing the value of ψ then, typically, b given by (7.9) will be a one-to-one function of v. This is so in particular for (m, d) exponential models.

Note also that for an (m, d) exponential model $\mathcal{M}$ with exponential representation (2.114) the log-likelihood derivatives $\ell_{r_1r_2}, \ell_{r_1r_2r_3}, \ldots$ are all linear functions of $s-\eta$ so that the construction of an approximate ancillary from these derivatives, discussed earlier, is in effect equivalent to the construction of the score ancillary (7.9).

We now turn to the alternative possibility of constructing an $(m-d)$-dimensional complementary ancillary b from directed likelihood statistics. This can be done as follows. We select a sequence of submodels $\mathcal{M}^0_{m-\nu}$, $\nu = 0, \ldots, m-d$, of $\mathcal{M}^0$ such that the parametric dimension of $\mathcal{M}^0_{m-\nu}$ is $m-\nu$ and such that $\mathcal{M}^0 = \mathcal{M}^0_m \supset \mathcal{M}^0_{m-1} \supset \cdots \supset \mathcal{M}^0_d = \mathcal{M}$. Let r_ν be the directed likelihood statistic for testing $\mathcal{M}^0_{m-\nu}$ against $\mathcal{M}^0_{m-\nu+1}$ and let

$$b = (r_1, \ldots, r_{m-d}). \tag{7.11}$$

In wide generality the distribution of b will be approximately standard $(m-d)$-dimensional normal, as desired. Improved approximation to standard normality can, in wide generality, be achieved by employing adjusted versions of the directed likelihoods, for instance, for each $\nu = 1, \ldots, m-d$, using the modified directed likelihood r^*_ν calculated with $(r^*_1, \ldots, r^*_{\nu-1}, a^0)$ as ancillary.

As is the case for the score-based ancillary the construct (7.11) is unique only when $d = m-1$, and it will therefore lead to different conditional models unless $\mathcal{M}$ is transformational or determined via a cut as discussed in connection with formula (7.10).

In cases of non-uniqueness and if there are no substantial reasons for choosing one of the versions of, say, the directed likelihood ancillary over the others, it may be preferable to work with inference procedures that do not require specification of an ancillary but do nevertheless satisfy the prescription of conditional inference in the sense of having, approximately, the relevant frequency properties conditional on any reasonable ancillary. We use the term **stable** to characterize procedures and quantities that have this kind of property. Many of the classical, first-order, limit statements of parametric inference are indeed stable in this sense; see, however, section 4.4. It is, however, often possible to achieve stability and at the same time improve on the classical limit results by one degree of asymptotic accuracy, and we shall consider procedures of this type in Chapter 8.

7.3 Jacobians of conditionality structures

Once a conditionality structure $(\hat{\theta}, a)$ has been established it is generally of interest to determine the Jacobian of the transformation from $(\hat{\theta}, a)$ to the sufficient statistic s, as a step towards finding an expression for the conditional distribution of $\hat{\theta}$ given a.

Here we shall discuss the Jacobians for the settings considered at the end of section 7.2. We will argue under the supposition that the conditional model for ζ given a^0 has already been found so that the problem consists in finding the Jacobian of the transformation $(\hat{\theta}, b) \to \zeta$, for fixed a^0.

First, suppose that b is the score ancillary, i.e. the statistic (7.9) which was constructed on the basis of the score vector for $\mathcal{M}^0$ at $\hat{\zeta}$.

We view ℓ as a function of ζ and (ζ, a^0), i.e. $\ell = \ell(\zeta; \zeta, a^0), \zeta$ as function of $(\hat{\theta}, b)$ and $\hat{\zeta}$ as the function $\zeta(\hat{\theta}, \psi_0)$. Coordinates of b are denoted by b_κ, b_λ, etc., the indices κ, λ, etc. running from 1 to $m - d$. The equations determining $\hat{\zeta}$ and b from ζ are

$$\hat{\ell}_i \hat{\zeta}^i_{/r} = 0 \qquad (r = 1, \ldots, d), \tag{7.12}$$

$$\hat{\ell}_i \hat{B}^i_\kappa = b_\kappa \qquad (\kappa = 1, \ldots, m - d). \tag{7.13}$$

Differentiating these with respect to $\hat{\theta}^s$ and b_λ, we obtain the equations

$$\hat{\ell}_{ij}\hat{\zeta}^i_{/r}\hat{\zeta}^j_{/s} + \hat{\ell}_{i;j}\hat{\zeta}^i_{/r}\zeta^j_{/s} + \hat{\ell}_i\hat{\zeta}^i_{/rs} = 0, \tag{7.14}$$

$$\hat{\ell}_{i;j}\hat{\zeta}^i_{/r}\zeta^j_{/\lambda} = 0, \tag{7.15}$$

$$\hat{\ell}_{ij}\hat{B}^i_\kappa\hat{\zeta}^j_{/s} + \hat{\ell}_{i;j}\hat{B}^i_\kappa\zeta^i_{/s} + \hat{\ell}_i\hat{B}^i_{\kappa/s} = 0, \tag{7.16}$$

$$\hat{\ell}_{i;j}\hat{B}^i_\kappa\zeta^j_{/\lambda} = \delta_{\kappa\lambda}, \tag{7.17}$$

where

$$\begin{aligned}
\zeta^j_{/s} &= \partial\zeta^j(\hat{\theta}, b)/\partial\hat{\theta}^s, \\
\hat{\zeta}^i_{/rs} &= \partial_r\partial_s\zeta(\hat{\theta}, \psi_0)/\partial\hat{\theta}^r\partial\hat{\theta}^s, \\
\zeta^j_{/\lambda} &= \partial\zeta^j(\hat{\theta}, b)/\partial b_\lambda, \\
\hat{B}^i_{\kappa/s} &= \partial B^i_\kappa(\hat{\theta}; \dot{a}^0)/\partial\hat{\theta}^s,
\end{aligned}$$

and $\delta_{\kappa\lambda}$ is the Kronecker delta. These equations may be written in matrix form as

$$\begin{bmatrix} \partial\hat{\zeta}/\partial\hat{\theta} \\ \hat{B} \end{bmatrix} \ell_{;}(\hat{\zeta}) \frac{\partial\zeta^{\mathrm{T}}}{\partial(\hat{\theta}, b)} = \begin{bmatrix} j(\hat{\theta}) & 0 \\ F & I \end{bmatrix}, \tag{7.18}$$

where we have used the fact that

$$\ell_{rs} = \ell_{ij}\zeta^i_{/r}\zeta^j_{/s} + \ell_i\zeta^j_{/rs}$$

and where F is the $(k-d)\times d$ matrix whose (κ, s)th element is $\hat{j}_{ij}\hat{B}^i_\kappa\hat{\zeta}^j_{/s} - \hat{\ell}_i\hat{B}^i_{\kappa/s}$. By the construction of B we have

$$\begin{bmatrix} \partial\zeta/\partial\theta \\ B \end{bmatrix} \jmath \begin{bmatrix} \partial\zeta/\partial\theta \\ B \end{bmatrix}^{\mathrm{T}} = \begin{bmatrix} j_0 & 0 \\ 0 & I \end{bmatrix},$$

where j_0 is the Riemannian metric on Ω induced from the observed information metric on $\mathscr{M}^0$, i.e.

$$j_{0rs} = \jmath_{ij}\zeta^i_{/r}\zeta^j_{/s}. \tag{7.19}$$

Consequently, the desired Jacobian is

$$|\partial\zeta/\partial(\hat{\theta}, b)^{\mathrm{T}}| = |j(\hat{\theta})||j_0(\hat{\theta})|^{-\frac{1}{2}}|\jmath(\hat{\zeta})|^{\frac{1}{2}}|\ell_{;}(\hat{\zeta})|^{-1}. \tag{7.20}$$

Note that (7.20) is independent of which choice one makes of the normalizing matrix B.

If $\mathscr{M}$ is an (m, d) exponential model (2.114) and $\mathscr{M}^0$ is the (m, m) model generated by $\mathscr{M}$ one finds on taking ζ to be η, the mean value parameter of $\mathcal{M}^0$, that $\zeta = s$ and $\ell_{;} = \Sigma^{-1}$, where $\Sigma = \text{var}(S;\theta)$. It follows that in this case (7.20) can be rewritten as

$$|\partial s/\partial(\hat{\theta}, b)^{\mathrm{T}}| = |j(\hat{\theta})||i(\hat{\theta})|^{-\frac{1}{2}}|\hat{\Sigma}|^{\frac{1}{2}}. \tag{7.21}$$

We now briefly consider the situations where the ancillary component b is given by directed likelihoods. If $d = m - 1$, the Jacobian has in fact been derived previously, see formulae (6.128)–(6.130). For general d an expression for the Jacobian may be found in Jensen (1992).

7.4 Derivation of $p^\dagger$ and p^* for curved exponential models

The p^*- and $p^\dagger$-formulae may be arrived at in a number of different ways, each of which illuminates some important aspect or aspects of the formulae. The most general, rigorous proof is due to Skovgaard (1990a). However, that proof does not, in an immediate way at least, reveal that p^* is exact for transformation models, as was shown to be the case in section 6.2. Here we shall outline a proof that is more direct but which, on the other hand, is restricted to curved exponential models.

For exponential models that are non-curved, i.e. of type (m, m), a proof that $p^\dagger$ and p^* are, under repeated sampling, correct to asymptotic order $O(n^{-3/2})$ and $O(n^{-1})$, respectively, is given by Barndorff-Nielsen and Cox (1989, sections 6.5 and 6.13); see also section 6.2 above. This result is instrumental in the following discussion of the general case of (m, d) exponential models.

We use the set-up and notations employed in section 2.10 and in the previous two sections in connection with the discussion of the score ancillary and the directed likelihood ancillary. The model $\mathcal{M}$ is an (m, d) curved exponential model

$$p(y;\theta) = A(\theta)b(y)\exp\{s^{\mathrm{T}}(y)\phi(\theta)\} \tag{7.22}$$

where $s(y)$ and $\phi(\theta)$ are vectors of dimension m and the parameter θ is of dimension d. We assume that the (m, m) exponential model $\mathcal{M}^0$ generated by (7.22) is a prime exponential model. Its model function is of the form

$$p(y;\phi) = a(\phi)b(y)\exp\{s^{\mathrm{T}}(y)\phi\} \tag{7.23}$$

and we have that $A(\theta) = a(\phi(\theta))$.

Writing, as usual, $k(\phi) = -\log a(\phi)$ for the cumulant function of

(7.23) we have that the p^*-formula corresponding to (7.23) is

$$\begin{aligned} p^*('\phi;\phi) &\doteq c(\phi)|j('\phi)|^{1/2}e^{\ell-'\ell} \\ &= \bar{c}(\phi)(2\pi)^{-m/2}|\nabla\nabla^{\mathrm{T}}k('\phi)|^{1/2}\exp\{s^{\mathrm{T}}(y)(\phi-'\phi)-k(\phi)+k('\phi)\} \end{aligned} \tag{7.24}$$

where $\bar{c}=c(2\pi)^{m/2}$ and where, in accordance with the notational conventions in sections 7.2 and 7.3, $'\phi$ is the maximum likelihood estimator of ϕ under $\mathscr{M}^0$. The corresponding $p^\dagger$-formula is obtained from (7.24) by deleting the factor $\bar{c}$.

We may now transform (7.24) to a joint density for $(\hat{\theta},a)$, where a is the score ancillary (7.9). The Jacobian required for this is given by (7.21) multiplied by $|'\phi_{/s}|=|'\Sigma|^{-1}$ and hence we obtain

$$p^*(\hat{\theta},a;\phi)=c(\phi)|'j|^{1/2}e^{\ell-'\ell}\{|\hat{j}|/|'\Sigma|\}\{|\hat{\Sigma}|/|\hat{\imath}|\}^{1/2}, \tag{7.25}$$

where $'j=j('\phi),\hat{j}=j(\hat{\theta}),'\Sigma=\Sigma('\phi),\hat{\Sigma}=\Sigma(\phi(\hat{\theta}))$, and $\hat{\imath}=i(\hat{\theta})$. Rewriting (7.25), using $'j='\Sigma$, and assuming now that we are in the submodel, where $\phi=\phi(\theta)$, we obtain

$$p^*(\hat{\theta},a;\theta)=q^*(\hat{\theta},a;\theta)p^*(\hat{\theta};\theta|a), \tag{7.26}$$

where $p^*(\hat{\theta};\theta|a)$ is the p^*-formula for the model $\mathscr{M}$ and where $q^*(\hat{\theta},a;\theta)$ is defined by

$$q^*(\hat{\theta},a;\theta)=\{c(\phi)/c(\theta,a)\}\{|\hat{j}|/|\hat{\imath}|\}^{1/2}\{|\hat{\Sigma}|/|'\Sigma|\}^{1/2}e^{\hat{\ell}-'\ell}. \tag{7.27}$$

In (7.27) $c(\theta,a)$ equals the norming constant of $p^*(\hat{\theta};\theta|a)$, and $\hat{\ell}=\ell(\theta;\hat{\theta},a)=\ell(\phi(\hat{\theta}))$, while $'\ell=\ell('\phi)$.

Using the likelihood expansions established in sections 5.3 and 5.4, it may be shown that each of the quantities $\{|\hat{j}|/|\hat{\imath}|\}^{1/2}$, $\{|\hat{\Sigma}|/|'\Sigma|\}^{1/2}$, and $e^{\hat{\ell}-'\ell}$, occurring in (7.27), possesses an expansion of the form $1+c_1(\theta,a)n^{-1/2}+O_p(n^{-1})$. Thus, to order $O(n^{-1})$ we have that $q^*(\hat{\theta},a;\theta)$ depends on a and θ only. This implies, in view of (7.26) and since by definition $p^*(\hat{\theta};\theta|a)$ integrates to 1 for fixed value of a, that to the same order we have

$$p(a;\theta)\stackrel{\dagger}{=}q^*(\hat{\theta},a;\theta), \tag{7.28}$$

$$p^*(\hat{\theta};\theta|a)\stackrel{\dagger}{=}p(\hat{\theta};\theta|a). \tag{7.29}$$

Furthermore, because we know from the above that the p^*-formula for $'\phi$ is valid to order $O(n^{-3/2})$, we have from formula (6.9) that $c(\phi)\stackrel{\dagger}{=}(2\pi)^{-m/2}$ and hence, defining $q^\dagger(\hat{\theta},a;\theta)$ by

$$q^\dagger(\hat{\theta},a;\theta)=\{c(\theta,a)/c(\phi)\}(2\pi)^{-(m-d)/2}q^*(\hat{\theta},a;\theta), \tag{7.30}$$

we find that (7.26) can be re-expressed as

$$p^*(\hat{\theta}, a; \theta) = q^\dagger(\hat{\theta}, a; \theta) p^\dagger(\hat{\theta}; \theta | a) \tag{7.31}$$

where, as usual,

$$p^\dagger(\hat{\theta}; \theta | a) = \bar{c}(\theta, a)^{-1} p^*(\hat{\theta}; \theta | a) = (2\pi)^{-d/2} |\hat{j}|^{1/2} e^{\ell - \hat{\ell}}.$$

Arguing as above we see that, in parallel with (7.29), we have

$$p^\dagger(\hat{\theta}; \theta | a) \overset{\dagger}{=} p(\hat{\theta}; \theta | a), \tag{7.32}$$

i.e. we have shown the postulated validity of $p^\dagger(\hat{\theta}; \theta | a)$ to order $O(n^{-1})$.

Equation (7.32) is, in fact, valid not only when a is the score statistic but also essentially for any other approximate ancillary which agrees with a to first order. In particular, this is the case for the (multivariate) directed likelihood r.

In contrast, we do not in general have that $p^*(\hat{\theta}; \theta | a) \overset{*}{=} p(\hat{\theta}; \theta | a)$. To obtain *-accuracy it is in general necessary to use an ancillary which agrees with the (multivariate) modified directed likelihood r^* to second order, i.e. with †-accuracy, and, in particular, we have that

$$p^*(\hat{\theta}; \theta | r^*) \overset{*}{=} p(\hat{\theta}; \theta | r^*). \tag{7.33}$$

For brevity we omit the verification of this latter statement. The necessary reasoning may be found in Barndorff-Nielsen (1984; 1986a).

7.5 The effect of sampling rules

In a variety of important problems, where data are subject to uninformative censoring or to sequential stopping, the likelihood function, and hence the values of functions directly calculated from it, do not depend on the precise censoring mechanism and, in particular, do not involve the potential censoring times of the individuals who are, in the data under analysis, uncensored.

In many of these situations the asymptotic chi-squared distribution of the likelihood ratio statistic w will hold, largely independent of the censoring or stopping rule.

In particular, the question then arises as to whether a Bartlett adjustment is possible and, if it is, as to its dependence on the sampling rule.

Parallel to this one may enquire to what extent p^* continues to constitute an approximation to the conditional distribution of the maximum likelihood estimator, and similarly for some of the quantities derived on the basis of p^*, for instance the modified directed likelihood r^*. Here, however, we mainly restrict the discussion to the Bartlett adjustment.

We address these questions by examining a number of special cases connected with Brownian motion and with Poisson processes and survival analysis: these lend themselves to fairly simple direct calculations. The examples are meant both to be of intrinsic interest and to point towards general conclusions. A key role is played by the observed information.

7.5.1 Point hypotheses concerning the drift of Brownian motion

We will discuss a number of problems concerning a Brownian motion $Y(t)$ with unknown drift μ and known variance parameter, taken, without loss of generality, to be one, i.e.

$$E\{Y(t)\} = \mu t, \text{var}\{Y(t)\} = t. \tag{7.34}$$

The null hypothesis to be tested is $\mu = \mu_0$. Suppose that we observe a full sample path ending, after uninformative stopping, at the point (t, y). The likelihood is obtained by dividing $(0, t)$ into a large number of small subintervals, each making an independent contribution, and there results, after rejection of contributions not depending on μ, the log-likelihood function

$$\ell(\mu) = y\mu - \tfrac{1}{2}t\mu^2, \tag{7.35}$$

in general a $(2, 1)$ curved exponential family. It follows that $\hat{\mu} = y/t$ and that the likelihood ratio test of $\mu = \mu_0$ is based on

$$w = (y - \mu_0 t)^2/t. \tag{7.36}$$

We consider the properties of the test statistic (7.36) for two different stopping rules.

Example 7.2 Fixed time

The simplest possibility is to observe $Y(\cdot)$ until a time t which is either fixed in advance or determined in a way that allows the time to be treated as an ancillary statistic held fixed for inference about μ. Then the terminal position is a random variable Y, normal with mean μt and variance t, and (7.36) has exactly a chi-squared distribution with 1 degree of freedom. □

Example 7.3 Inverse sampling

Suppose next that $\mu_0 > 0$ and that observation continues until first passage to the position $y_0 > 0$. Then the random time T has the inverse Gaussian distribution with density under the null hypothesis

$$\phi^-(t) = \frac{y_0}{\sqrt{(2\pi)}t^{3/2}} \exp - \left\{ \frac{(y_0 - \mu_0 t)^2}{2t} \right\}. \tag{7.37}$$

The distribution of the random variable w of (7.36),

$$w = (y_0 - \mu_0 T)^2/T,$$

is known again to be exactly chi-squared with 1 degree of freedom. An elementary proof is by transformation of variables. Denote by $t_1(w) > t_2(w)$ the roots of the quadratic equation in t,

$$wt = (y_0 - \mu_0 t)^2. \tag{7.38}$$

It is then easily shown from (7.37) that

$$[\phi^-(t)/(dw/dt)]_{t=t_i(w)} = \frac{y_0 q_1(w)}{\mu_0 t_i(w) + y_0}, \tag{7.39}$$

where $q_1(w) = e^{-\frac{1}{2}w}/\sqrt{(2\pi w)}$ is the density of the chi-squared distribution with 1 degree of freedom. Each term (7.39) has to be taken with appropriate sign. Then the density of w is the sum of two contributions, one from $t_1(w)$ and one from $t_2(w)$, and on combination these yield the required chi-squared form, $q_1(w)$.

The separate contributions are, however, not of identical form. That is, $\operatorname{sgn}(y_0 - \mu_0 t)\sqrt{w}$ does not have a standard normal distribution. The statistical implication is that while the two-sided test based on the asymptotic theory (7.36) has exactly the correct properties, the corresponding statistically more interesting one-sided tests based on a normal distribution hold only as approximations, valid as $y_0 \to \infty$. □

If $Y^*(t) = Y(t) + kt$, where k is a constant, then $Y^*(\cdot)$ is a Brownian motion with drift $\mu + k$ and variance parameter 1 and the stopping boundary $y(t) = y_0$ becomes $y^*(t) = y_0 + kt$. Thus Example 7.3 applies whenever there is a single linear boundary, the mean path corresponding to the null hypothesis intersecting the boundary in the region $t > 0$.

7.5.2 *Point hypotheses concerning a single Poisson process*

We now consider a variety of problems connected with a Poisson process of rate ρ, the null hypothesis being that $\rho = \rho_0$. If in total time of observation t there are observed n points, the log-likelihood is

$$\ell(\rho) = n \log \rho - \rho t, \tag{7.40}$$

with $\hat{\rho} = n/t$ and

$$w = 2[n \log\{n/(\rho_0 t)\} - (n - \rho_0 t)]. \tag{7.41}$$

The model is a (2, 1) exponential family with statistic (n, t), unless the sampling scheme is such as to fix either n or t.

Example 7.4 Sampling for a fixed time or until a fixed number of events

If t is preassigned, then n is the observed value of a Poisson random variable of mean $\rho_0 t$, and, on series expansion,

$$E(w) = 1 + \tfrac{1}{6}(\rho_0 t)^{-1} + \cdots. \tag{7.42}$$

Correspondingly, if n is preassigned, so that $2\rho_0 t$ has a chi-squared distribution with $2n$ degrees of freedom,

$$E(w) = 1 + \tfrac{1}{6}n^{-1} + \cdots. \tag{7.43}$$

Now under the null hypothesis $n/\rho t$ tends to unity in probability under both sampling rules, so that (7.42) and (7.43) give equivalent Bartlett adjustments. We disregard complications in (7.42) arising from the discrete nature of the Poisson distribution. □

Example 7.5 Censored life-testing: simple schemes

We now consider a situation related to, but not identical with, the observation of a single Poisson process. Suppose that m individuals are observed, their failure times $T_1, \ldots, T_m$ being independently exponentially distributed with parameter ρ, mean $1/\rho$. If all individuals are fully observed, the situation is equivalent to that of Example 7.4 with $m = n$, fixed, $T_1 + \cdots + T_m$ being the total time of observation, T. Consider, however, various forms of uninformative right censoring in which observation of some individuals is stopped before failure, their time under test being observed.

(a) *Type I censoring.* At a preassigned time z_0, all individuals who have not failed are censored. Thus the number n of failures is binomially distributed with index m and parameter $1 - e^{-\rho_0 z_0}$, whereas, given m, the total time 'at risk' is $T = (m - n)z_0 + U_n$, where U_n is the sum of n independent and identically distributed random variables with the truncated exponential density $\rho_0 e^{-\rho_0 u}/(1 - e^{-\rho_0 z_0})$ $(u < z_0)$.

(b) *Type II censoring.* After a preassigned number n of failures, the $m - n$ individuals remaining are censored. The total time on test, T, is $(2\rho_0)^{-1}$ times a chi-squared random variable with $2n$ degrees of freedom, and the discussion of Example 7.4 applies. □

Example 7.6 Censored life-testing: general scheme

To represent more general schemes of censoring, such as arise in clinical trials for instance, suppose first that the ith individual has, as above, failure time S_i, and in addition a censoring time Z_i. Initially we treat the Z_i as either preassigned constants or, if random, as generated by a mechanism independent of the S_i and of the parameter ρ. We observe the random variable $X_i = \min(S_i, Z_i)$ and an indicator of failure or censoring. The failure times of censored individuals are not observed; the censoring times of individuals who fail, sometimes called potential censoring times, may or may not be known but are in any case generally though to be irrelevant.

Now an individual failing at x_i contributes $\rho e^{-\rho x_i}$ to the likelihood, whereas an individual censored at x_i contributes $e^{-\rho x_i}$. Thus the full log-likelihood is exactly (7.40) with n the number of failures and $T = \sum X_i$ the total time at risk. The maximum likelihood estimate $\hat{\rho}$ and the likelihood ratio statistic w have the form discussed earlier; no further specification of the censoring mechanism is needed. Note that this discussion covers type I censoring but not type II censoring, although the same likelihood function applies.

Further, because the expected information contributed by any individual is bounded, it is easily shown that, provided only that the $\{Z_i\}$ and ρ are such that as $n \to \infty$ the total expected information also tends to infinity, $\hat{\rho}$ is asymptotically normal and w is asymptotically chi-squared with 1 degree of freedom.

It would be possible, given all the $\{Z_i\}$, including the potential censoring times, to examine more refined approximations to the distribution of w. Such a study is to some extent eased by supposing the Z_i to correspond to random variables, possible models being that these are:

(a) independently distributed with densities $g_i(\cdot)$ and distribution functions $G_i(\cdot)$;
(b) independently and identically distributed with density $g(\cdot)$ and distribution function $G(\cdot)$;
(c) independently and identically distributed in an exponential distribution of parameter γ.

Under (a), for example, the full log-likelihood has an additional term $\log g_i(t_i)$ from a censored individual and $\log\{1 - G_i(t_i)\}$ from a failed individual. Expression (7.40) is now a partial log-likelihood; cf. section 8.6; $\hat{\rho}$ and w are unaltered.

We work first with the exponential censoring model (c). According to this, in (7.41) the random variable n is binomially distributed with index m and probability of 'success' $\rho_0(\rho_0 + \gamma)^{-1}$, and $2(\rho_0 + \gamma)t$ has a chi-squared distribution with $2m$ degrees of freedom independently of n; of course we are now dealing with a composite null hypothesis, γ being a nuisance parameter.

It follows on series expansion that

$$E(w) = 1 + \frac{\rho_0 + \gamma}{6\rho_0 m} + \cdots. \qquad (7.44)$$

Now n/m converges in probability to $\rho_0(\rho_0 + \gamma)^{-1}$, so that we may write the right-hand side of (7.44) as

$$1 + \tfrac{1}{6}n^{-1} + \cdots, \qquad (7.45)$$

showing that the same Bartlett adjustment is involved as in the earlier examples.

Unfortunately this simple conclusion does not hold in general. An elementary but tedious calculation for the fixed censoring model gives, to order $\{E(n)\}^{-1}$,

$$E(w) = 1 + \frac{1}{E(n)}\left\{\frac{1}{6} + \frac{(\pi_2 - 2\pi_1)(1 - \pi_0) + 2\pi_1^2}{(1 - \pi_0)^2}\right\}, \qquad (7.46)$$

where

$$\pi_k = \frac{1}{m}\sum(\rho z_i)^k e^{-\rho z_i} \qquad (k = 0, 1, 2). \qquad (7.47)$$

Note that $E(n) = m(1 - \pi_0)$.

In particular, in type I censoring with $z_i = z_0$ $(i = 1, \ldots, m)$,

$$E(w) = 1 + \frac{1}{E(n)}\{\tfrac{1}{6} + a(z_0\rho)\}, \tag{7.48}$$

where

$$a(x) = xe^{-x}\{(x-2) + (x+2)e^{-x}\}(1 - e^{-x})^{-2}. \tag{7.49}$$

The function $a(x)$ is small for large x and has a maximum of about 0.212 near $x = 2.6$, corresponding to about 10% censoring.

If we know the potential censoring times of all individuals, that is including those individuals who fail rather than being censored, then the right-hand side of (7.46) can be calculated and a Bartlett adjustment found. If only the observed censoring times are available, π_k can be consistently estimated by

$$\tilde{\pi}_k = \frac{1}{m}\sum{}^{(z)}\rho^k z_i^k, \tag{7.50}$$

where $\Sigma^{(z)}$ denotes a sum with respect to i over the censored individuals. Note that a procedure depending on $\tilde{\pi}_1, \tilde{\pi}_2$ is not in conflict with sufficiency if the unobserved z_i are regarded as parameters.

Censoring does raise some conceptually puzzling features. Thus it is not clear on general grounds that the potential censoring times of individuals who in fact fail are relevant to inference about ρ. □

7.5.3 Composite hypotheses

We now look at two-sample versions of the situations sketched above.

Example 7.7 Comparison of Brownian motions
Suppose we have two independent Brownian motions $Y_1(t)$ and $Y_2(t)$ of unit variance and with drifts μ_1 and μ_2, it being required to test the hypothesis $\mu_1 = \mu_2$. Under this hypothesis we write μ for the common value of μ_1 and μ_2. The log-likelihood corresponding to stopping at $(t_i, y_i)(i = 1, 2)$ is the sum of two terms of the form (7.35) and the likelihood ratio statistic is

$$w = (y_1/t_1 - y_2/t_2)^2(1/t_1 + 1/t_2)^{-1}.$$

It is now feasible to analyse a number of stopping rules including: (a) fixed–fixed, i.e. t_1, t_2 fixed, when w is exactly chi-squared with 1 degree of freedom; (b) inverse–inverse, i.e. y_1, y_2 fixed, when the same distributional result holds: (c) inverse–fixed, i.e. y_1 and t_2 fixed; (d) inverse–derived, in which y_1 is fixed; (e) first hit of circle. We consider cases (c)–(e) in more detail.

(c) *Inverse–fixed sampling.* Suppose $Y_1(\cdot)$ is observed till first hit of level $c > 0$ at the random time T_1 while, on the other hand, t_2 is fixed. Under the null hypothesis and assuming that $\mu > 0$, we find

$$E(w) = 2 + \mu c^2 + \mu^2 t_2 - \{t_2 + (c + \mu t_2)^2\} E\{(T_1 + t_2)^{-1}\}.$$

For the limiting operation where c and t_2 tend simultaneously to infinity in such a way that c/t_2 has a limiting value q in $(0, \infty)$, we obtain, on writing

$$(T_1 + t_2)^{-1} = \mu(c + \mu t_2)^{-1}\left[1 + \frac{c}{c + \mu t_2}\left\{\frac{T_1}{E(T_1)} - 1\right\}\right]^{-1}$$

and expanding, that

$$E(w) = 1 + 2q(q + \mu)^{-3} t_2^{-1} + O(t_2^{-\frac{3}{2}}). \qquad (7.51)$$

In particular, then, the mean value is not 1, in contrast to the cases (a) and (b).

(d) *Inverse–derived sampling.* Let the first process $Y_1(\cdot)$ be observed until it reaches level $c > 0$ and let the observation period for the second process be equal to that of the first, that is $T_2 = T_1 = T$, say, follows the inverse Gaussian distribution $\mathrm{IG}(c^2, \mu^2)$ and the conditional distribution of $Y_2(T)$ given $T = t$ is $\mathrm{N}(\mu t, t)$. Paraphrased, we are observing the two-dimensional Brownian motion until it first hits a vertical barrier at $Y_1 = c$. A calculation shows that in the present situation W is exactly distributed as χ_1^2.

(e) *Sampling till first hit of circle.* In this case $(Y_1(t), Y_2(t))$ is observed until the time T when the process first hits the circle with centre $(0, 0)$ and radius r_0. Then we have

$$w = \tfrac{1}{2} r_0^2 t^{-1}\{1 - \sin(2v)\},$$

where v is determined by $(y_1(t), y_2(t)) = r_0(\cos v, \sin v)$.

The stopping time T and the angle V are independent, with V following a von Mises distribution,

$$V \sim \{2\pi I_0(r_0\rho)\}^{-1} \exp\{r_0 \rho \cos(v - \psi)\}, \qquad (7.52)$$

while

$$T \sim I_0(r_0\rho)e^{\frac{1}{2}\rho^2 t}p_0(t); \tag{7.53}$$

here ρ and ψ are determined by $(\mu_1, \mu_2) = \rho(\cos\psi, \sin\psi)$ and $p_0(t)$ is the probability density function having Laplace transform

$$\int_0^\infty e^{-ut}p_0(t)dt = I_0(r_0\sqrt{(2u)})^{-1}.$$

To see this one simply has to note that if $P_{(\mu_1,\mu_2)}$ denotes the probability measure for the process $(Y_1(\cdot), Y_2(\cdot))$ and if we write P_0 for $P_{(0,0)}$ then by the Cameron–Martin–Girsanov theorem, as formulated, for instance, by Sørensen (1983), we have

$$\begin{aligned} P_{(\mu_1,\mu_2)}\{T\in dt, V\in dv\} &= \exp\{\ell(\mu_1,\mu_2) - \ell(0,0)\}P_0\{T\in dt, V\in dv\} \\ &= e^{-\frac{1}{2}\rho^2 t}P_0\{t\in dt\}\exp\{r_0\rho\cos(r-\psi)\}(2\pi)^{-1}dv, \end{aligned} \tag{7.54}$$

it being obvious that T and V are independent under P_0, with V following a uniform distribution. It follows that, under the null hypothesis $\mu_1 = \mu_2$,

$$\begin{aligned} E(w) &= \tfrac{1}{2}r_0^2\{2\pi I_0(r_0\rho)\}^{-1}\int_{-\pi}^{\pi}\exp(r_0\rho\cos v)\sin^2 v\,dv\,E(T^{-1}) \\ &= r_0^2\{2r_0\rho I_0(r_0\rho)\}^{-1}I_1(r_0\rho)E(T^{-1}). \end{aligned}$$

It does not seem possible to find an explicit expression for $E(T^{-1})$, but for $\rho > 0$ an asymptotic expansion of this quantity may be obtained by letting $\tau = E(T)$ and $\Delta = T - \tau$ and expanding $T^{-1} = \tau^{-1}(1 + \Delta\tau^{-1})^{-1}$. This yields an expansion for $E(T^{-1})$ in powers of τ^{-1}, the coefficients of which are determined by the cumulants of T. The cumulant transform $\kappa(\lambda)$ for T is given by

$$\kappa(\lambda) = -\log I_0(r_0\sqrt{(-2\lambda)});$$

see (7.53). Thus we find, for $r_0 \to \infty$,

$$E(w) = 1 + \rho^{-1}r_0^{-1} + O(r_0^{-2}), \tag{7.55}$$

for which we have used also the asymptotic expansion for the Bessel functions $I_\nu(x)$, as $x \to \infty$:

$$I_\nu(x) = (2\pi)^{-\frac{1}{2}}x^{-\frac{1}{2}}e^x\left\{1 - \frac{4\nu^2 - 1}{8x} + \cdots\right\}.$$

One notes that, in particular, there is a non-trivial Bartlett factor under the present sampling rule. □

Next we consider two independent Poisson processes having intensities ρ_1 and ρ_2, respectively. The first process is observed over the interval $[0, t_1]$, and the second over the interval $[0, t_2]$, the corresponding counts being n_1 and n_2. Assuming that the joint stopping rule applied is non-informative, we have that the log-likelihood function for (ρ_1, ρ_2) is

$$\ell(\rho_1, \rho_2) = -\rho_1 t_1 + n_1 \log \rho_1 - \rho_2 t_2 + n_2 \log \rho_2.$$

The hypthesis to be tested is $\rho_1 = \rho_2$, and for this the log-likelihood ratio statistic w is given by

$$w = 2\{n_1 \log(n_1/t_1) + n_2 \log(n_2/t_2) - n_+ \log(n_+/t_+)\},$$

where $n_+ = n_1 + n_2$, $t_+ = t_1 + t_2$. The common value of ρ_1 and ρ_2 under the null hypothesis will be denoted by ρ.

Example 7.8 Two Poisson processes

As in Example 7.7 we consider a number of different sampling rules.

(a) *Fixed–fixed sampling.* When t_1 and t_2 are fixed in advance we have

$$E(w) = 1 + \tfrac{1}{6}(t_1^{-1} + t_2^{-1} - t_+^{-1})\rho^{-1} + O(t_{\min}^{-3/2}),$$

where $t_{\min} = \min(t_1, t_2)$. This expression may be rewritten as

$$E(w) = 1 + \tfrac{1}{6}(n_1^{-1} + n_2^{-1} - n_+^{-1}) + O(n_{\min}^{-3/2}),$$

with $n_{\min} = \min(n_1, n_2)$.

(b) *Inverse–inverse sampling.* When, instead, the counts n_1 and n_2 are fixed, we find

$$E(w) = 1 + \tfrac{1}{6}(n_1^{-1} + n_2^{-1} - n_+^{-1}) + O(n_{\min}^{-3/2}),$$

the same structure as under (a).

(c) *Inverse–fixed sampling.* Suppose n_1 and t_2 are taken as fixed. Then

$$E(w) = 1 + \tfrac{1}{6}\{n_1^{-1} + (\rho t_2)^{-1} - (n_1 + \rho t_2)^{-1}\} + \frac{2n_1 \rho t_2}{(n_1 + \rho t_2)^3} + O(\{\min(n_1, t_2)\}^{-3/2}).$$

In comparison with (a) and (b) there is here an additional term $2n_1\rho t_2(n_1+\rho t_2)^{-3}$.

(d) *Inverse–derived sampling.* Let the first process be observed till n_1 points have occurred and let the observation period for the second process be equal to that of the first, i.e. we take T_2 to be equal to the first passage time T_1. In this case

$$E(w)=1+\tfrac{1}{4}n_1^{-1}+O(n_1^{-3/2})$$

or, equivalently,

$$E(w)=1+\tfrac{1}{6}(n_1^{-1}+n_2^{-1}-n_+^{-1})+O(n_{\min}^{-3/2}),$$

as in (a) and (b), since $n_2/n_1, n_+/n_1$ tend respectively to 1 and $\frac{1}{2}$.

(e) *Fixed–derived sampling.* Observing the first process over the fixed interval $[0,t_1]$ and the second process till the count equals that of the first, that is $n_2=n_1$, yields

$$E(w)=1+\tfrac{1}{4}(\rho t_1)^{-1}+O(t_1^{-3/2}),$$

as in (a), (b) and (d). □

The concurrence of (a), (b), (d) and (e) in the above example is similar to that of (a), (b) and (d) in Example 7.7.

Finally, for $i=1,2$, let $S_{i1},\dots,S_{in_i}$ be independent random variables following the exponential distribution with mean ρ_i^{-1}. We think of these as failure times and suppose that the observation of the failure times is subject to uninformative censoring. Letting r_i denote the number of observed failures and with t_i as the total time on test for failure group i we have that the log-likelihood ratio statistic for testing $\rho_1=\rho_2$ is

$$w=2\{r_1\log(r_1/t_1)+r_2\log(r_2/t_2)-r_+\log(r_+/t_+)\},$$

where $r_+=r_1+r_2$, $t_+=t_1+t_2$. In the special case of no censoring the situation is, in essence, the same as in Example 7.8 (b).

Example 7.9 Comparison of censored failure times

Suppose the two failure groups are subjected to independent exponential censoring, so that the observation resulting from failure of item (i,j) is $Y_{ij}=\min(S_{ij},Z_{ij})$, together with the information of whether Y_{ij} is a failure time or a censoring time; here the Zs are assumed to be mutually independent and independent of the Ss, and with Z_{ij} following an exponential distribution of mean γ_i^{-1}. Then

R_i and T_i are independent, R_i has a binomial distribution with index n_i and probability parameter $\rho_i/(\rho_i+\gamma_i)$, and the distribution of $2(\rho_i+\gamma_i)T_i$ is χ^2 with $2n_i$ degrees of freedom. Consequently, under the null hypothesis $\rho_1=\rho_2=\rho$,

$$E(w)=1+\tfrac{1}{6}(n_1^{-1}+n_2^{-1}-n_+^{-1})+\frac{2n_1n_2}{n+\phi^3}(\psi_1-\psi_2)^2+O(n_{\min}^{-3/2}),$$

where

$$\psi_i=\rho/(\rho+\gamma_i),$$
$$\phi=(n_1\psi_1+n_2\psi_2)/(n_1+n_2).$$

Thus the Bartlett adjustment is the same as for the case of no censoring if and only if the two degrees of censoring γ_1 and γ_2 are equal. □

In view of the examples discussed here and results indicated in *Further results and exercises* 7.7–7.9, the qualitative position appears to be as follows. If the boundary is too irregular or if the stopping rule is too data-dependent, neither will maximum likelihood estimates be asymptotically normal nor will w be close to χ^2. In suitably smooth situations, the 'usual' limiting forms will hold and some further smoothness, e.g. approximate linearity of a stopping boundary, will ensure the existence of a Bartlett adjustment factor for w as well as the corresponding accuracy of the p^* formula. The Bartlett adjustment factor will in general depend on the sampling rule although interesting families of sampling rules related in a simple way may lead to the same value of the adjustment.

In general, the directed likelihood statistic is more sensitive to the stopping rule than is w.

7.6 Independence of second order

In this section, to avoid some notational difficulties, we diverge from the standard practice of indicating random variables by capital letters and their values by small letters and we use lower case in both instance. Let x be a random variable of dimension m, values of the generic coordinates of which we denote by x^i, x^j, etc. $(i,j=1,\ldots,m)$. As in Barndorff-Nielsen and Cox (1989), we denote the cumulants of x by $\kappa^i,\kappa^{i,j},\kappa^{ij},\kappa^{i,jk}$, etc., where $\kappa^i=E(x^i)$, $\kappa^{i,j}=\mathrm{cov}(x^i,x^j)$, $\kappa^{ij}=$

$E(x^i x^j)$, $\kappa^{i,jk} = \text{cov}(x^i, x^j x^k)$. We assume that $\kappa^i = 0$ and we write κ for the covariance matrix $[\kappa^{i,j}]$.

Now, suppose x is partitioned into two components, of dimensions q and $m - q$, respectively. We shall indicate generic coordinates of these two components by $x^a, x^b, \ldots$ and by $x^r, x^s, \ldots$, where $a, b, \ldots$ run from 1 to q and $r, s, \ldots$ run from $q + 1$ to m. If the joint cumulant matrix $[\kappa^{a,r}]$ equals 0 then x^a and x^r are asymptotically independent to order $O(n^{-\frac{1}{2}})$ assuming a standard repeated sampling scheme. When the third-order joint cumulant (array) $[\kappa^{a,b,r}]$ is also 0 this independence statement may be sharpened to asymptotic independence to order $O(n^{-1})$ of x^r and the residual of x^a after quadratic regression on x^r. And similarly if $[\kappa^{a,r,s}]$ is 0.

To show this, we first recall that the Edgeworth expansion for x, including the first-order correction term only, is

$$\phi_m(x; \kappa)\{1 + \tfrac{1}{6}\kappa^{i,j,k} h_{ijk}(x; \kappa)\}, \tag{7.56}$$

where $h_{ijk}(x; \kappa)$ is the third-order tensorial Hermite polynomial in x derived from the $N(0, \kappa)$ distribution.

We now assume that the two components are orthogonal, i.e.

$$\kappa^{a,r} = 0. \tag{7.57}$$

Then, writing $\tilde{y}^a$ for the residual of x^a after quadratic regression on the vector variate x^r and $\tilde{z}^r$ for the residual of x^r after quadratic regression on the vector variate x^a, we have that with error $O(n^{-1})$, under repeated sampling,

$$\tilde{y}^a \doteq y^a, \tilde{z}^r \doteq z^r \tag{7.58}$$

where

$$y^a = x^a - \tfrac{1}{2}\kappa^a_{r,s}(x^r x^s - \kappa^{r,s}), \tag{7.59}$$

$$z^r = x^r - \tfrac{1}{2}\kappa^r_{a,b}(x^a x^b - \kappa^{a,b}). \tag{7.60}$$

Here $\kappa^a_{r,s} = \kappa^{a,t,u}\kappa_{rt}\kappa_{su}$ and $\kappa^r_{a,b} = \kappa^{r,c,d}\kappa_{ac}\kappa_{bd}$, $[\kappa_{rs}]$ denoting the inverse matrix of $[\kappa^{r,s}]$. The linear part in x^r is missing from (7.59) due to (7.57), and similarly for (7.60). The asymptotic equivalences (7.58) rely furthermore on the fact that, to order $O(n^{-1/2})$, we have that the variance matrix for the quadratic components satisfies

$$\kappa^{rs,tu} \doteq \kappa^{r,t}\kappa^{s,u} + \kappa^{r,u}\kappa^{s,t} \tag{7.61}$$

and similarly for $\kappa^{ab,cd}$, these latter relations being exactly valid under the limiting normal distribution. Basically, the quadratic component of the regression of x^a on x^r is of the form

$$\sum_{\substack{r \leqslant s \\ t \leqslant u}} \kappa^{a,t,u} \lambda_{rs,tu}(x^r x^s - \kappa^{r,s})$$

where we have dropped the summation convention to take account of the fact that $x^r x^s = x^s x^r$ and where $\lambda_{rs,tu}$ is the inverse of the covariance matrix $\kappa^{rs,tu}$ of the set of variables $x^r x^s, r \leqslant s$. Instead of $\lambda_{rs,tu}$ we may use the approximation obtained via (7.61), but it then remains to prove that the result can be re-expressed as in (7.59) where we are summing freely over the indices. The simplest way to do this is first to verify the result on the assumption that the coordinates x^r are uncorrelated and then to note that both the exact residual $\tilde{y}^a$ and its approximation y^a are invariant under linear transformations of the vector x^r.

To the asymptotic order concerned, the Edgeworth approximation (7.56) may now, due to the factorization property of Hermite polynomials (cf. *Further notes and exercises* 7.10) be factorized as

$$g(y)h(z),$$

where

$$g(y) = \phi(y; [\kappa^{a,b}])\{1 + \tfrac{1}{6}\kappa^{a,b,c} h_{abc}(y; [\kappa^{a,b}])\} \tag{7.62}$$

with a similar expression for $h(z)$. The Edgeworth expansion for the joint distribution of y and z is therefore of the form

$$g(y)h(z)J(y,z), \tag{7.63}$$

where $J(y,z)$ is the inverse of the Jacobian determinant of the transformation from x to (y,z).

It follows that if either $\kappa^{a,r,s} = 0$ for all r and s or $\kappa^{a,b,r} = 0$ for all a and b, in which cases $J(y,z) = 1$, then y^a and z^r are independent to one order higher than are x^a and x^r. Furthermore, if $\kappa^{a,r,s} = 0$ we have $y^a \doteq x^a$ to order $O(n^{-1})$ and hence x^a is independent of z^r to that same order. Note that $\kappa^{a,r,s}$ equals 0 in particular if the conditional mean of the vector variate x^a given the vector variate x^r is 0. A similar result obtains if $\kappa^{a,b,r} = 0$.

This result is, in particular, of some interest in connection with likelihood analysis for a certain type of parametric models, the orthogeodesic models (Barndorff-Nielsen and Blæsild, 1993b). The parameter θ in those models is partitioned into two components, χ

and ψ, that are orthogonal with respect to expected information. Furthermore, the defining properties of the orthogeodesic models imply that the joint cumulants of log-likelihood derivatives of the form $\lambda_{a,r,s} = K\{\ell_a, \ell_r, \ell_s\}$, where a is a general suffix for coordinates of χ and r and s are general suffices for coordinates of ψ, are all 0. For such models we therefore have that ℓ_χ, i.e. the score vector for χ, is independent of the vector

$$\ell_\psi - \tfrac{1}{2}\lambda_{\psi,\chi,\chi}\lambda^{\chi\chi}\lambda^{\chi\chi}(\ell_\chi \ell_\chi - \lambda_{\chi,\chi}) \tag{7.64}$$

to order $O(n^{-1})$, under ordinary repeated sampling. Here λ indicates cumulants of log-likelihood derivatives and $\lambda^{\chi\chi}$ is the inverse of $\lambda_{\chi,\chi}$. With a straightforward interpretation, the notation used in (7.64) applies to the case of multivariate ψ and χ.

Example 7.10 A special type of orthogeodesic model
Suppose for simplicity that ψ is one-dimensional, and consider an exponential model of the form

$$\exp[\psi\{r(x) + \beta^{\mathrm{T}}(\chi)s(x)\} - \alpha(\psi) - h(x)], \tag{7.65}$$

where β and s are vector functions. Such models constitute a subclass of the orthogeodesic models and hence $\lambda_{\psi,\psi,\chi} = 0$.

There are many instances of this type of model. For example, the von Mises–Fisher models are of this kind, with ψ being the precision parameter and χ determining the mean direction. □

Example 7.11 Location–scale model
Consider a sample $Y_1, \ldots, Y_n$ from a location–scale family

$$\sigma^{-1} f\left(\frac{y-\mu}{\sigma}\right)$$

for which the known density f is symmetric about 0. Then $\lambda_{\mu,\sigma,\sigma} = 0$, and ℓ_μ is second-order independent of

$$\ell_\sigma - \tfrac{1}{2}\lambda_{\sigma,\mu,\mu}\lambda^{\mu\mu}\lambda^{\mu\mu}(\ell_\mu^2 - \lambda_{\mu,\mu}). \tag{7.66}$$

In particular, if f is the standard normal density we have that

$$\ell_\mu = n\sigma^{-2}(\hat{\mu} - \mu),$$
$$\ell_\sigma = n\sigma^{-1}\{-1 + \hat{\sigma}^2/\sigma^2 + (\hat{\mu} - \mu)^2/\sigma^2\}.$$

The two score components ℓ_μ and ℓ_σ are uncorrelated but not independent, in spite of the fact that $\hat{\mu}$ and $\hat{\sigma}$ are independent.

However, in this case the expression (7.66) equals

$$(n-1)\sigma^{-3}(s^2-\sigma^2), \tag{7.67}$$

where s is the usual estimate of σ. The quantity (7.67) is not only second-order asymptotically independent but in fact completely independent of ℓ_μ.

Note that while for arbitrary location-scale models with symmetric density f we have $\lambda_{\mu,\sigma,\sigma}=0$, it is not true in general that $\lambda_{\mu,\mu,\sigma}=0$.

If, more generally, the density is of the form $f(x-\mu;\sigma)$ with $f(\cdot;\sigma)$ symmetric about 0 for all fixed σ, but σ not necessarily a scale parameter, we again obtain that ℓ_μ is second-order independent of (7.66), because $\lambda_{\mu,\sigma,\sigma}=0$. □

Example 7.12 Inverse Gaussian distribution
For the inverse Gaussian distribution

$$\frac{\sqrt{\lambda}}{\sqrt{(2\pi)}}\exp\{\sqrt{(\lambda\phi)}\}y^{-\frac{3}{2}}\exp\{-\tfrac{1}{2}(\lambda y^{-1}+\phi y)\},$$

the mean value of which is $\mu=\sqrt{(\lambda/\phi)}$, we find that the score vector relative to parameterization by (μ,λ) may be written as

$$\ell_\mu = n\lambda^{-2}\mu^2(\hat{\mu}-\mu),$$

$$\ell_\lambda = \frac{n}{2}\lambda^{-1}\{1-\lambda/\hat{\lambda}-\lambda^{-1}\hat{\mu}^{-1}(\hat{\mu}-\mu)^2\},$$

where $\hat{\mu}=\bar{Y}$ and $\hat{\lambda}^{-1}=n^{-1}\sum(y_i^{-1}-\bar{y}^{-1})$. The two components ℓ_μ and ℓ_λ are uncorrelated but not independent, whereas $\hat{\mu}$ and $\hat{\lambda}$ are independent.

We have here another instance of an orthogeodesic model and with the interpretations $\psi=\lambda$ and $\chi=\mu$ we find that (7.64) takes the form

$$\frac{n}{2}\lambda^{-1}\{1-\lambda/\hat{\lambda}-\lambda^{-2}\hat{\mu}^{-1}(\hat{\mu}-\mu)^3\}$$

which is indeed second-order asymptotically independent of ℓ_μ. □

Further results and exercises

7.1 In relation to the constructions of ancillary statistics discussed previously in section 7.2, and earlier in section 2.5, discuss cases based on multinomial distributions with cell probabilities specified in terms of a low-dimensional parameter and those based on random sample sizes of known distribution.

[Section 7.2]

7.2 For a model with a single parameter the quantity

$$\gamma^2 = i^{-2}v_{2,2} - i^{-3}v_{2,1} - 1$$

is called the **Efron curvature** of the model. It constitutes the first-order approximation of $i^{-2}\,\mathrm{var}(\hat{j} - \hat{\imath})$ or, equivalently, of $\mathrm{var}(\hat{j}/\hat{\imath} - 1)$, the latter form showing that γ^2 is parameterization invariant. The Efron curvature is thus of geometric nature and has an interpretation as a measure of the deviation of the model from the (1, 1) exponential family form, with γ being equal to 0 for such families.

[Efron, 1975; Section 7.2]

7.3 For a translation family with θ as the translation parameter we have

$$\frac{\mathrm{var}(\hat{\theta}|a) - \hat{j}^{-1}}{\mathrm{var}(\hat{\theta}|a) - \hat{\imath}^{-1}} = O_p(n^{-\frac{1}{2}})$$

where a denotes the Efron–Hinkley ancillary. Thus $\hat{j}^{-1}$ is a better approximation to the conditional variance of $\hat{\theta}$ than $\hat{\imath}^{-1}$.

[Efron and Hinkley, 1978; Section 7.2]

7.4 Let (Y_i, Z_i) $(i = 1, \ldots, n)$, be independent and identically distributed such that $Z_i \sim \mathrm{N}(\theta_1, 1)$ and $Y_i = \theta_2 Z_i + U_i$ with $U_i \sim \mathrm{N}(0, 1)$. Show that

$$a = \sqrt{(2n)}\{\textstyle\sum(Z_i - \bar{Z})^2 - n\}$$

is exactly ancillary, i.e. a is exactly distribution constant and $(\hat{\theta}, a)$ is sufficient, where $\theta = (\theta_1, \theta_2)$. Verify also that a is of the type (7.9) and that with error $O(n^{-1})$ the conditional distribution of $\hat{\theta}$ given a is normal with mean 0 and variance $(\hat{j}^{-1} + i - \hat{\imath})^{-1}$.

[Hinkley, 1980; Skovgaard, 1985a; Section 7.2]

7.5 Consider the (2, 1) exponential model determined by observation of two independent Brownian motions with common unknown drift parameter $\mu > 0$ and common known diffusion coefficient σ^2, the first process being observed till it first hits a fixed level $x_0 > 0$ and the second being observed for a fixed time period $[0, t]$. Show that if the directed likelihood r for testing this model against the ambient (2, 2) exponential model is taken as the approximate ancillary for the p^*-formula then that formula is accurate to order $O(n^{-1})$ only, whereas $O(n^{-3/2})$ holds if instead the modified directed likelihood r^* is adopted.

[Barndorff-Nielsen, 1984; Section 7.4]

7.6 For a process of diffusion type which starts at $y_0 > 0$ and satisfies a stochastic differential equation of the form

$$dy(t) = \beta a_t^2(y(\cdot))dt + a_t(y(\cdot))db(t)$$

for some known non-anticipating functional a_t and with b denoting standard Brownian motion, show that under suitable conditions on a_t and on the stopping time t, the log-likelihood function is

$$\ell = -\tfrac{1}{2}\beta^2 \int_0^t a_s^2(y(\cdot))ds + \beta\{y(t) - y_0\},$$

so that $w = \hat{j}(\hat{\beta} - \beta)^2$, where

$$\hat{j} = \int_0^t a_s^2(y(\cdot))ds$$

is the observed information. Furthermore, under fairly weak additional conditions on a_t, the statistic w follows a χ_1^2 distribution exactly, both under sampling till the observed information $\hat{j}$ reaches a preassigned level, and under sampling till the process first hits a horizontal barrier at $y' > y_0'$ where in the latter case β is supposed to be positive. It suffices that a_t is strictly positive and satisfies a Lipschitz condition.

[Sørensen, 1983; Section 7.5]

7.7 If the stopping rules of Examples 7.2 and 7.3 are combined, observation stops as soon as either $t \geqslant t_0$ or $y \geqslant y_0$. Show that

$$w = \begin{cases} (y_0 - t\mu_0)^2/t & (t \leqslant t_0) \\ (y - t_0\mu_0)^2/t_0 & \text{otherwise,} \end{cases}$$

where y denotes the value of the process at t_0 if the process is stopped at that time. Show that the density of t is given by $\phi^-(t)$ over

$0 \leqslant t \leqslant t_0$ and that of y is

$$\frac{1}{\sqrt{(2\pi t_0)}}\left[\exp\left\{-\frac{(y-\mu_0 t_0)^2}{2t_0}\right\}-\exp\left\{2\mu_0 a_0-\frac{(y-2y_0-\mu_0 t_0)^2}{2t_0}\right\}\right]$$

for $-\infty < y < y_0$. The second term arises from the image 'sink' at $2y_0$ to correct for non-absorption by t_0.

Detailed analysis shows that the limiting distribution of w is chi-squared with 1 degree of freedom but that the next term in the asymptotic expansion is too large and complex in form to be captured by a Bartlett adjustment. Equivalently, the asymptotic distribution of the maximum likelihood estimate is normal but for $\mu = \mu_0$ not given by the p^*-formula.

[Barndorff-Nielsen and Cox, 1984b; Section 7.5]

7.8 Show that in the corresponding problem with two parallel boundaries the distribution of the likelihood ratio statistic can in principle be found from the diffusion approximation to the distribution at time t conditional on nonabsorption (Cox and Miller, 1965, p. 222). The nearer the situation to the single-boundary from the closer will be the approximation by chi-squared as in the previous exercise. Show that in the 'worst case' of two symmetrical boundaries parallel to the direction of motion the density of w is

$$\frac{\pi}{2w^2}(q - 3q^9 + 5q^{25} - \cdots),$$

where $q = \exp\{-\pi^2/(8w)\}$, which can be expressed in terms of theta functions. Show that for small w, for example for $w < 1$, the first term is an excellent approximation and that it tends to zero as $w \to 0$, in contrast with the chi-squared density $q_1(w)$, reflecting that the stopping rule inhibits the occurrence of small values of w. Show further that for large w, corresponding to small values of the stopping time t, the above density is very close to $2q_1(w)$; give also a probabilistic proof by relating the two-barrier problem to two separate single-barrier problems.

Thus the 'usual' asymptotic theory does not apply in this context, unless the probability is concentrated overwhelmingly on one boundary, but the significance level calculated from the chi-squared distribution is wrong by a factor of at most 2.

[Barndorff-Nielsen and Cox, 1984b; Section 7.5]

7.9 Consider the two-dimensional Brownian motion with independent components of unit variance per unit time and drift $\rho_0(\cos\psi, \sin\psi)$, where ρ_0 is known. Suppose that, according to a certain optional stopping rule, the process is stopped at time t and at point (x, y).

Show that the log-likelihood is

$$\rho_0(x\cos\psi + y\sin\psi) - \tfrac{1}{2}t\rho_0^2$$

and that the likelihood ratio statistic for testing $\psi = \psi_0$ is $w = 2\rho_0 r\{1 - \cos(\hat{\psi} - \psi_0)\}$, where $r^2 = x^2 + y^2$. For the three stopping rules (a) fixed time, (b) first passage to a circle centre the origin, (c) first passage to a vertical barrier, it can be shown that $\hat{\psi}$ is asymptotically normal and the likelihood ratio statistic asymptotically chi-squared with 1 degree of freedom. Further, a Bartlett adjustment is available and can be determined either by evaluation of $E(w)$ or, more directly, using (6.78) together with the respective expressions for the normalizing constant of p^*.

[Barndorff-Nielsen and Cox, 1984b; Section 7.5]

7.10 *Factorization property of Hermite polynomials*
Consider the tensorial Hermite polynomials (Barndorff-Nielsen and Cox, 1989, Section 5.7)

$$h^{i_1 \cdots i_n}(x; \kappa)$$

in a variable $x = (x_1, \ldots, x_m)$ with associated covariance matrix $\kappa = [\kappa^{i,j}]$. Let x be partitioned into two subvectors x_1 and x_2, of dimensions q and $m - q$, and denote generic coordinates of these subvectors by x^a and x^r, respectively. Show that if the subvectors are uncorrelated, i.e. $[\kappa^{a,r}] = 0$, then

$$h^{a_1 \cdots a_\nu r_1 \cdots r_{n-\nu}}(x; \kappa) = h^{a_1 \cdots a_\nu}(x_1; [\kappa_1]) h^{r_1 \cdots r_{n-\nu}}(x_2; [\kappa_2])$$

where $\kappa_1 = [\kappa^{a,b}]$ and $\kappa_2 = [\kappa^{r,s}]$ are the covariance matrices of x_1 and x_2. Similarly for the covariant versions $h_{i_1 \cdots i_n}$ of the Hermite polynomials.

[Section 7.6]

Bibliographic notes

The material on approximate ancillaries in section 7.2 is based on Efron and Hinkley (1978), Skovgaard (1985a), Barndorff-Nielsen (1983), Pedersen (1981) and Barndorff-Nielsen (1988a, Chapter 4), while the discussion in sections 7.3 and 7.4 draws on Barndorff-Nielsen (1980; 1983; 1986a; 1988a). For a different approach to local second-order ancillaries, see Cox (1980) and McCullagh (1984a). Section 7.5 reproduces parts of Barndorff-Nielsen and Cox (1984b), and the results of section 7.6 are from Barndorff-Nielsen and Blæsild (1992).

There are connections between section 7.5 and the work of Liptser and Shiryayev (1978, section 17.5), Grambsch (1983) and Lai and Siegmund (1983) on schemes for autoregressive and other processes in which sampling continues until a preassigned observed information is achieved, although we shall not develop here this connection or these particular applications. Woodroofe (1982) has studied in depth estimation and associated problems connected with sequential sampling, concentrating on the 'square-root' boundary connected with repeated significance tests. The statement containing formula (7.54) combines results discussed by Gordon and Hudson (1977), Stern (1977) and Wendel (1980).

CHAPTER 8

Various notions of pseudo-likelihood and higher-order theory

8.1 Introduction

In most of this chapter we think of the parameter θ as partitioned, $\theta = (\psi, \chi)$, into a parameter of interest ψ and a nuisance parameter χ, the dimensions of these parameters being denoted by d_ψ and d_χ. We shall, in particular, pay attention to how some of the quantities studied behave under interest-respecting reparameterizations as defined earlier, i.e. under reparameterizations of the form

$$(\psi, \chi) \to (\phi, \lambda)$$

where

$$\phi = \phi(\psi), \quad \lambda = \lambda(\psi, \chi), \tag{8.1}$$

the distinguishing feature being that ϕ depends on ψ only.

For this kind of reparameterization we have the important fact that $i_{\chi\chi}$ and $i^{\psi\psi}$, i.e. the $\chi\chi$-block of the expected information matrix for (ψ, χ) and the $\psi\psi$-block of the inverse of that matrix, both behave tensorially in the sense that

$$i_{\chi\chi}(\psi, \chi) = \lambda_{/\chi} i_{\lambda\lambda}(\phi, \lambda) \lambda_{/\chi}^{\mathrm{T}}, \tag{8.2}$$

$$i^{\psi\psi}(\psi, \chi) = \psi_{/\phi} i^{\phi\phi}(\phi, \lambda) \psi_{/\phi}^{\mathrm{T}}, \tag{8.3}$$

where we have used the shorthand notation for Jacobian matrices introduced in section 2.2. The same is not true under reparameterization generally. The analogous properties for the observed information j, as defined at the end of section 5.2, hold as well.

As discussed in section 3.4, the profile likelihood $L_{\mathrm{P}}(\psi)$ for a parameter of interest ψ can fairly often be thought of as if it were a genuine likelihood, and, in particular, it may be used to test hypotheses and obtain confidence regions. Inference on ψ based on treating $L_P(\psi)$ as if it were a proper likelihood is, however, sometimes quite misleading, especially, but not only, if the data contain insufficient information about χ. This is particularly likely to happen if the dimension of χ is large.

Various adjusted versions of profile likelihood have been introduced as a remedy for this type of problem. We shall discuss several such versions here.

The first is the **modified profile likelihood** L_{MP} which in general, for its calculation, requires explicit specification of an ancillary statistic. In a range of important cases this causes no problem; in particular, such specification is not called for if the model is of the (m, m) exponential family type, since then the maximum likelihood estimator is already minimal sufficient, nor if the model is a transformation model, for then the maximal invariant may serve as the ancillary.

For situations where the ancillary is not more or less readily at hand it is desirable to have available an alternative adjusted version $L_A^\dagger$ of the profile likelihood L_P, that does not require specification of an ancillary but which nevertheless has the following properties:

1. under ordinary repeated sampling $L_A^\dagger$ is equivalent to L_{MP} to order $O(n^{-1})$;
2. statement 1 holds conditionally on any reasonable ancillary statistic;
3. $L_A^\dagger$ is, like L_{MP}, invariant under interest-respecting reparameterizations.

In section 8.3 we shall show the existence of such a combinant $L_A^\dagger$, which is termed the **stably adjusted profile likelihood**. It requires in general the calculation of expected and observed information matrices and certain derivatives of the expected information with respect to parameters.

In general we call a statistical quantity **stable** if it respects the conditionality principle relative to any reasonable ancillary without requiring specification of the ancillary.

Prior to the discussion, in section 8.3, of the stably adjusted profile likelihood $L_A^\dagger$ we consider, at the end of section 8.2, the role of parameter orthogonality in connection with the modified profile likelihood L_{MP}. Under parameter orthogonality $L_{MP}(\psi)$ can be approximated by another adjusted profile likelihood, the quantity $L_A(\psi)$ which was discussed briefly in section 4.4. The latter adjusted version is like $L_A^\dagger(\psi)$ in that it does not require specification of an ancillary for its calculation and it also has properties 1 and 2 above. However, it is not invariant under interest-respecting reparameterizations.

Implications of the results of section 8.3 for the directed likelihood are considered in section 8.4.

The behaviour of key likelihood quantities when the number of nuisance parameters is large is of particular interest, and some discussion of this issue is given in section 8.5.

The final section 8.6 briefly reviews a number of other pseudo-likelihoods, in particular the partial likelihood.

8.2 Modified profile likelihood

To obtain the modified profile likelihood $L_{MP}(\psi)$ for a parameter of interest ψ we argue as follows. Let $M(\psi)$ be a modulating factor applied to the profile likelihood $L_P(\psi)$ and write

$$L_{MP}(\psi) = M(\psi)L_P(\psi). \tag{8.4}$$

There are several cases where there are strong arguments for a particular $M(\psi)$ and it turns out that these can be put in common form, which we therefore adopt for general use. In discussing this we suppose, as in the discussion in Chapters 5 and 6, that the data can be replaced in the likelihood by $(\hat{\psi}, \hat{\chi}, a)$, where a is ancillary, so that the full likelihood is $L(\psi, \chi; \hat{\psi}, \hat{\chi}, a)$.

The first particular case is where the joint distribution of $(\hat{\psi}, \hat{\chi})$ given a can be factorized in the form

$$p(\hat{\psi}, \hat{\chi}; \psi, \chi \mid a) = p(\hat{\psi}; \psi \mid a)p(\hat{\chi}; \psi, \chi \mid \hat{\psi}, a), \tag{8.5}$$

where it is appealing to use the marginal distribution of $\hat{\psi}$ (given a) as the base for inference about ψ. Then the requirement is that the factor $M(\psi)$ is such as to produce, at least approximately, the likelihood corresponding to $p(\hat{\psi}; \psi \mid a)$.

The second particular case is where the joint distribution of $(\hat{\psi}, \hat{\chi})$ given a takes the form

$$p(\hat{\psi}, \hat{\chi}; \psi, \chi \mid a) = p(\hat{\chi}; \psi, \chi \mid a)p(\hat{\psi}; \psi \mid \hat{\chi}, a), \tag{8.6}$$

where it is appealing to use the conditional distribution of $\hat{\psi}$ given $\hat{\chi}$ (and a) for inference about ψ, and the factor $M(\psi)$ is to be chosen accordingly. Important examples of (8.6) arise within the prime exponential family where ψ is a component of the canonical parameter and conditioning on the remaining components of the canonical statistic leads to 'exact' inference. It turns out that in both cases (8.5) and (8.6) the aim is achieved by the choice

$$M(\psi) = |\ell_{\chi;\hat{\chi}}(\psi, \hat{\chi}_\psi; \hat{\psi}, \hat{\chi}, a)|^{-1} |j_{\chi\chi}(\psi, \hat{\chi}_\psi; \hat{\psi}, \hat{\chi}, a)|^{\frac{1}{2}}, \tag{8.7}$$

where $\ell_{\chi;\hat{\chi}}$ is the $d_\chi \times d_\chi$ matrix of mixed second-order derivatives of the log-likelihood function ℓ. Note that, like L_{P}, the modified profile likelihood L_{MP} is parameterization invariant, in the sense that under interest-respecting reparameterizations L_{MP} changes at most by a factor depending on the observations only. This follows directly from (8.7).

Example 8.1 Normal distribution
Let $Y_1, \ldots, Y_n$ be a random sample from the normal distribution $\mathrm{N}(\mu, \sigma^2)$. The modified profile likelihood for σ^2 is

$$L_{\mathrm{MP}}(\sigma^2) = \sigma^{-(n-1)/2} \exp\{-\tfrac{1}{2}\sigma^{-2} \textstyle\sum (y_j - \bar{y})^2\}. \tag{8.8}$$

The estimate of σ^2 obtained by maximizing $L_{\mathrm{MP}}(\sigma^2)$ is the usual estimate $s^2 = (n-1)^{-1} \sum (y_i - \bar{y})^2$. In fact, $L_{\mathrm{MP}}(\sigma^2)$ equals the marginal likelihood function $L(\sigma^2; s^2)$ and also the conditional likelihood function $L(\sigma^2; s^2 \mid \bar{y})$. □

Example 8.2 Normal theory linear regression
Let $Y_1, \ldots, Y_n$ be independent and such that $Y_j \sim \mathrm{N}(\mu + z_j^{\mathrm{T}}\beta, \sigma^2)$ where the known covariate vectors z_j are of dimension q and satisfy $z_1 + \cdots + z_n = 0$ and $\mathrm{rank}[z_1, \ldots, z_n] = q$. The modified profile likelihood for the mean effect μ, as determined by the specifications $\chi = (\beta, \sigma^2)$ and $\psi = \mu$ may be written

$$L_{\mathrm{MP}}(\mu) = \{1 + t^2/(n-1-q)\}^{-n/2 + q/2 + 1}, \tag{8.9}$$

where

$$t = \sqrt{n}(\bar{Y} - \mu)/\sqrt{\{\textstyle\sum (Y_j - \hat{\mu} - z_j^{\mathrm{T}}\hat{\beta})^2/(n-q-1)\}}$$

is the usual Student test statistic for μ. Note that, disregarding a proportionality factor, (8.9) differs from the probability density function for t only in that the latter function has exponent $-n/2 + q/2$, rather than $-n/2 + q/2 + 1$. The expression for $L_{\mathrm{MP}}(\mu)$ should also be compared to that of the profile likelihood for μ which is

$$L_P(\mu) = \{1 + t^2/(n-1-q)\}^{-n/2},$$

so that if q/n is not small there is an appreciable difference from the usual result. □

For prime exponential models simple explicit formulae for the modified profile likelihood are available.

Consider first a prime exponential family with partitioned variables $s = (s^{(1)}, s^{(2)})$, $\phi = (\phi^{(1)}, \phi^{(2)})$, $\eta = E_\phi(S) = (\eta^{(1)}, \eta^{(2)})$, and mixed parameter $(\eta^{(1)}, \phi^{(2)})$, and let Σ and Δ be the variance matrix of the canonical statistic S and the inverse of that matrix, respectively. Then

$$L_{MP}(\phi^{(2)}) = |\Sigma_{11}(\hat{\eta}^{(1)}, \phi^{(2)})|^{\frac{1}{2}} L_P(\phi^{(2)}). \tag{8.10}$$

Under an arbitrary parameterization (ψ, χ) of the prime exponential model we have that the two matrices entering the modulation factor (8.7) of the modified profile likelihood for ψ are of the form

$$j_{\chi\chi}(\psi, \hat{\chi}_\psi) = \tilde{\phi}_{/\chi}^{\mathrm{T}} \tilde{\Sigma} \tilde{\phi}_{/\chi} - (s - \tilde{\eta})^{\mathrm{T}} \tilde{\phi}_{/\chi\chi}, \tag{8.11}$$

$$\ell_{\chi;\hat{\chi}}(\psi, \hat{\chi}_\psi) = \hat{\phi}_{/\chi}^{\mathrm{T}} \tilde{\Sigma} \tilde{\phi}_{/\chi}, \tag{8.12}$$

where, as usual, a circumflex and a tilde indicate evaluation at $(\hat{\psi}, \hat{\chi})$ and $(\psi, \hat{\chi}_\psi)$, respectively.

Example 8.3 Gamma mean

For a sample $Y_1, \ldots, Y_n$ from the gamma model and with the mean $\mu = \lambda/\alpha$ as the parameter of interest, we find, from (8.11),

$$L_{MP}(\mu) = \zeta(\hat{\lambda}_\mu)^{\frac{1}{2}} L_P(\mu), \tag{8.13}$$

where

$$\zeta(\lambda) = \Psi'(\lambda) - \lambda^{-1} \tag{8.14}$$

and $\Psi'(\lambda)$ is the digamma function of Example 6.21, i.e. $d\log\Gamma(\lambda)/d\lambda$. □

We now turn to the justification of (8.7) starting first from the factorization (8.5). When that factorization holds we may apply the p^*-formula separately to $p(\hat{\psi}, \hat{\chi}; \psi, \chi | a)$ and to $p(\hat{\chi}; \psi, \chi | \hat{\psi}, a)$, in the latter instance for fixed ψ. The details of the derivation are as follows.

We have

$$p^*(\hat{\psi}, \hat{\chi}; \psi, \chi | a) = c(\psi, \chi, a) | j(\hat{\psi}, \hat{\chi})|^{1/2} e^{\ell - \hat{\ell}}. \tag{8.15}$$

Furthermore,

$$p(\hat{\chi}; \psi, \chi | \hat{\psi}, a) = p(\hat{\chi}_\psi; \psi, \chi | \hat{\psi}, a) | \hat{\chi}_{\psi/\hat{\chi}} |, \tag{8.16}$$

$$p^*(\hat{\chi}_\psi; \psi, \chi | \hat{\psi}, a) = c(\chi | \psi, \hat{\psi}, a) | j_{\chi\chi}(\psi, \hat{\chi}_\psi) |^{1/2} e^{\ell - \ell_P}, \tag{8.17}$$

where the same version ℓ of the full log-likelihood function is employed in (8.15) and (8.17), and where $\hat{\ell} = \ell(\hat{\psi}, \hat{\chi})$ and $\ell_P = \ell(\psi, \hat{\chi}_\psi)$, the profile log-likelihood for ψ. Moreover, $\hat{\chi}_{\psi/\hat{\chi}}$ is the matrix of partial derivatives of $\hat{\chi}_\psi$ with respect to $\hat{\chi}$, when $\hat{\chi}_\psi$ is considered as a function of $(\hat{\psi}, \hat{\chi}, a)$ and ψ, and $c(\chi | \psi, \hat{\psi}, a)$ denotes the norming constant of the p^*-formula when this is applied to the model function $p(\hat{\chi}_\psi; \psi, \chi | \hat{\psi}, a)$ for fixed ψ. It is an underlying assumption that $\hat{\chi}_{\psi/\hat{\chi}}$ is non-singular.

Note that the fact that $p^*(\hat{\chi}_\psi; \psi, \chi | \hat{\psi}, a)$ can be given the form (8.17), with ℓ the same as in (8.15), hinges on the assumption that the marginal distribution for $\hat{\psi}$ does not depend on χ; this implies namely that for fixed ψ the combinant $(\hat{\chi}_\psi, \hat{\psi}, a)$ is sufficient and the statistic $(\hat{\psi}, a)$ is ancillary, so that p^* can be applied to the submodel determined by fixing ψ and using the log-likelihood function $\ell = \ell(\psi, \chi; \hat{\psi}, \hat{\chi}, a)$ from the original model, the latter conceived, however, as a function of χ and $(\hat{\psi}, \hat{\chi}_\psi, a)$.

Inserting equations (8.15) and (8.17) in (8.5) we find, recalling the use of the symbols °, † and * introduced in section 6.1,

$$\begin{aligned} p(\hat{\psi}; \psi | a) &\doteq \frac{c(\psi, \chi, a)}{c(\chi | \psi, \hat{\psi}, a)} \frac{|j(\hat{\psi}, \hat{\chi})|^{1/2}}{|j_{\chi\chi}(\psi, \hat{\chi}_\psi)|^{1/2}} |\hat{\chi}_{\psi/\hat{\chi}}|^{-1} e^{\ell_P - \hat{\ell}} \\ &= \frac{c(\psi, \chi, a)}{c(\chi | \psi, \hat{\psi}, a)} |\hat{j}^{\psi\psi}|^{-1/2} C(\psi) e^{\ell_P - \hat{\ell}} \end{aligned} \tag{8.18}$$

where

$$C(\psi) = \frac{|j_{\chi\chi}(\hat{\psi}, \hat{\chi})|^{1/2} |j_{\chi\chi}(\psi, \hat{\chi}_\psi)|^{1/2}}{|\ell_{\chi;\hat{\chi}}(\psi, \hat{\chi}_\psi; \hat{\psi}, \hat{\chi}, a)|} \tag{8.19}$$

and where we have used the fact, shown below, that

$$|\hat{\chi}_{\psi/\hat{\chi}}|^{-1} = D(\psi) \tag{8.20}$$

with

$$D(\psi) = |j_{\chi\chi}(\psi, \hat{\chi}_\psi; \hat{\psi}, \hat{\chi}, a)| / |\ell_{\chi;\hat{\chi}}(\psi, \hat{\psi}_\psi; \hat{\psi}, \hat{\chi}, a)|. \tag{8.21}$$

The final expression in (8.18) constitutes an approximation, of order $O(n^{-3/2})$, under ordinary repeated sampling, to the model function for $\hat{\psi}$ given a; the exact form of that model function may not be accessible.

If we are interested only in the marginal likelihood function for ψ based on $\hat{\psi}$ we may discard factors in (8.18) that depend on the observations only and also the two norming constants (cf. formula (6.9)), and this leads to the modified profile likelihood $L_{\text{MP}}(\psi)$ as defined by (8.7).

When (8.5) holds it is sensible to look for a version of $L_{\text{MP}}(\psi)$ which depends on the observations through $(\hat{\psi}, a)$ only, because this is the case for the likelihood function determined by $p(\hat{\psi}; \psi \mid a)$. Such a version exists, at least in all the main instances of (8.5) and, in particular, for composite transformation models with ψ as the maximal invariant parameter (Barndorff-Nielsen, 1985).

Next, starting from (8.6) one obtains the same expression $L_{\text{MP}}(\psi)$ now as an approximation to the conditional likelihood for ψ given $\hat{\chi}$ and a, this time applying p^* to $p(\hat{\psi}, \hat{\chi}; \psi, \chi \mid a)$ and $p(\hat{\chi}; \psi, \chi \mid a)$ separately, in the latter case combined with a transformation from $p(\hat{\chi}; \psi, \chi \mid a)$ to $p(\hat{\chi}_\psi; \psi, \chi \mid a)$. It is essential here that, as follows from the particular character of the factorization (8.6), $(\hat{\chi}_\psi, a)$ is minimal sufficient with a ancillary when ψ is considered as fixed and that

$$\begin{aligned} &\log p(\hat{\chi}_\psi; \psi, \chi \mid a) - \log p(\hat{\chi}_\psi; \psi, \hat{\chi}_\psi \mid a) \\ &= \log p(\hat{\chi}; \psi, \chi \mid a) - \log p(\hat{\chi}; \psi, \hat{\chi}_\psi \mid a) \\ &= \ell - \ell_{\text{P}}, \end{aligned}$$

where the first log-likelihood difference is that entering in the p^* expression for $p(\hat{\chi}_\psi; \psi, \chi \mid a)$, whereas the ℓ of the last difference is the log-likelihood function of the original model.

As a third possibility, it may happen that the joint distribution of $\hat{\psi}$ and $\hat{\chi}_\psi$ factorizes as

$$p(\hat{\psi}, \hat{\chi}_\psi; \psi, \chi \mid a) = p(\hat{\chi}_\psi; \psi, \chi \mid a) p(\hat{\psi}; \psi \mid \hat{\chi}_\psi, a), \tag{8.22}$$

whereas (8.6) does not hold. Applying p^* to $p(\hat{\psi}, \hat{\chi}_\psi; \psi, \chi \mid a)$, via the transformation $(\hat{\psi}, \hat{\chi}) \leftrightarrow (\hat{\psi}, \hat{\chi}_\psi)$, and to $p(\hat{\chi}_\psi; \psi, \chi \mid a)$, the modified profile likelihood L_{MP} again emerges, now as an approximation to $p(\hat{\psi}; \psi \mid \hat{\chi}_\psi, a)$ considered as a pseudo-likelihood for ψ.

Note, incidentally, that if (8.6) holds then so does (8.22) and $p(\hat{\psi}, \psi \mid \hat{\chi}, a) = p(\hat{\psi}; \psi \mid \hat{\chi}_\psi, a)$.

Note also that in all three of the above cases the modified profile likelihood approximates the target likelihood to order $O(n^{-3/2})$ in regions of the form $\{\psi : |\psi - \hat{\psi}| \leqslant cn^{-1/2}\}$.

We still have to derive (8.20). For this we consider the likelihood equation for $\hat{\chi}_\psi$, i.e.

$$\ell_\chi(\psi, \hat{\chi}_\psi; \hat{\psi}, \hat{\chi}, a) = 0,$$

viewed as an identity. Differentiation with respect to $\hat{\chi}$ yields

$$\ell_{\chi\chi}(\psi, \hat{\chi}_\psi; \hat{\psi}, \hat{\chi}, a)\hat{\chi}_{\psi/\hat{\chi}} + \ell_{\chi;\hat{\chi}}(\psi, \hat{\chi}_\psi; \hat{\psi}, \hat{\chi}, a) = 0, \tag{8.23}$$

from which (8.20) follows.

Modified profile likelihood may also be of interest in model checking. For instance, suppose that we are considering a location–scale model whose error law has probability density function f and that we wish to investigate various forms of deviation from f, the relevant type of deviation being embodied as a family of alternative error laws f_λ, where λ is a parameter indexing these laws. To draw inference on λ one may study the marginal likelihood for λ based on the configuration statistic u, and this is expressible as

$$L(\lambda; u) = \iint \sigma^{-n} \prod_{i=1}^{n} f_\lambda((x_i - \mu)/\sigma)\sigma^{-1} d\sigma d\mu; \tag{8.24}$$

cf. formula (2.104). Alternatively, we may adopt the modified profile likelihood $L_{\mathrm{MP}}(\lambda)$ for λ as a pseudo-likelihood. From a computational viewpoint this has the advantage over (8.24) of not involving a multiple integral.

The general expression for $L_{\mathrm{MP}}(\psi)$, given by (8.4) and (8.7), may be recast as

$$L_{\mathrm{MP}}(\psi) = D(\psi)|j_{\chi\chi}(\psi, \hat{\chi}_\psi)|^{-1/2} L_{\mathrm{P}}(\psi), \tag{8.25}$$

where $D(\psi)$ is defined by (8.21).

Note that if $\hat{\chi}_\psi$ does not depend on ψ, i.e. if $\hat{\chi}_\psi = \hat{\chi}$, then $D(\psi) = 1$ and

$$L_{\mathrm{MP}}(\psi) = |j_{\chi\chi}(\psi, \hat{\chi}_\psi)|^{-1/2} L_{\mathrm{P}}(\psi), \tag{8.26}$$

as is apparent from (8.25).

Formula (8.10) is an illustration of (8.26), because $\hat{\eta}^{(1)} = \hat{\eta}^{(1)}_{\phi^{(2)}} = s^{(1)}$. It turns out that $D(\psi) = 1$, so that (8.26) holds, as it also does in some cases where $\hat{\chi}_\psi \neq \hat{\chi}$.

Example 8.4 Location–scale model

Writing $g = -\log f$, where f denotes the known density function, we have

$$\begin{aligned}\ell(\mu,\sigma) &= -n\log\sigma - \sum g(\sigma^{-1}(y_i-\mu))\\ &= -n\log\sigma - \sum g(\hat\sigma/\sigma(a_i + (\hat\mu-\mu)/\hat\sigma))\\ &= \ell(\mu,\sigma;\hat\mu,\hat\sigma,a),\end{aligned} \tag{8.27}$$

where $a = (a_1,\dots,a_n)$ and $a_i = (y_i - \hat\mu)/\hat\sigma$.

With σ as interest parameter we find

$$\ell_{\mu\mu}(\mu,\sigma) = -\sigma^{-2}\sum g''(\sigma^{-1}(y_i-\mu)), \tag{8.28}$$

$$\ell_{\mu;\hat\mu}(\mu,\sigma) = \sigma^{-2}\sum g''(\sigma^{-1}(y_i-\mu)). \tag{8.29}$$

Hence, in this case we even have

$$\ell_{\mu;\hat\mu}(\mu,\sigma) = j_{\mu\mu}(\mu,\sigma) \tag{8.30}$$

which is, in fact, stronger than $D(\psi) = 1$. □

The right-hand side of (8.26) equals the adjusted profile likelihood $L_A(\psi)$ discussed in Chapter 4, i.e.

$$L_A(\psi) = |j_{\chi\chi}(\psi,\hat\chi)|^{-1/2} L_P(\psi). \tag{8.31}$$

When ψ and χ are orthogonal we have $\hat\chi_\psi \stackrel{\dagger}{=} \hat\chi$ and hence, by (8.20), $D(\psi) \stackrel{\dagger}{=} 1$ and

$$L_{MP}(\psi) \stackrel{\dagger}{=} L_A(\psi). \tag{8.32}$$

The adjusted profile likelihood L_A is simpler to calculate than L_{MP} because it does not involve the factor $D(\psi)$ which requires specification of the ancillary a for its determination. However, this simplification is at the cost of loss of exact parameterization invariance, i.e. L_A is not invariant under interest-respecting reparameterizations and, furthermore, to calculate L_A one generally has to determine an orthogonal parameterization, and this is often difficult, though possible in principle if ψ is of dimension 1. Further, when (8.5) or (8.6) holds, L_{MP} is of accuracy $O(n^{-3/2})$ in moderate-deviation regions, whereas the accuracy of L_A is generally $O(n^{-1})$.

Note that in the location–scale problem considered in Example 8.4 μ and σ are not expected orthogonal unless the density f is symmetric or, slightly more generally, unless $\int y g''(y) f(y) dy = 0$.

8.3 Stable adjustments of profile likelihood

As discussed in section 8.1, particularly for cases where the ancillary is not explicitly available, it is useful to have adjusted versions of the profile likelihood L_P that, in contrast to the modified profile likelihood L_{MP}, do not require specification of an ancillary but which to some extent share other key properties with L_{MP}. To be specific, let $L_0^\dagger$ denote an adjusted profile likelihood. Then it is desirable, in particular, that $L_0^\dagger$ sould have the following properties:

1. $L_0^\dagger$ equals L_{MP} to order $O(n^{-1})$ under ordinary repeated sampling.
2. $L_0^\dagger$ is invariant under interest-respecting reparameterizations.
3. When one of the three types of factorization (8.5), (8.6) or (8.22) holds, the approximation of $L_0^\dagger$ to the relevant marginal or conditional likelihood is of accuracy $O(n^{-1})$ conditionally on any reasonable ancillary, that is, $L_0^\dagger$ is stable in the sense indicated in section 8.1.

We now derive and discuss two adjusted versions of L_P which meet these three requirements, in particular the stably adjusted profile likelihood $L_A^\dagger$. For simplicity, we consider only one-dimensional interest parameters ψ.

The modulating factor $M(\psi)$, which is given by (8.7) and which turns the profile likelihood $L_P(\psi)$ into the modified profile likelihood $L_{MP}(\psi)$, may be expressed as

$$M(\psi) = G(\psi)\{|i_{\chi\chi}(\psi, \hat{\chi}_\psi)|/|j_{\chi\chi}(\psi, \hat{\chi}_\psi)|\}^{1/2}, \tag{8.33}$$

where

$$G(\psi) = |j_{\chi\chi}(\psi, \hat{\chi}_\psi)|/\{|\ell_{\chi;\hat{\chi}}(\psi, \hat{\chi}_\psi)||i_{\chi\chi}(\psi, \hat{\chi}_\psi)|^{1/2}\}. \tag{8.34}$$

The ratio between information determinants in braces in (8.33) is invariant under interest-respecting reparameterizations, and under such reparameterizations $G(\psi)$ changes at most by a factor depending on the observations alone, that is to say, $G(\psi)$ is invariant in the sense of likelihood calculations.

Letting $g(\psi) = \log G(\psi)$, we shall show later that

$$g(\psi) - g(\hat{\psi}) \mathrel{\dot{=}} g_0(\psi), \tag{8.35}$$

where

$$g_0(\psi) = \int_{\hat{\psi}}^{\psi} h(\psi, \hat{\chi}_\psi)d\psi, \tag{8.36}$$

with

$$h(\psi, \chi) = \mathrm{tr}\{i_{\chi\chi}^{-1}(\psi, \chi)H_{\chi\chi\psi}\}, \tag{8.37}$$

$$H_{\chi\chi\psi} = i_{\chi\psi/\chi} - i_{\chi\chi/\chi}i_{\chi\chi}^{-1}i_{\chi\psi} - \tfrac{1}{2}(i_{\chi\chi/\psi} - i_{\chi\chi/\chi}i_{\chi\chi}^{-1}i_{\chi\psi}). \tag{8.38}$$

In (8.38) at the first place where $i_{\chi\chi/\chi}$ occurs the matrix multiplication involves the second χ in $i_{\chi\chi/\chi}$ whereas at the second place the multiplication is relative to the third χ. It is possible to express $H_{\chi\chi\psi}$ in terms of second- and third-order moments of first- and second-order log-likelihood derivatives; see formulae (8.55) and (8.56) below.

A direct calculation shows that $H_{\chi\chi\psi}$ behaves tensorially under interest-respecting reparameterizations. This also follows because, in the language of differential geometry, $H_{\chi\chi\psi}$ equals the Riemannian embedding curvature, also known as the Euler–Schouten curvature or the second fundamental form, of the submanifold of Ω determined by fixing ψ and under the metric given by expected information i. In consequence, $h(\psi, \chi)$ is invariant under interest-respecting reparameterizations, and hence $g_0(\psi)$ is parameterization invariant.

We now define the **stably adjusted profile likelihood** $L_A^\dagger$ by

$$L_A^\dagger(\psi) = \{|i_{\chi\chi}(\psi, \hat{\chi}_\psi)|/|j_{\chi\chi}(\psi, \hat{\chi}_\psi)|\}^{1/2} \exp\{g_0(\psi)\} L_P(\psi). \quad (8.39)$$

The crucial steps in its calculation in any specific problem are the evaluation of the observed and expected information matrices and the derivatives of the expected information entering $H_{\chi\chi\psi}$ in (8.38). Evaluation of the integral in $g_0(\psi)$, cf. formula (8.36), can be avoided; see *Further results and exercises* 8.6.

Example 8.5 Normal location–scale problem
Consider, again, a sample of size n from the normal location–scale model with mean μ and variance σ^2, and let $\psi = \sigma^2$ and $\chi = \mu$. Then

$$\begin{aligned} \ell &= -\tfrac{1}{2}n\{\log \sigma^2 + \hat{\sigma}^2/\sigma^2 + (\hat{\mu} - \mu)^2/\sigma^2\} \\ &= -\tfrac{1}{2}n\{\log \psi + \hat{\psi}/\psi + (\hat{\chi} - \chi)^2/\psi\}, \end{aligned} \quad (8.40)$$

$$\begin{aligned} \ell_\chi &= n(\hat{\chi} - \chi)/\psi, \quad \ell_\psi = -\tfrac{1}{2}n\{\psi^{-1} - \hat{\psi}/\psi^2 - (\hat{\chi} - \chi)^2/\psi^2\} \\ \ell_{\chi\chi} &= -n/\psi, \quad \ell_{\chi\psi} = -n(\hat{\chi} - \chi)/\psi^2, \end{aligned} \quad (8.41)$$

and hence

$$i_{\chi\chi} = n/\psi, \quad i_{\chi\psi} = 0.$$

It follows that

$$g_0(\psi) = \tfrac{1}{2}\log(\psi/\hat{\psi}) = g(\psi), \quad (8.42)$$

i.e. in this case $L_A^\dagger(\psi) = L_{MP}(\psi) = L(\sigma^2; s^2)$ exactly. □

Example 8.6 Linear-nonlinear regression
Let Y_i, $i = 1, \ldots, n$, be independent normal with variance σ^2 and mean

$$E(Y_i) = \beta g(z_i; \gamma),$$

where β and γ are parameters and the z_i are covariates. We assume for simplicity that σ^2 is known. The log-likelihood function is then

$$\ell(\beta,\gamma) = -\tfrac{1}{2}\sigma^{-2}\sum\{y_i - \beta g(z_i;\gamma)\}^2$$

and we find

$$\begin{aligned}
&i_{\beta\beta} = \sigma^{-2}\sum g(z_i;\gamma)^2, \quad i_{\beta\gamma} = \sigma^{-2}\beta\sum g(z_i;\gamma)g_{/\gamma}(z_i;\psi),\\
&i_{\gamma\gamma} = \sigma^{-2}\beta^2\sum g^2_{/\gamma}(z_i;\gamma)\\
&j_{\beta\beta} = i_{\beta\beta}, \quad j_{\beta\gamma} = i_{\beta\gamma} - \sigma^{-2}\sum\{y_i - \beta g(x_i;\gamma)\}g_{/\gamma}(z_i;\gamma),\\
&j_{\gamma\gamma} = i_{\gamma\gamma} - \sigma^{-2}\beta\sum\{y_i - \beta g(z_i;\gamma)\}g_{/\gamma\gamma}(z_i;\gamma).
\end{aligned}$$

Suppose first that γ is the parameter of interest. We then find that the embedding curvature $H_{\beta\beta\gamma}$, calculated from (8.38), is identically zero, i.e. $\exp\{g_0(\psi)\} = 1$. The first adjustment factor in (8.39) is also 1 and hence $L^\dagger_A(\gamma)$ is the profile likelihood for γ.

If, on the other hand, the interest parameter is β the first adjustment factor is

$$1 - \beta^{-1}\{\sum g^2_{/\psi}\}^{-1}\sum(y_i - \beta g)g_{/\psi\psi} \tag{8.43}$$

while the second is determined form

$$h(\beta,\gamma) = \beta^{-1}\{\sum g^2_{/\psi}\}^{-2}\{\sum g^2_{/\psi}\sum gg_{/\psi\psi} - \sum g_{/\psi}gg_{/\psi\psi}\sum gg_{/\psi}\} \tag{8.44}$$

via (8.36)–(8.38). In (8.43), (8.44) g is short for $g(z_i;\gamma)$, etc.

□

To prove (8.36) we note that first-order Taylor expansion of $g(\psi)$ in ψ around $\hat{\psi}$ yields the approximation

$$\begin{aligned}
g(\psi) - g(\hat{\psi}) \mathrel{\mathring{=}} (\psi - \hat{\psi})[&\mathrm{tr}\{j_{\chi\chi}(\hat{\psi},\hat{\chi})^{-1}k(\hat{\psi},\hat{\chi})\}\\
&- \tfrac{1}{2}\mathrm{tr}\{\hat{\imath}^{-1}_{\chi\chi}(\hat{\imath}_{\chi\chi/\psi} + \hat{\imath}_{\chi\chi/\chi}\hat{\chi}_{\psi/\psi}(\hat{\psi}))\}],
\end{aligned} \tag{8.45}$$

where

$$k(\hat{\psi},\hat{\chi}) = -\{\hat{\ell}_{\chi\chi\psi} + \hat{\ell}_{\chi\chi\chi}\hat{\chi}_{\psi/\psi}(\hat{\psi}) + \hat{\ell}_{\chi\psi;\hat{\chi}} + \hat{\ell}_{\chi\chi;\hat{\chi}}\hat{\chi}_{\psi/\psi}(\hat{\psi})\} \tag{8.46}$$

and where we have used formula (5.109), which in the present notation may be written as

$$\{\log|A(\psi)|\}_{/\psi} = \mathrm{tr}\{A(\psi)^{-1}A(\psi)_{/\psi}\}.$$

To eliminate $\hat{\ell}_{\chi\psi;\hat{\chi}}$ and $\hat{\ell}_{\chi\chi;\hat{\chi}}$ from (8.46) we note that

$$(\hat{\ell}_{\chi\psi})_{/\hat{\chi}} = \hat{\ell}_{\chi\chi\psi} + \hat{\ell}_{\chi\psi;\hat{\chi}}, \quad (\hat{\ell}_{\chi\chi})_{/\hat{\chi}} = \hat{\ell}_{\chi\chi\chi} + \hat{\ell}_{\chi\chi;\hat{\chi}}. \tag{8.47}$$

As in section 5.2 we use ν to indicate joint moments of log-likelihood derivatives, writing $\nu_{\chi,\chi,\psi}$ for $E(\ell_\chi \ell_\chi \ell_\psi)$, $\nu_{\chi\chi,\psi}$ for $E(\ell_{\chi\chi}\ell_\psi)$, etc. Since

$$\hat{\ell}_{\chi\psi} \stackrel{\circ}{=} -i_{\chi\psi}, \quad \hat{\ell}_{\chi\chi} \stackrel{\circ}{=} -i_{\chi\chi},$$
$$\hat{\ell}_{\chi\chi\psi} \stackrel{\circ}{=} \nu_{\chi\chi\psi}, \quad \hat{\ell}_{\chi\chi\chi} \stackrel{\circ}{=} \nu_{\chi\chi\chi}, \tag{8.48}$$

it follows from (8.47) that, to first asymptotic order,

$$\hat{\ell}_{\chi\psi;\hat{\chi}} \stackrel{\circ}{=} -i_{\chi\psi/\chi} - \nu_{\chi\chi\psi}, \quad \hat{\ell}_{\chi\chi;\hat{\chi}} \stackrel{\circ}{=} -i_{\chi\chi/\chi} - \nu_{\chi\chi\chi}. \tag{8.49}$$

Further, differentiation with respect to ψ of the likelihood equation for $\hat{\chi}_\psi$ shows that

$$\hat{\chi}_{\psi/\psi} = -\ell_{\psi\chi}(\psi, \hat{\chi}_\psi)\ell_{\chi\chi}(\psi, \hat{\chi}_\psi)^{-1} \tag{8.50}$$

and hence

$$\hat{\chi}_{\psi/\psi}(\hat{\psi}) \stackrel{\circ}{=} -i_{\psi\chi} i_{\chi\chi}^{-1}. \tag{8.51}$$

Inserting (8.48)–(8.50) into (8.46) gives

$$k(\hat{\psi}, \hat{\chi}) \stackrel{\circ}{=} i_{\chi\psi/\chi} - i_{\chi\chi/\chi} i_{\chi\chi}^{-1} i_{\chi\psi}, \tag{8.52}$$

and this combined with (8.45) yields (8.35–38).

Since

$$i_{\chi\psi/\chi} = -\nu_{\chi\chi\psi} - \nu_{\chi\psi,\chi}, \quad i_{\chi\chi/\chi} = -\nu_{\chi\chi\chi} - \nu_{\chi\chi,\chi} \tag{8.53}$$

we may, using the balance relations

$$\nu_{\chi\chi\psi} + 2\nu_{\chi\psi,\chi} + \nu_{\chi\chi,\psi} + \nu_{\chi,\chi,\psi} = 0,$$
$$\nu_{\chi\chi\chi} + 3\nu_{\chi\chi,\chi} + \nu_{\chi,\chi,\chi} = 0, \tag{8.54}$$

re-express $H_{\chi\chi\psi}$ as

$$H_{\chi\chi\psi} = \tau_{\chi\chi,\psi} + \tfrac{1}{2}\tau_{\chi,\chi,\psi}, \tag{8.55}$$

where

$$\tau_{\chi\chi,\psi} = \nu_{\chi\chi,\psi} - \nu_{\chi\chi,\chi} i_{\chi\chi}^{-1} i_{\chi\psi},$$
$$\tau_{\chi,\chi,\psi} = \nu_{\chi,\chi,\psi} - \nu_{\chi,\chi,\chi} i_{\chi\chi}^{-1} i_{\chi\psi}. \tag{8.56}$$

Note that $\tau_{\chi\chi,\psi}$ is the joint cumulant (or moment) of $\ell_{\chi\chi}$ and of the residual of the score component ℓ_ψ after regression on the complementary component ℓ_χ, and similarly for $\tau_{\chi,\chi,\psi}$. Both $\tau_{\chi\chi,\psi}$ and $\tau_{\chi,\chi,\psi}$ behave as tensors under interest-respecting reparameterizations.

A calculation similar to that given above shows that

$$L_A^\dagger(\psi) \stackrel{\dagger}{=} \exp\{b_0(\psi)\} L_P(\psi), \tag{8.57}$$

where

$$b_0(\psi) = \frac{1}{2}\int_{\hat{\psi}}^{\psi} b(\psi, \hat{\chi}_\psi)d\psi \tag{8.58}$$

$$b(\psi, \chi) = \mathrm{tr}\{i_{\chi\chi}^{-1}(\psi, \chi)(\tau_{\chi\chi.\psi} + \tau_{\chi,\chi.\psi})\} \tag{8.59}$$

and where, in consequence of the tensorial behaviour of $\tau_{\chi\chi.\psi}$ and $\tau_{\chi,\chi.\psi}$, we have that $b_0(\psi)$ is invariant under interest-respecting reparameterizations. Thus $L_B^\dagger$ defined by

$$L_B^\dagger(\psi) = \exp\{b_0(\psi)\}L_P(\psi) \tag{8.60}$$

constitutes a stable and invariant adjusted profile likelihood which, like $L_A^\dagger$, is †-equivalent to the modified profile likelihood L_{MP}.

A comparison with Bartlett (1955) shows that $-\frac{1}{2}b(\psi, \chi)$ equals the integral of a bias correction to the profile score. Otherwise expressed,

$$E[\{\ell_P(\psi)\}_{/\psi}] \stackrel{\dagger}{=} -\tfrac{1}{2}b(\psi, \chi). \tag{8.61}$$

8.4 Stable adjustments of directed likelihood

The stably adjusted profile likelihood $L_A^\dagger$ discussed in section 8.3 and given by formula (8.39) may be used to define an associated stably adjusted version of the directed likelihood r for a one-dimensional parameter of interest ψ.

In the case where no nuisance parameters are present $L_A^\dagger$ does not, however, enter and the stably adjusted version of the directed likelihood is defined by

$$r_A^\dagger = r + r^{-1}\log(u_A/r), \tag{8.62}$$

where

$$u_A = (\hat{\imath}/i)^{1/2}\hat{j}^{-1/2}\ell_\psi. \tag{8.63}$$

The combinant u_A, which is a test statistic of the score type, is invariant under interest-respecting reparameterizations. One notes the similarity of (8.62) to the definition (6.102) of the modified directed likelihood statistic. If the model is a location model one has, in fact, $r_A^\dagger = r^*$. It can be proved that $r_A^\dagger$ is standard normally distributed to accuracy $O(n^{-1})$, conditionally on an ancillary statistic and hence also unconditonally.

Example 8.7 Inverse Gaussian model
Consider the inverse Gaussian distribution

$$\frac{\sqrt{\lambda}}{\sqrt{(2\pi)}} e^{\sqrt{(\lambda\psi)}} y^{-3/2} \exp\{-\tfrac{1}{2}(\lambda y^{-1} + \psi y)\}$$

with λ taken as known. In this case we have

$$\begin{aligned} r &= \lambda^{1/4}\hat{\psi}^{-1/4}(\hat{\psi}^{1/2} - \psi^{1/2}), \\ u &= \tfrac{1}{2}\lambda^{1/4}\hat{\psi}^{-3/4}(\hat{\psi} - \psi), \\ u_A &= \lambda^{1/4}\psi^{3/4}(\psi^{-1/2} - \hat{\psi}^{-1/2}). \end{aligned}$$

Table 8.1 compares the approximations $\Phi(r)$, $\Phi(r_A^\dagger)$, and $\Phi(r^*)$ to the exact tail probability of this distribution for $\lambda = 9$, $\psi = 0.1$ and 1 and selected values of y. □

The above definitions of u_A and $r_A^\dagger$ may be extended to the situation where nuisance parameters are present by generalizing (8.62) to

$$r_A^\dagger = r_A + r_A^{-1}\log(u_A/r_A), \tag{8.64}$$

Table 8.1 *Exact and approximate tail probabilities (per cent) for the inverse Gaussian distribution when $\lambda = 9$, $\psi = 0.1$ and 1*

y	Exact	$\Phi(r)$	$\Phi(r_A^\dagger)$	$\Phi(r^*)$
$\psi = 0.1$				
1	0.675	0.364	1.18	0.689
3	19.36	11.81	24.25	20.30
5	39.79	26.29	44.83	41.71
10	30.84	47.95	28.62	28.39
20	10.50	22.86	10.66	9.07
30	4.27	11.81	4.74	3.57
50	0.898	3.50	1.16	0.729
75	0.159	0.837	0.237	0.128
$\psi = 1.0$				
1	3.55	2.28	4.22	3.61
2	32.18	23.98	33.71	32.66
3	39.27	50.00	38.48	38.48
5	11.56	18.56	11.90	11.20
10	0.548	1.343	0.647	0.519
20	0.0^2174	0.0^2718	0.0^2255	0.0^2163

where r_A is the directed likelihood for ψ calculated from $\ell_A^\dagger = \log L_A^\dagger$, i.e.

$$r_A = \text{sgn}('\psi - \psi)[2\{\ell_A^\dagger('\psi) - \ell_A^\dagger(\psi)\}]^{1/2}, \tag{8.65}$$

$'\psi$ denoting the maximum point of $\ell_A^\dagger$. Furthermore, u_A is defined by

$$u_A = \{\bar{\imath}^{\psi\psi}/'i^{\psi\psi}\}^{1/2}{'j}_A^{\dagger\,-1/2}\ell_{A/\psi}^\dagger. \tag{8.66}$$

Here $\bar{\imath}^{\psi\psi}$ is the $\psi\psi$-part of the expected formation matrix evaluated at $(\psi, \hat{\chi}_\psi)$, $j_A^\dagger(\psi) = -\ell_{A/\psi\psi}^\dagger$ is the corresponding 'observed information' derived from $\ell_A^\dagger$, and a prime indicates evaluation at the values $'\psi$ and $\hat{\chi}_{'\psi}$ of ψ and χ. In this general case, too, $r_A^\dagger$ is stable, invariant, and standard normally distributed with †-accuracy, i.e. error $O(n^{-1})$.

Calculation of (8.64) requires determination of the maximum point $'\psi$ of $\ell_A^\dagger$ and calculation of the derivative $\ell_{A/\psi}^\dagger$. This can be avoided by working with an alternative stable and parameterization invariant adjusted directed likelihood defined by

$$r_0^\dagger = r + r^{-1}\log(u_0/r), \tag{8.67}$$

where

$$u_0 = \tilde{C}(\psi)\{\tilde{\imath}^{\psi\psi}/\hat{\imath}^{\psi\psi}\}^{1/2}j_P(\hat{\psi})^{-1/2}\ell_{P\psi}, \tag{8.68}$$

$$\tilde{C}(\psi) = \{|\hat{\imath}_{\chi\chi}||\tilde{j}_{\chi\chi}|/|\tilde{\imath}_{\chi\chi}||\hat{j}_{\chi\chi}|\}^{1/2}e^{-g_0(\psi)}. \tag{8.69}$$

Here, as usual, a tilde and a circumflex indicate evaluation at $(\psi, \hat{\chi}_\psi)$ and $(\hat{\psi}, \hat{\chi})$, $\ell_{P\psi} = d\ell_P(\psi)/d\psi$ is the profile score and $g_0(\psi)$ is defined by (8.36)–(8.38).

We present two simple examples. In the first example the statistics $r^*, r_0^\dagger$ and $r_A^\dagger$ all reduce to the same quantity, while the second enables one to compare the process of simply adjusting the directed profile likelihood to the alternative of adjusting first the likelihood function and then the corresponding directed likelihood.

Example 8.8 Exponential regression

Let Y_i $(i = 1, \ldots, n)$ be independent and exponentially distributed random variables where Y_i has mean $\lambda e^{-\psi z_i}$ and $\sum z_i = 0$ and the z_i are covariates. We take ψ as the parameter of interest and note that λ is expected orthogonal to ψ. The log-likelihood function for (ψ, λ) can be written

$$\ell(\psi, \lambda) = \ell(\psi, \lambda; \hat{\psi}, \hat{\lambda}, a) = -n\log\lambda - n\hat{\lambda}_\psi\lambda^{-1} \tag{8.70}$$

where a is exactly ancillary with $a_i = \log y_i - \log\hat{\lambda} + \psi z_i$, $\hat{\psi}$ satisfies the equation $\sum z_i y_i e^{\psi z_i} = 0$ and

$$\hat{\lambda}_\psi = \frac{1}{n}\sum y_i e^{\psi z_i} = \frac{1}{n}\sum \hat{\lambda}\exp\{a_i + (\psi - \hat{\psi})z_i\}. \tag{8.71}$$

Some useful relations and quantities for this example that are derivable from (8.70) and (8.71) are as follows:

$$\sum e^{a_i} = n, \quad \Sigma z_i e^{a_i} = 0, \tag{8.72}$$

$$i_{\psi\psi} = \sum z_i^2, \quad i_{\psi\lambda} = 0, \quad i_{\lambda\lambda} = ny^{-2}, \tag{8.73}$$

$$\ell_{;\hat{\psi}} = \lambda^{-1}\sum \hat{\lambda} z_i \exp\{a_i + (\psi - \hat{\psi})z_i\}, \quad \ell_{;\hat{\lambda}} = -n\hat{\lambda}_\psi(\lambda\hat{\lambda})^{-1}, \tag{8.74}$$

$$\ell_{\lambda;\hat{\psi}} = -\lambda^{-1}\ell_{;\hat{\psi}}, \quad \ell_{\lambda;\hat{\lambda}} = n\hat{\lambda}_\psi\hat{\lambda}^{-1}\lambda^{-2}, \tag{8.75}$$

$$\ell_{\psi\psi} = -\lambda^{-1}\sum \hat{\lambda} z_i^2 \exp\{a_i + (\psi - \hat{\psi})z_i\}, \quad \ell_{\lambda\lambda} = n(\lambda^{-2} - 2\lambda^{-3}\hat{\lambda}_\psi), \tag{8.76}$$

$$\ell_{\lambda\psi} = -\ell_{\lambda;\hat{\psi}}. \tag{8.77}$$

From (8.75) and (8.76), the factor $D(\psi)$ of (8.21) is

$$D(\psi) = \hat{\lambda}\hat{\lambda}_\psi^{-1} = \frac{\hat{\lambda}}{\sqrt{n}}|j_{\lambda\lambda}(\psi, \hat{\lambda}_\psi)|^{1/2} \tag{8.78}$$

where by (8.73) and (8.76)

$$j_{\lambda\lambda}(\psi, \hat{\lambda}_\psi) = i_{\lambda\lambda}(\psi, \hat{\lambda}_\psi) = n\hat{\lambda}_\psi^{-2}. \tag{8.79}$$

In view of (8.73) the quantity $h(\psi, \lambda)$ of (8.37) satisfies $h(\psi, \lambda) = 0$ and hence $g_0(\psi)$ of (8.36) is

$$g_0(\psi) = 0. \tag{8.80}$$

Using (8.25), (8.39) and (8.78)–(8.80) it follows that

$$L_{\mathrm{MP}}(\psi) = L_{\mathrm{A}}^{\dagger}(\psi) = L_{\mathrm{P}}(\psi) = (\hat{\lambda}/\hat{\lambda}_\psi)^n. \tag{8.81}$$

The profile score $\ell_{\mathrm{P}\psi}$ is

$$\ell_{\mathrm{P}\psi} = -n\hat{\lambda}_\psi^{-1}(\hat{\lambda}_\psi)_{/\psi} = -\ell_{;\hat{\psi}}(\psi, \hat{\lambda}_\psi) \tag{8.82}$$

where $\ell_{;\hat{\psi}}(\psi, \lambda)$ is given in (8.74). Furthermore,

$$\ell_{;\hat{\lambda}}(\hat{\psi}, \hat{\lambda}) - \ell_{;\hat{\lambda}}(\psi, \hat{\lambda}_\psi) = \ell_{;\hat{\psi}}(\hat{\psi}, \hat{\lambda}) = 0,$$

and using (8.76) we have that

$$|\ell_{;\hat{\theta}}(\hat{\theta}) - \ell_{;\hat{\theta}}(\hat{\theta}_\psi) \quad \ell_{\lambda;\hat{\theta}}(\hat{\theta}_\psi)| = n(\hat{\lambda}\hat{\lambda}_\psi)^{-1}\ell_{\mathrm{P}\psi} \tag{8.83}$$

with $\theta = (\psi, \lambda)$. Combining the identity

$$|j(\hat{\theta})| = |j_{\lambda\lambda}(\hat{\theta})|\,|j_{\mathrm{P}}(\hat{\psi})| \tag{8.84}$$

with (8.79) and (8.83), the quantity u of (6.108) reduces to

$$u = j_{\mathrm{P}}(\hat{\psi})^{-1/2}\ell_{\mathrm{P}\psi}(\psi) = -\left(\sum z_i^2 a_i\right)^{-1/2}\left(\frac{\hat{\lambda}}{\hat{\lambda}_\psi}\right)\sum z_i \exp\{a_i + (\psi - \hat{\psi})z_i\}, \tag{8.85}$$

where the profile information function $j_{\mathrm{P}}(\psi)$ evaluated at $\hat{\psi}$ is

$$j_{\mathrm{P}}(\hat{\psi}) = \sum z_i^2 a_i. \tag{8.86}$$

Moreover, from (8.79), (8.81) and the fact that $i^{\psi\psi}(\theta) = (\sum z_i^2)^{-1}$ does not depend on θ, the quantities u_{A} and u_0 of (8.66) and (8.68) required to calculate $r_{\mathrm{A}}^\dagger$ and $r_0^\dagger$ also reduce to

$$u_{\mathrm{A}} = u_0 = u = j_{\mathrm{P}}(\hat{\psi})^{-1/2}\ell_{\mathrm{P}\psi}(\psi), \tag{8.87}$$

and hence $r^* = r_0^\dagger = r_{\mathrm{A}}^\dagger$. □

Example 8.9 Inverse Gaussian model

Let Y_i $(i = 1, \ldots, n)$ be independent Gaussian observations $\mathrm{IG}(\phi, \gamma)$ each with probability density function

$$p(y; \phi, \gamma) = \frac{\sqrt{\phi}}{\sqrt{2\pi}} e^{\sqrt{(\phi\gamma)}} y^{-3/2} \exp\{-\tfrac{1}{2}(\phi y^{-1} + \gamma y)\}.$$

We take ϕ as the parameter of interest, with γ as the nuisance parameter. The log-likelihood function of $\theta = (\phi, \gamma)$ is

$$\ell(\theta) = \frac{n}{2}\log\phi + n(\phi\gamma)^{1/2} - \tfrac{1}{2}\phi\sum y_i^{-1} - \tfrac{1}{2}\gamma\sum y_i \tag{8.88}$$

with score components

$$\ell_\phi = \frac{n}{2}\{\phi^{-1} + (\gamma/\phi)^{1/2} - (\textstyle\sum y_i^{-1}/n)\}, \tag{8.89}$$

$$\ell_\gamma = \frac{n}{2}\{(\phi/\gamma)^{1/2} - (\textstyle\sum y_i/n)\}. \tag{8.90}$$

Using the likelihood equations the expressions (8.88), (8.89) and (8.90) can be written

$$\ell = \frac{n}{2}\log\phi + n(\phi\gamma)^{1/2} - \frac{n}{2}[\phi\{\hat{\phi}^{-1} + (\hat{\gamma}/\hat{\phi})^{1/2}\} + \gamma(\hat{\phi}/\hat{\gamma})^{1/2}],$$

$$\ell_\phi = \frac{n}{2}[\phi^{-1} + (\gamma/\phi)^{1/2} - \{\hat{\phi}^{-1} + (\hat{\gamma}/\hat{\phi})^{1/2}\}],$$

$$\ell_\gamma = \frac{n}{2}\{(\phi/\gamma)^{1/2} - (\hat{\phi}/\hat{\gamma})^{1/2}\}.$$

The maximum likelihood estimate of γ for known ϕ is

$$\hat{\gamma}_\phi = \hat{\gamma}\phi/\hat{\phi} = \phi\bar{y}^{-2},$$

while the profile log-likelihood function of ϕ is

$$\ell_P(\phi) = \frac{n}{2}(\log\phi - \phi\hat{\phi}^{-1}). \tag{8.91}$$

The modified profile likelihood function of ϕ is easily calculated using the mixed parameterization (ϕ, λ) with $\lambda = E(Y_i) = \sqrt{(\phi/\gamma)}$. Since $\hat{\lambda}_\phi = \hat{\lambda}$ we have $D(\phi) = 1$ and we find

$$L_{MP}(\phi) = L_A^\dagger(\phi) = \phi^{(n-1)/2}\exp(-\tfrac{1}{2}n\phi/\hat{\phi}). \tag{8.92}$$

To determine the various adjusted and modified directed likelihoods we also need the quantities

$$\ell_{;\hat{\gamma}} = \frac{n}{4}\{\gamma\hat{\gamma}^{-1}(\hat{\phi}/\hat{\gamma})^{1/2} - \phi(\hat{\phi}\hat{\gamma})^{-1/2}\},$$

$$\ell_{;\hat{\phi}} = \frac{n}{4}\{2\phi\hat{\phi}^{-2} + \phi\hat{\phi}^{-1}(\hat{\gamma}/\hat{\phi})^{1/2} - \gamma(\hat{\phi}\hat{\gamma})^{-1/2}\},$$

$$\ell_{\gamma;\hat{\gamma}} = \frac{n}{4}\hat{\gamma}^{-1}(\hat{\phi}/\hat{\gamma})^{1/2},$$

$$j_{\gamma\phi} = i_{\gamma\phi} = \frac{n}{4}(\phi\gamma)^{-1/2},$$

$$j_{\phi\phi} = i_{\phi\phi} = \frac{n}{4}\{2\phi^{-2} + \phi^{-1}(\gamma/\phi)^{1/2}\},$$

$$\tilde{\ell}_\phi(\phi) = -\frac{n}{2}\hat{\phi}^{-1}(1 - \phi\hat{\phi}^{-1}),$$

$$j_P(\phi) = \frac{n}{2}\phi^{-2}.$$

We then obtain

$$\ell_{;\hat{y}}(\hat{\theta}) - \ell_{;\hat{y}}(\hat{\theta}_\phi) = 0,$$

$$\ell_{;\hat{\phi}}(\hat{\theta}) - \ell_{;\hat{\phi}}(\hat{\theta}_\phi) = \frac{n}{2}\hat{\phi}^{-1}(1 - \phi\hat{\phi}^{-1})$$

and hence the statistic u of (6.108) is

$$u = \{n\phi\hat{\phi}^{-1}/2\}^{1/2}(1 - \phi\hat{\phi}^{-1}). \tag{8.93}$$

Furthermore, the combinant u_0 of (8.68)–(8.69) reduces to $u_0 = u$ and hence

$$r^* = r_0^\dagger. \tag{8.94}$$

The directed likelihood itself is

$$r = \mathrm{sgn}(\hat{\phi} - \phi)\,[n\{\log(\hat{\phi}/\phi) + \phi\hat{\phi}^{-1} - 1\}]^{1/2}.$$

To calculate $r_A^\dagger$ we note that, by (8.92), the adjusted log-likelihood, score and information functions are, respectively,

$$\ell_A^\dagger(\phi) = \frac{n-1}{2}\left\{\log\phi - \frac{n}{n-1}\phi\hat{\phi}^{-1}\right\}, \tag{8.95}$$

$$\ell_{A\phi}^\dagger(\phi) = \frac{n-1}{2}\left\{\phi^{-1} - \frac{n}{n-1}\hat{\phi}^{-1}\right\}, \tag{8.96}$$

$$j_A^\dagger(\phi) = \frac{n-1}{2}\phi^{-2}. \tag{8.97}$$

The point $'\phi$ which maximizes (8.95) can be obtained as the root of the adjusted score equation (8.96), and is

$$'\phi = \frac{n-1}{n}\hat{\phi} = (n-1)\{\textstyle\sum(y_i^{-1} - \bar{y}^{-1})\}^{-1}. \tag{8.98}$$

All in all we find

$$r_A = \mathrm{sgn}('\phi - \phi)\{(n-1)(\log('\phi/\phi) - \phi/'\phi + 1)\}^{1/2}, \tag{8.99}$$

$$u_A = \{(n-1)/2\}^{1/2}(1 - \phi'\phi^{-1}). \tag{8.100}$$

which are the quantities needed for determining $r_A^\dagger$ as defined by (8.64).

We can assess and compare the standard normal approximation

Table 8.2 *Inverse Gaussian model with ϕ as the interest parameter. Exact and approximate tail probabilities, as percent; $n = 2$*

x	Exact	$\Phi(r)$	$\Phi(r_A)$	$\Phi(r_0^\dagger)$	$\Phi(r_A^\dagger)$
$0.0^4 5$	0.505	$0.0^3 468$	0.126	0.202	0.551
$0.0^2 1$	2.52	0.0140	0.753	1.280	2.73
0.01	7.97	0.167	2.86	4.82	8.44
0.25	38.29	6.03	21.25	30.4	38.75
1.0	31.73	73.29	50.0	38.74	31.92
3.0	8.33	33.18	17.12	11.08	8.55
5.0	2.53	13.40	6.1	3.53	2.64
8.0	0.468	3.62	1.33	0.679	0.493
20.0	$0.0^3 774$	1.26	$0.0^2 316$	$0.0^2 122$	$0.0^3 841$

to the distributions of the statistics $r_A^\dagger$ and $r_0^\dagger = r^*$ using the distributional result $n\phi\hat{\phi}^{-1} \sim \chi^2_{n-1}$.

Two features of Table 8.2 are noteworthy. The tail probabilities generated by $\Phi(r_A^\dagger)$ provide very accurate approximation to the actual tail probabilities and a substantial improvement over that provided by $\Phi(r_0^\dagger)$. This is surprising in this instance since we have only established that $\Phi(r_A^\dagger)$ is correct to order $O(n^{-1})$ while $\Phi(r_0^\dagger)$, being equal in this example to $\Phi(r^*)$, is correct to order $O(n^{-3/2})$. Secondly, the tail probabilities generated by $\Phi(r_A)$, although inaccurate, are a great improvement over those provided by $\Phi(r)$. □

The explanation of the features observed in Table 8.2 is connected to a special structure, which in fact both the models of Examples 8.8 and 8.9 possess. We pursue this point below since it may be helpful in suggesting a means to choose between $r_A^\dagger$ and $r_0^\dagger$ in specific instances.

Because $\sum(y_i^{-1} - \bar{y}^{-1}) \sim \Gamma((n-1)/2,\ \phi/2)$ and is stochastically independent of $\bar{y} \sim \mathrm{IG}(n\phi, n\gamma)$, we have that $'\phi$ is stochastically independent of $\hat{\gamma}_\phi$. The exact density function of $'\phi$ given $\hat{\gamma}_\phi$ is

$$p('\phi; \phi \mid \hat{\gamma}_\phi) = p('\phi; \phi) = \Gamma\{(n-1)/2)\}^{-1}$$
$$\{(n-1)/2\}^{(n-1)/2}\phi^{-1}(\phi{}'\phi^{-1})^{(n+1)/2}\exp\{-(n-1)\phi{}'\phi^{-1}/2\}. \quad (8.101)$$

Again due to the special character of (8.101), it produces a marginal likelihood for ϕ that is the same as the adjusted profile likelihood (8.92). We can apply the single-parameter version of r^* to $p('\phi; \phi)$ of (8.101) to obtain a statistic $r_m^* = r_m + r_m^{-1}\log(u_m/r_m)$. To determine

u_{m} we must express (8.95) as

$$\ell_{\mathrm{A}}^{\dagger} = \frac{n-1}{2}\{\log\phi - \phi'\phi^{-1}\} \tag{8.102}$$

and calculate

$$\ell_{\mathrm{A};'\phi}^{\dagger} = \frac{n-1}{2}\phi\hat{\phi}^{-2} \tag{8.103}$$

yielding

$$u_{\mathrm{m}} = u_{\mathrm{A}}, \tag{8.104}$$

where u_{A} is given in (8.100). Upon noting that $r_{\mathrm{m}} = r_{\mathrm{A}}$ of (8.99), one obtains

$$r_{\mathrm{m}}^{*} = r_{\mathrm{A}}^{\dagger}. \tag{8.105}$$

The numerical accuracy of the tail probabilities generated by $\Phi(r_{\mathrm{A}}^{\dagger})$ in Table 8.2 is then accounted for by the fact that $\Phi(r_{\mathrm{A}}^{\dagger}) = \Phi(r_{\mathrm{m}}^{*})$, where r_{m}^{*} is normally distributed to order $O(n^{-3/2})$. Although the large-deviation property of r^{*} is the same whether or not there are nuisance parameters present, one would naturally expect the actual numerical performance of r^{*} to deteriorate as the number of nuisance parameters increases while n is kept fixed. The higher accuracy of $\Phi(r_{\mathrm{A}}^{\dagger})$ over $\Phi(r_{0}^{\dagger})$ may be because $r_{\mathrm{A}}^{\dagger}$ is equal to the r^{*}-statistic from a one-dimensional model whereas $r_{0}^{\dagger}$ is equal to an r^{*}-statistic from a two-dimensional model. In terms of their score functions, the difference between profile likelihood and modified profile likelihood is that the latter has, to order $O(n^{-1})$, been corrected for bias (cf. sections 8.2, 8.3, in particular (8.61)) and this may account for the observed improvement of the approximation to the tail probabilities generated by $\Phi(r_{\mathrm{A}})$ over those generated by $\Phi(r)$.

Examples 8.7–8.9 relate to ordinary repeated sampling and illustrate the numerical accuracy of $r_{\mathrm{A}}^{\dagger}$ and $r_{0}^{\dagger}$. However, as is the case for the modified directed likelihood, the main interest in these quantities is likely to concern situations with a large number of nuisance parameters.

8.5 Large number of nuisance parameters

Suppose that we have m samples $Y_{i1},\ldots,Y_{ik_i}$ $(i=1,\ldots,m)$ from a model of the same form, the parameters for the ith sample being ψ and χ_i, i.e. χ_i varies from sample to sample while ψ is the same throughout. The maximum likelihood estimate of ψ, which of course equals the estimate obtained from the profile likelihood $L_P(\psi)$, may be inefficient or even inconsistent. The two cases of normal variates Y_{ij} with either a common mean μ or a common variance σ^2 illustrate this; see section 4.3. It is therefore natural to explore the use of adjusted profile likelihood and adjusted directed likelihood for inference on ψ.

We do not undertake a systematic investigation of this point here, but will discuss some examples.

Example 8.10 Precision of duplicate measurements
We discuss in more detail and generality the problem introduced in Example 4.2. Duplicate measurements are taken of m objects using the same measurement device. This results in observations Y_{ij} where $i = 1, \ldots, m$, and $j = 1, 2$. It is assumed that Y_{ij} follows a location-scale model, i.e. $Y_{ij} \sim \sigma^{-1} f((y_{ij} - \mu_i)/\sigma)$ with f a known probability density function on the real line. Letting a denote the configuration ancillary,

$$a = ((y_{11} - \hat{\mu}_1)/\hat{\sigma}, (y_{12} - \hat{\mu}_1)/\hat{\sigma}, \ldots, (y_{m1} - \hat{\mu}_m)/\hat{\sigma}, (y_{m2} - \hat{\mu}_m/\hat{\sigma}),$$

we find the modified profile likelihood for σ to be

$$L_{MP}(\sigma) = \sigma^{-m} \prod_i (g''(y_{i1} - \hat{\mu}_{i\sigma})/\sigma)$$
$$+ g''((y_{i2} - \hat{\mu}_{i\sigma})/\sigma)\}^{-1/2} \sum_{i,j} f((y_{ij} - \hat{\mu}_{i\sigma})/\sigma), \quad (8.106)$$

where $g = -\log f$. In the present case the factor $|\hat{\chi}_{\psi/\hat{\chi}}|$ equals 1 because for fixed a, σ and $\hat{\sigma}$ the estimator $\hat{\mu}_{i\sigma}$ is a translation of $\hat{\mu}_i$.

In particular, for the normal error law we have

$$L_{MP}(\sigma) = \sigma^{-m} \exp\{-\sigma^{-2} \sum_i (y_{i1} - y_{i2})^2/4\}.$$

This equals the marginal likelihood for σ based on s^2, and it yields the usual estimate $s^2 = m^{-1} \sum_i (y_{i1} - y_{i2})^2$ for σ^2, whereas the full maximum likelihood estimate $\hat{\sigma}^2 = \frac{1}{2}s^2$ is not even consistent.

In general, the modified profile likelihood $L_{MP}(\sigma)$ may be compared to the marginal likelihood for σ which, on account of (2.104), is given by

$$L(\sigma; a_0) = \sigma^{-m} \sum_{i=1}^{m} \int_{-\infty}^{\infty} f(y_{i1}/\sigma - \zeta) f(y_{i2}/\sigma - \zeta) d\zeta, \quad (8.107)$$

where $a_0 = (y_{11} - y_{12}, \ldots, y_{m1} - y_{m2})$. While these two pseudo-likelihoods do generally differ, in the normal case they coincide, and the same is true for the extreme value distribution $f(y) = \exp(-y)\exp\{-\exp(-y)\}$. In the latter case we have

$$L_{\text{MP}}(\sigma) = L(\sigma; a_0) = \sigma^{-m} e^{-y_{++}/\sigma} \prod_i \left\{ \sum_j e^{-y_{ij}/\sigma} \right\}^{-2},$$

where $y_{++} = \sum_{i,j} y_{ij}$. □

Suppose now that the m samples are from a prime exponential model

$$a(\psi, \chi) b(y) \exp(u^{\text{T}} \psi + v^{\text{T}} \chi) \tag{8.108}$$

with partitioned canonical parameter $\phi = (\psi, \chi)$, partitioned canonical statistic $s = (u, v)$, and with (ψ, λ_i) as the corresponding mixed parameter for the ith sample. Using formula (8.10), we find

$$L_{\text{MP}}(\psi) = \prod_{i=1}^{m} [|\Sigma_{22}(\psi, \hat{\chi}_i)|^{1/2} a^{n_i}(\psi, \hat{\chi}_i) \exp\{\psi^{\text{T}} u_{i+} + \lambda(\psi, \hat{\chi}_i)^{\text{T}} v_{i+}\}], \tag{8.109}$$

where

$$u_{i+} = u(y_{i1}) + \cdots + u(y_{ik_i}), \quad v_{i+} = v(y_{i1}) + \cdots + v(y_{ik_i}),$$

and $\hat{\chi}_i = \bar{v}_i = k_i^{-1} v_{i+}$.

Example 8.11 Inverse Gaussian samples

When the prime exponential model is the inverse Gaussian

$$\frac{\sqrt{\psi}}{\sqrt{(2\pi)}} y^{-3/2} \exp\{-\tfrac{1}{2}(y - \chi)^2/(y\chi^2)\} \tag{8.110}$$

and if we let $k_1 = \cdots = k_m = k$, formula (8.109) takes the form

$$L_{\text{MP}}(\psi) = \psi^{m(k-1)/2} e^{-\chi w_+/2} \tag{8.111}$$

where $w_+ = \sum w_i$ and $w_i = \sum(y_{ij}^{-1} - \bar{y}_{i+}^{-1})$. The modified likelihood (8.111) coincides with the marginal likelihood for ψ based on $(w_1, \ldots, w_m)$, i.e.

$$L_{\text{MP}}(\psi) = L(\psi; w_1, \ldots, w_m). \tag{8.112}$$

Furthermore, the maximum modified profile likelihood estimate

$\hat{\psi}_{MP}$ for ψ is consistent for m and/or k tending toward infinity. On the other hand the unconditional maximum estimate of ψ is

$$\hat{\psi} = mk/w_+ \tag{8.113}$$

and this is not consistent for $m \to \infty$ and k fixed. □

Examples 8.10 and 8.11 and similar examples show that in some, highly regular, cases with increasing number of nuisance parameters the modified profile likelihood yields optimal estimates of the parameters of interest. In general, the estimate $\hat{\psi}_{MP}$ derived from $L_{MP}(\psi)$ will not be consistent for $m \to \infty$ and bounded k_i. However, it will generally be more nearly consistent than $\hat{\psi}$. Otherwise expressed, in a situation where the k_i increase slowly to infinity with m, both $\hat{\psi}_{MP}$ and $\hat{\psi}$ are consistent but $\hat{\psi}_{MP}$ tends to ψ considerably faster than $\hat{\psi}$. To illustrate this, we consider the case of m matched pairs.

Example 8.12 Matched binary pairs

We now apply modified profile likelihood to matched binary pairs, shown in Example 4.6 to be a situation in which unconditional maximum likelihood gives inconsistent answers. Here Y_{j1}, Y_{j2} are independent binary (0, 1) variables with $P(Y_{j1} = 1) = e^{\chi_j}/(1 + e^{\chi_j})$, $P(Y_{j2} = 1) = e^{\chi_j + \psi}/(1 + e^{\chi_j + \psi})$. We consider for the present purpose only pairs for which $y_{j1} + y_{j2} = 1$; let s_{01}, s_{10} denote the number of (0, 1) and (1, 0) responses. Then the modified profile likelihood for ψ is

$$L_{MP}(\psi) = \exp\{\tfrac{1}{2}\psi(s_{01} - s_{10})\}\{\exp(\tfrac{1}{4}\psi) + \exp(-\tfrac{1}{4}\psi)\}^{-3(s_{10}+s_{01})}. \tag{8.114}$$

Let $\rho = e^{\psi}$, so that ρ is the odds ratio. Then as $m \to \infty$, the unconditional maximum likelihood estimate tends to ρ^2, as shown in Example 4.6, while $\hat{\rho}_{MP}$ tends to $\{(5\rho + 1)/(\rho + 5)\}^2$. Table 8.3 compares these values. Of course in this instance exact conditional

Table 8.3 *Matched pairs. Odds ratio ρ and corresponding infinite sample values of the maximum modified profile likelihood estimator $\hat{\rho}_{MP}$ and of the ordinary maximum likelihood estimator $\hat{\rho}$*

ρ	1	1.5	2	3	5	10	15	20
$\hat{\rho}_{MP}$	1	1.71	2.47	4.00	6.76	11.56	14.44	16.32
$\hat{\rho}$	1	2.25	4	9	25	100	225	400

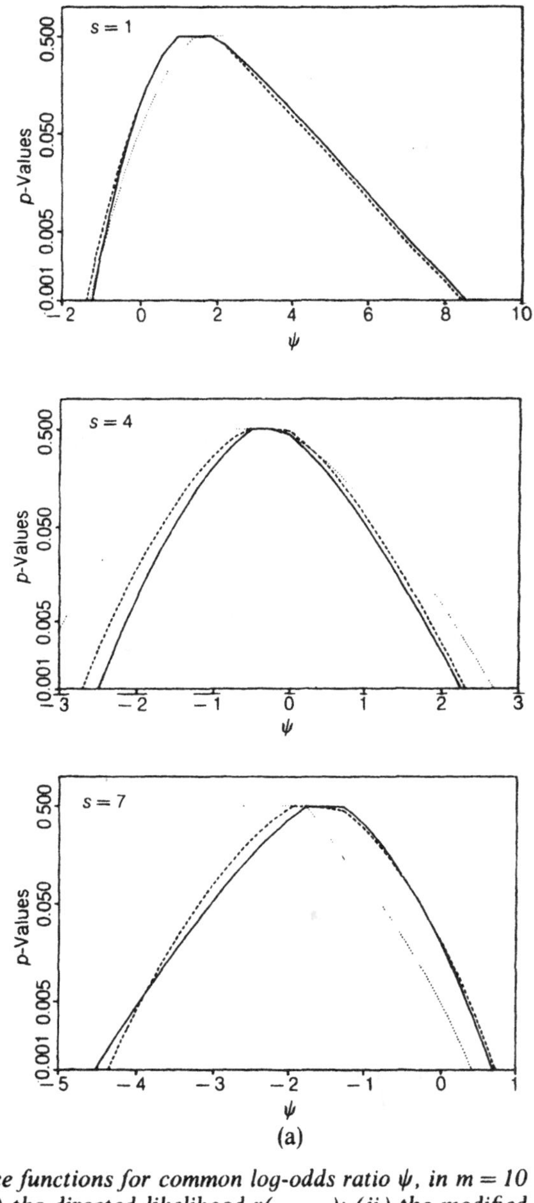

Figure 8.1 *(a) Confidence functions for common log-odds ratio ψ, in $m = 10$ 2×2 tables, based on (i) the directed likelihood r(......); (ii) the modified directed likelihood r^*(– – –); (iii) the conditional directed likelihood r^c (——). (b) Corresponding pseudo-likelihoods: (i) profile likelihood (......); (ii) modified profile likelihood (– – – –); (iii) conditional likelihood (——).*

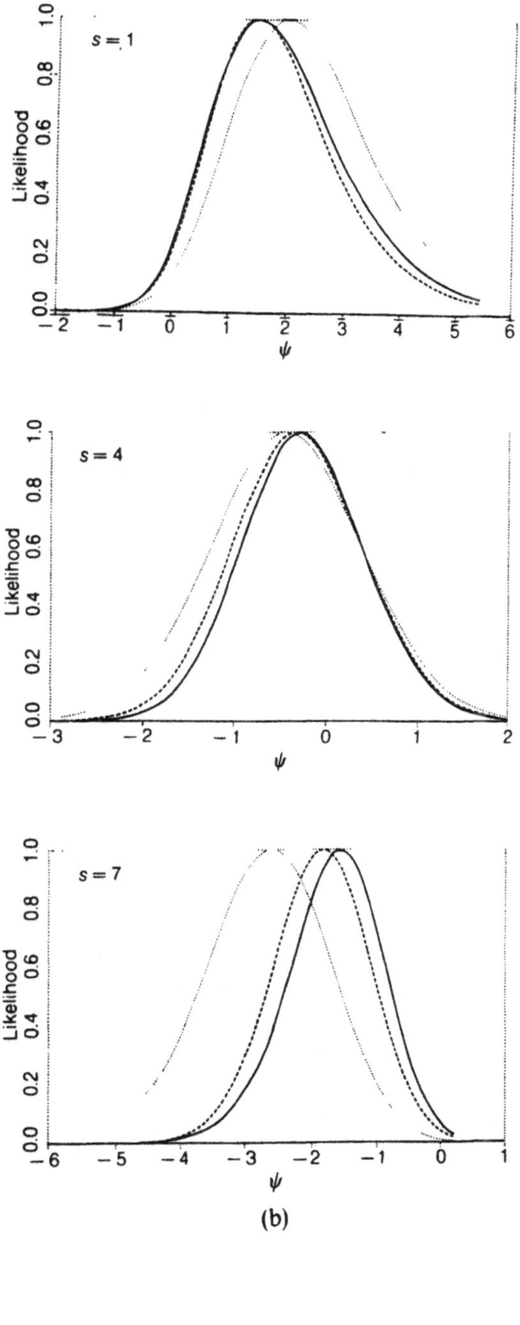

(b)

inference for ψ is straightforward, based on the binomial distribution, but the example demonstrates that modified profile likelihood gives a substantial improvement over the unconditional maximum likelihood estimate, particularly in the region of values of ρ of greatest practical interest. □

As a final illustration we consider an application of the modified directed likelihood to a matched pairs setting, much like that of the previous example. The discussion below also throws some further light on the modified profile likelihood and on the nature of the decomposition of the adjustment term of r^*, considered in section 6.6.4.

Example 8.13 Common odds ratio in 2 × 2 tables
Let Y_{jk} $(j = 1, \ldots, m;\ k = 1, 2)$ be independent binomial random variables, $Y_{jk} \sim \text{bin}(n_{jk}, \pi_{jk})$ with $\pi_{j1} = e^{\chi_j}/(1 + e^{\chi_j})$ and $\pi_{j2} = e^{\chi_j + \psi}/(1 + e^{\chi_j + \psi})$. As in Example 8.12 we consider inference about the common log-odds ratio ψ in the presence of the m nuisance parameters $\chi_1, \ldots, \chi_m$. But here we assume that $n_{j1} = 1$ and $n_{j2} = 2$ and that $Y_{j+} = 1$ for all j. Under the conditional distribution given $(Y_{1+}, \ldots, Y_{m+})$ we have that the row sum $s = Y_{+1}$ is minimal sufficient.

Figure 8.1 shows, for $m = 10$ and several values of s, the confidence functions for ψ based on (1) the directed likelihood r treated as standard normal; (ii) the modified directed likelihood r^* treated as standard normal; (iii) the conditional directed likelihood r^c (or, equivalently, any other monotonic function of Y_{+1}) with its exact conditional distribution, given $(Y_{1+}, \ldots, Y_{m+}) = (1, \ldots, 1)$. Each confidence function determines, as its level sets, the nested family of confidence limits corresponding to the associated procedure (i), (ii) or (iii). Thus, for instance, in Fig. 8.1 the confidence function corresponding to (ii) with $s = 4$ determines the 0.95 confidence interval $(-2.3, 1.8)$ for ψ. For the confidence functions the correction (6.137) has been used. Plots of the (normalized) profile likelihood, modified profile likelihood and conditional likelihood functions for ψ are also given in Fig. 8.1. The information correction z_{INF} is nearly 0, in agreement with the discussion in section 6.6.4. One notes in particular that all three functions are nearly of normal shape, the differences being mainly in location and spread.

Other closely related numerical and graphical illustrations have been given in Pierce and Peters (1992).

8.6 Further definitions of pseudo-likelihood

The initial definition of likelihood corresponds directly to the probability density of the full data y as specified by the model under study. Subsequently we have used marginal and conditional likelihoods. These are 'ordinary' likelihoods attached to derived random systems, for instance in the former case to a random system in which only a particular statistic is considered. The only possible difficulty in their use concerns whether the part of the full likelihood omitted carries useful information about the parameter of interest.

It is, however, possible to introduce further forms of likelihood that do not correspond directly to the observation of random subsystems but which may have sufficient of the properties of likelihood (or of associated functions such as the score function) to make use of the term likelihood sensible. The advantages of such generalizations are partly concerned with difficult problems with many nuisance parameters and partly with applications in which calculating the full likelihood in useful form is difficult. Profile likelihood and its adjustments are instances of this.

In some ways, however, the most direct of these extensions is **partial likelihood**, defined as follows. Suppose that the full data can be represented by a sequence

$$\{w_1, y_1, w_2, y_2, \ldots, w_n, y_n\} \tag{8.115}$$

which is such that given $\{w_1, y_1, \ldots, w_r\}$ the random variable Y_r has a distribution depending only on the parameter of interest ψ, whereas the conditional distribution of W_r given $\{w_1, y_1, \ldots, y_{r-1}\}$ either is not available in useful form or is so dependent on nuisance parameters that it contains no information about ψ. Then from the full likelihood

$$\prod_{j=1}^{n} p_{W_j|W_1,Y_1,\ldots,Y_{j-1}} \prod_{j=1}^{n} p_{Y_j|W_1,Y_1,\ldots,W_j} \tag{8.116}$$

it is useful for inference about ψ to extract the second factor, called the partial likelihood for ψ, based on the particular decomposition (8.116): we write

$$\ell_{\mathrm{PT}}(\psi) = \sum_{j=1}^{n} \log p_{Y_j|W_1,Y_1,\ldots,W_j}. \tag{8.117}$$

A number of issues arise. First how can a suitable decomposition (8.116) be found? Next how can it be verified that the omitted factor indeed contains little or no information about ψ? Finally to what extent do maximum partial likelihood estimates and other statistics obtained from $\ell_{PT}(\psi)$ have the 'usual' properties, at least to the first order of asymptotic theory?

So far as we are aware, no general constructive methods are available to answer the first question. The issues raised in discarding the omitted factor are largely those involved in the various extended notions of ancillarity; see section 2.5. In particular, the omitted factor may not depend on ψ. Finally, the key first-order properties associated with $\ell_{PT}(\psi)$ follow from the martingale structure of the gradient of (8.117), as outlined in section 3.3. That is, taking the case of scalar ψ for simplicity, $d\ell_{PT}(\psi)/d\psi$ is a sum of uncorrelated terms of zero mean and thus has as variance a sum of contributions from each term and hence estimated via an 'observed information'

$$[-d^2\ell_{PT}(\psi)/d\psi^2]_{\psi=\hat{\psi}_{PT}}, \tag{8.118}$$

where $\hat{\psi}_{PT}$ is the value of ψ that maximizes $\ell_{PT}(\psi)$.

In some cases the partial log-likelihood may be the same as the profile log-likelihood. When there are many nuisance parameters there is, however, no guarantee that the profile log-likelihood has good properties, so that there is a conceptual advantage to the partial log-likelihood.

Example 8.14 Stochastic process with a special state.
Consider a discrete-time stochastic process with discrete states $\{0, 1, \ldots\}$, the state 0 having special properties. Suppose that, conditionally on state 0 being occupied at time n, and independently of the previous history of the process, the probability that it will be occupied at time $n+1$ is $p_n(\psi)$, where ψ is an unknown parameter. Suppose, further, that on leaving state 0 the system executes 'tours' determined by a probability mechanism of complicated and possibly unknown form. If large-sample arguments are to be used, we suppose that the number of distinct sojourns in state 0 is large.

The decomposition (8.116) is as follows: W_1 gives the path up to and including the first entry to state 0; Y_1 specifies the number of subsequent periods up to the next exit from state 0, but does not specify the state to which the system then moves; W_2 gives the state entered on leaving 0 and the subsequent path up to and including the second entry to state 0, and so on. The partial likelihood based on Y in the sequence $\{w_j, y_j\}$ is then

$$\prod\left[\{1 - p_{e_j+r_j}(\psi)\}\prod_{k=0}^{r_j-1} p_{e_j+k}(\psi)\right], \tag{8.119}$$

where the jth sojourn in state 0 starts at entry time e_j and ends at time $e_j + r_j$.

In particular, if the probability of remaining in state 0 is always θ, this is $\Pi_j\{(1-\psi)\psi^{r_j-1}\}$ and is the same as the marginal likelihood of the sojourns in state 0.

This example can be generalized in various ways. For instance, in a semi-Markov process it may be convenient to take W_j to be the time between $(j-1)$th and jth transitions and Y_j to be state occupied after the jth transition. □

Example 8.15 Censored failure data

In Example 2.9, we gave in (2.32) the form of the log-likelihood for censored failure data under a so-called proportional hazards model involving a base-line hazard function $\rho_0(x)$ and a vector β of regression coefficients. If $\rho_0(x)$ is specified in parameter form with a fairly small number of unknown parameters then there is no difficulty in calculating, for example, a profile likelihood for components of β of particular interest. If, however, $\rho_0(x)$ is left arbitrary, so that the specification becomes semiparametric, a partial likelihood can be obtained via the following decomposition. Let $t_{(1)} < t_{(2)} < \cdots < t_{(n)}$ be the times at which failures occur. For simplicity we suppose that there are no ties.

Denote by Y_j an identifier of the individual who fails at time $t_{(j)}$ and let W_j denote a complete specification of the behaviour of the system, for example including cases of censoring, from $t_{(j-1)} + 0$ up to and including the occurrence of a failure at $t_{(j)}$; W_j does, however, not identify the individual failing. Then the contribution to the likelihood from Y_j is

$$\frac{e^{\beta^{\mathrm{T}} z_j}}{\sum_{\ell \in \mathscr{R}_j} e^{\beta^{\mathrm{T}} z_\ell}},$$

where $\mathscr{R}_j$ is the risk set at $t_{(j)}$, i.e. the set of individuals who are at risk of failure at $t_{(j)}$, or in other words who have neither failed nor been censored before then. The partial likelihood for ψ is consequently

$$L_{\mathrm{PT}}(\psi) = \prod_{j=1}^{n} \frac{e^{\beta^{\mathrm{T}} z_j}}{\sum_{\ell \in \mathscr{R}_j} e^{\beta^{\mathrm{T}} z_\ell}}, \tag{8.120}$$

□

There are a considerable number of further forms of pseudo-likelihood, some but not all generalizations of partial likelihood. An immediate generalization is that ℓ_{PT} as defined in (8.117) could still contain a nuisance parameter χ, so that, say, a profile likelihood would be calculated from ℓ_{PT}. This would still be valuable if χ were of small dimension, as contrasted with a nuisance parameter of large dimension in the full likelihood.

A more radical modification, starting again from the sequential structure (8.115), is to use instead of (8.117) the simpler form

$$\ell_{\mathrm{SPT}}(\psi) = \sum_{j=1}^{n} \log p_{Y_j | W_j}, \tag{8.121}$$

assuming that the conditional distributions concerned depend only on ψ. While the expectation of the gradient of $\ell_{\mathrm{SPT}}(\psi)$ is zero, successive terms are not in general uncorrelated. Thus (8.121) typically gives on maximization an asymptotically normal estimate of ψ, but there will be some loss of efficiency if (8.117) is also available.

Equation (8.121) is related to a sequence of Markov likelihoods that can be defined quite generally but which are especially relevant for time-series models. The full log-likelihood, based on a sequence of observations $Y_1, \ldots, Y_n$, is given by the sum of conditional terms, in which Y_r, say, is taken conditionally on $Y_1, \ldots, Y_{r-1}$. As a series of approximations to this we may let

$$\ell_{M_0}(\theta) = \sum_{j=1}^{n} \log p_{Y_j}(y_j; \theta)$$

$$\ell_{M_1}(\theta) = \sum_{j=1}^{n} \log p_{Y_j | Y_{j-1}}(y_j | y_{j-1}; \theta),$$

$$\ell_{M_2}(\theta) = \sum_{j=1}^{n} \log p_{Y_j | Y_{j-1}, Y_{j-2}}(y_j | y_{j-1}, y_{j-2}; \theta), \tag{8.122}$$

etc. If the process is k-dependent Markov, then ℓ_{M_k} will be the standard log-likelihood. In general, estimation of all parameter components will be possible only via ℓ_{M_k} for sufficiently large k. Thus for stationary Gaussian processes we need $k \geqslant \dim(\theta) - 2$, the efficiency increasing with increasing k towards that of the full log-likelihood (Azzalini, 1983).

A yet more general method is to associate with each Y_j a conditioning set C_j consisting of suitable other $Y_k (k \neq j)$ and possibly other variables and then to consider the pseudo-likelihood

$$\ell_{\text{PS}}(\theta) = \sum_{j=1}^{n} \log p_{Y_j | C_j}(y_j | c_j; \theta). \tag{8.123}$$

The gradient again has expectation zero and under rather general conditions will be asymptotically normal, but evaluation of its variance requires construction of the covariance of different terms.

Example 8.16 Spatial process
Consider a set of normally distributed random variables $\{Y_{i,j}\}$ defined on the lattice $(i, j = 0, \ldots, m)$ and following the Gaussian autoregressive scheme in which, given all other random variables,

$$\begin{aligned} E(Y_{i,j}) &= \mu + \beta(y_{i+1,j} + y_{i-1,j} + y_{i,j+1} + y_{i,j-1} - 4\mu) \\ &= \nu + \beta(y_{i+1,j} + y_{i-1,j} + y_{i,j+1} + y_{i,j-1}), \end{aligned}$$

say, with conditional variance σ^2. Then by taking C_j to be the set of the four nearest neighbours of (i, j) we can write

$$\begin{aligned} \ell_{\text{PS}}(\nu, \beta) = -n \log \sigma - \sum \{ y_{i,j} - \nu - \beta(y_{i+1,j} \\ + y_{i-1,j} + y_{i,j+1} + y_{i,j-1}) \}^2 / (2\sigma^2). \end{aligned}$$

We assume for simplicity that the sum is taken over those lattice points for which all four neighbouring values are available. Thus, at the cost of some loss of efficiency, estimation is reduced to a standard least squares problem. □

Among other existing types of pseudo-likelihood are the following:

1. *Empirical likelihood* (Owen, 1988; 1991). This, in essence, is the concept of profile likelihood extended to nonparametric settings. It involves consideration of multinomial distributions with support on the data values.
2. *Quasi-likelihood* (Wedderburn, 1974; McCullagh and Nelder, 1989). Here distributional assumptions are replaced by low-order moment assumptions; for further discussion see section 9.2 below.

3. *Penalized likelihood* cf., for instance, Green (1987).
4. *Extended likelihood* (Barndorff-Nielsen, 1993). This concept associates a 'likelihood' function with certain types of combinant.

There also exist various definitions of predictive likelihood connected with inference about as yet unobserved random variables rather than with unknown parameters; see Butler (1986; 1989).

Further results and exercises

8.1 Show that the modified profile score is approximately unbiased in the sense that

$$E\{\ell_{\mathrm{MP}}(\psi)_{/\psi}\} \stackrel{\dagger}{=} 0.$$

[Liang, 1987; McCullagh and Tibshirani, 1990; Ferguson, Reid and Cox, 1991; Barndorff-Nielsen, 1992; Section 8.2–8.3]

8.2 Let $d(\psi) = \log D(\psi)$, with $D(\psi)$ as defined by (8.21). Show that $d(\psi) \stackrel{\dagger}{=} 0$, and hence that formula (8.32) holds if and only if

$$\{(\hat{j}_{\chi\chi})^{rs}\hat{j}_{ra}\}_{/s} = O_{\mathrm{p}}(n^{-1/2})$$

or, equivalently, if and only if

$$\{(i_{\chi\chi})^{rs}i_{ra}\}_{/s} = O(n^{-1/2}).$$

Here indices r and s refer to coordinates of χ, while a indicates a coordinate of ψ, and $/s$ indicates differentiation with respect to χ^s.

[Barndorff-Nielsen and McCullagh, 1993; Section 8.2]

8.3 If the statistical model is a composite transformation model with ψ as the maximal invariant parameter and with χ as the group parameter, then orthogonality of ψ and χ implies that, to order $O(n^{-1})$ and under regularity assumptions, the modified profile likelihood $L_{\mathrm{MP}}(\psi)$ equals the Laplace approximation to the integral representation (2.104) of the marginal likelihood function for ψ.

[Section 8.2]

8.4 Consider the linear functional relationship model for independent observations Y_i and Z_i $(i = 1, \ldots, n)$ and specified by

$$Y_i \sim \mathrm{N}(\eta_i, \sigma^2), \quad Z_i \sim \mathrm{N}(\alpha + \beta\eta_i, \sigma^2).$$

We assume that σ^2 is known, the unknown parameters thus being α, β and $\eta_1, \ldots, \eta_n$.

Show that if the interest and nuisance parameters are $\psi = \alpha$ and $\chi = (\beta, \eta_1, \ldots, \eta_n)$, respectively, then both $h(\psi, \chi)$ and $H_{\chi\chi\psi}$, given by the formulae (8.37) and (8.38), are equal to 0. Verify also that if, instead, we take $\psi = \beta$ and $\chi = (\alpha, \eta_1, \ldots, \eta_n)$ then $H_{\chi\chi\psi} \neq 0$ but, again, $h(\psi, \chi) = 0$.

[Salamin, 1993; Section 8.3]

8.5 Suppose that ψ is a scalar parameter of interest and λ a scalar nuisance parameter. Show, by noting that

$$\frac{\partial \log j_{\lambda\lambda}(\psi, \hat{\lambda}_\psi)}{\partial \psi} = \frac{1}{\bar{j}(\psi, \hat{\lambda}_\psi)} \left\{ \frac{\partial \bar{j}_{\lambda\lambda}(\psi, \hat{\lambda}_\psi)}{\partial \psi} + \frac{\partial \bar{j}_{\lambda\lambda}}{\partial \lambda} \frac{\partial \hat{\lambda}_\psi}{\partial \psi} \right\},$$

where $\bar{j}$ is the observed information per observation, that the left-hand side tends in probability to zero if and only if

$$\nu_{\psi\lambda\lambda} = E(\partial^3 \ell / \partial\psi \partial\lambda^2) = 0.$$

More generally, if λ is a vector parameter the corresponding condition is that

$$\operatorname{tr}(i_{\lambda\lambda}^{-1} \nu_{\psi\lambda\lambda}) = 0.$$

Suppose that under a (2, 2) exponential family, (ψ, λ) is a mixed orthogonal parameterization in which one component is a canonical parameter and the other is the complementary mean parameter. Show that $\nu_{\psi\lambda\lambda} = 0$ if ψ is the mean parameter, whereas in general $\nu_{\psi\lambda\lambda} \neq 0$ if ψ is the canonical component. Hence 'explain' why, in linear normal theory problems in which the parameter of interest is a regression coefficient, adjustment of the likelihood function may be of minor importance, whereas if the parameter of interest is a variance (in the presence of a large number of regression parameters regarded as nuisance parameters), adjustment of the likelihood function is essential.

[Cox and Reid, 1992; Section 8.2–8.3]

8.6 The function $g_0(\psi)$ from (8.36), which forms part of the stably adjusted profile likelihood $L_A^\dagger(\psi)$, may be approximated as $g_0(\psi) \doteq h_0(\psi)$ where

$$h_0(\psi) \doteq r_\psi \hat{\imath}^{\psi\psi} h(\hat{\psi}, \hat{\chi}).$$

Here r_ψ is the directed likelihood for ψ and $h(\psi, \chi)$ is defined by (8.37)–(8.38).

Show that $h_0(\psi)$ is invariant under interest-respecting reparameterizations and discuss the properties of the alternatives to $L_A^\dagger(\psi)$ and to the adjusted directed likelihoods $r_A^\dagger$ and $r_0^\dagger$ (formulae (8.62) and (8.67)) determined by substituting $h_0(\psi)$ for $g_0(\psi)$ in the expression (8.39) for $L_A^\dagger(\psi)$.

[Sections 8.3–8.4]

8.7 Consider a prime exponential model (2.105) with the canonical parameter ϕ and the canonical statistic s partitioned as $\phi = (\psi, \chi)$ and $s = (u, v)$ and with cumulant function $k(\phi) = k(\psi, \chi)$. The modified profile likelihood $L_{MP}(\psi)$ approximates the conditional likelihood $L(\psi; u|v)$ for ψ given v, to order $O(n^{-1})$ for large deviations and to order $O(n^{-3/2})$ for moderate deviations. This approximation immediately yields the following approximation for the conditional cumulant function:

$$k(\psi|v) \doteq k(\psi, \hat{\chi}_\psi) - \hat{\chi}_\psi^T v + \tfrac{1}{2}\log|k_{/\chi\chi}(\psi, \hat{\chi}_\psi)|.$$

Use this relation to derive approximations to the conditional cumulants of u given v and show that these approximations are valid to order $O(n^{-1})$ at least, and that they agree with the approximations obtainable from the mixed, or direct-tilted, expansion (Barndorff-Nielsen and Cox, 1989, section 7.5) for the conditional density. Compare also to the expansions for conditional cumulants given by McCullagh (1987, section 5.6).

[Pace and Salvan, 1992; Section 8.2]

8.8 Compare the conditional properties, given a local second-order ancillary a, of the two versions of the maximum likelihood test corresponding to standardization by the observed information and by the expected information, respectively. Show that, to order $O(n^{-1})$, the conditional test level is equal to the nominal level if observed information is used, but that this is in general not the case if expected information is used. Similar properties hold for the score test.

[Cox, 1980; Severini, 1990]

Bibliographic notes

The modified profile likelihood was introduced by Barndorff-Nielsen (1983) and further discussed by Barndorff-Nielsen (1985; 1988a) and Barndorff-Nielsen and McCullagh (1993).

The material in sections 8.3 and 8.4 is based on Barndorff-Nielsen (1993) and Barndorff-Nielsen and Chamberlin (1991; 1994), respectively, and that of section 8.5 is from Barndorff-Nielsen (1983), Pierce and Peters (1992) and joint work in progress at the time of writing by O.E. Barndorff-Nielsen, D.A. Pierce and D. Peters. Haberman (1977) gave a deep discussion of the behaviour of ordinary maximum likelihood estimates in cases of many nuisance parameters, with particular reference to the Rasch model.

The partial likelihood L_{PT} was introduced in the context of survival analysis by Cox (1972) and in its general form by Cox (1975). There is now an extensive literature on properties and applications of partial likelihood; see for instance Cox and Oakes (1984), Andersen and Borgan (1985) and Andersen, Borgan, Gill and Keiding (1993). The general type of pseudo-likelihood given by (8.123) can be traced back to Ogawara (1951) for the analysis of time series but has been discussed systematically by Oakes (1988), in the context of survival analysis, and by Lindsay (1988). In spatial statistics it is an important tool in connection with Markov random field models; see Besag (1974; 1977; 1978; 1986) and Ripley (1988), cf. also Barndorff-Nielsen (1990d, section 10.5.3).

CHAPTER 9

Further aspects

9.1 Pivots

As the first of a miscellaneous series of topics, each of which will be discussed only briefly, we now consider pivots, which are a useful device for handling certain special problems and often lead to concise statement of the conclusions about a parameter of interest. We consider for simplicity a scalar parameter ψ of interest, with nuisance parameters χ. Then a **pivot** for ψ is a function $p(Y, \psi)$ of the observed random variables Y and the parameter of interest ψ such that

1. the distribution of $p(Y, \psi)$, evaluated under the distribution of Y when the parameter value is (ψ, χ), is the same for all (ψ, χ);
2. the function $p(y, \psi)$ is monotonic in ψ for almost all y.

We call the distribution of $p(Y, \psi)$ the **pivotal distribution**, and denote its upper quantiles by p_ε^*, i.e.

$$P\{p(Y, \psi) < p_\varepsilon^*; \psi, \chi\} = 1 - \varepsilon. \tag{9.1}$$

We shall deal only with the case where the pivotal distribution is continuous.

Note that if $p(Y, \psi)$ is a pivot, so is any one-to-one function of p; we usually choose that function for which the pivotal distribution has an especially convenient form. It would be possible via a probability integral transformation to arrange that the pivotal distribution is uniform on (0, 1) or indeed to achieve any other fixed distribution, but this is not particularly helpful.

Simple examples are that if $Y_1, \ldots, Y_n$ are independent and identically distributed $N(\mu, \sigma^2)$ and $\bar{Y}, s^2$ are the usual estimators of mean and variance, then if σ^2 is known, equal to σ_0^2, say, we have that

$$(\bar{Y} - \mu)/(\sigma_0/\sqrt{n}) \tag{9.2}$$

is a pivot for μ, and if σ^2 is unknown

$$(\bar{Y}-\mu)/(s/\sqrt{n}) \tag{9.3}$$

is a pivot for μ, the pivotal distributions being respectively the standard normal distribution and the Student t distribution with $n-1$ degrees of freedom. In both cases the pivots have been chosen so that the pivotal distribution is of standard form.

More generally, let $\tilde{Y}$ be any location statistic and $\tilde{s}$ any estimate of scale, i.e.

$$\begin{aligned} \tilde{Y}(y_1+b,\ldots,y_n+b) &= \tilde{Y}(y_1,\ldots,y_n)+b, \\ \tilde{s}(cy_1+b,\ldots,cy_n+b) &= c\tilde{s}(y_1,\ldots,y_n), \end{aligned} \tag{9.4}$$

for all b and all $c>0$. Then (9.2) and (9.3) can be replaced by

$$(\tilde{Y}-\mu)/(\sigma_0/\sqrt{n}), \quad (\tilde{Y}-\mu)/(\tilde{s}/\sqrt{n}), \tag{9.5}$$

the pivotal distributions depending on the particular choice of statistics. Furthermore, (9.5) are pivots for any given scale and location family $\sigma^{-1}h\{(y-\mu)/\sigma\}$.

If ancillary statistics, A, are available it would be natural to consider pivotal distributions conditionally on $A=a$, the observed value. If the definition of the pivot can absorb a in such a way that the pivotal distribution does not depend on a, i.e. so that the pivot and A are independent, then conditional and unconditional distributions of the pivot coincide. When possible this is an effective way of dealing with ancillaries, but the device is, for example, not available for the general location problem.

We can similarly define quantities that are pivotal to a particular order of asymptotic theory, i.e. for which (9.1) holds with error $O(n^{-k/2})$ for some given k. An example is the directed likelihood statistic

$$p(Y,\psi)=r=\operatorname{sgn}(\hat{\psi}-\psi)[2\{\ell_P(\hat{\psi})-\ell_P(\psi)\}]^{1/2}, \tag{9.6}$$

in which the asymptotic pivotal distribution is N(0, 1); for this and refinements, see section 6.6.

The required monotonicity 2 in ψ is immediate for the pivots (9.2), (9.3) and (9.5) but requires for (9.6) that $\ell_P(\psi)$ has a single maximum (in the range of interest). From the monotonicity it is easy to obtain confidence limits for ψ at any required level. For example, from (9.3) we have that

$$P\left(\frac{\bar{Y}-\mu}{s/\sqrt{n}} < t^*_{n-1,\varepsilon}; \mu, \sigma\right) = 1 - \varepsilon$$

and, because of the monotonicity, this is inverted into

$$P\left\{\mu > \bar{Y} - t^*_{n-1,\varepsilon}\frac{s}{\sqrt{n}}; \mu, \sigma\right\} = 1 - \varepsilon; \tag{9.7}$$

here $t^*_{n-1,\varepsilon}$ is the upper ε point of the Student t distribution with $n-1$ degrees of freedom. This gives a lower $1-\varepsilon$ confidence limit for μ, and an upper limit is obtained similarly.

The point of this as a presentational device is that all that is necessary is to specify $\bar{Y}, s$ and the form of the pivotal distribution in order for confidence limits at any level to be defined implicitly. Preferably, of course, the pivotal distribution should be tabulated and of a simple familiar form.

Barnard (1974; 1977b) has developed an approach to inferential problems in which pivots play a primary role. For this it is necessary to define a basis for choosing between different pivots on the basis of their 'selectivity'. Pivots based on sufficient statistics and closely related to log-likelihood functions are to be preferred.

A pivotal treatment can also be given for some prediction problems. Here the pivot is a function not of Y and ψ but of Y and Z, the as yet unobserved random variable whose value is to be predicted. The simplest instance is that of predicting a new observation Z from $N(\mu, \sigma_0^2)$ on the basis of a set $Y_1, \ldots, Y_n$ of independent identically distributed values from that distribution, of mean $\bar{Y}$. The natural pivot is

$$(Z - \bar{Y})/\{\sigma_0\sqrt{(1 + 1/n)}\}, \tag{9.8}$$

the pivotal distribution being the standard normal and the required monotonicity in Z being clear; note the parallel with (9.2).

As noted in section 9.4 below, an asymptotic theory of prediction takes different forms depending on the relation between the sample size of the data and the number of observations involved in the aspect to be predicted.

The first possibility is that these are comparable; for example, it may be required to predict the mean of a new sample of size similar to that observed. A pivot can be constructed fairly directly and, in principle at least, an Edgeworth expansion found for it. Let θ be the unknown parameter vector and Z the value to be predicted, having an asymptotic normal distribution with mean $\tilde{m}(\theta)$ and variance $\tilde{v}(\theta)/\tilde{n}$. If from the observed random variables we construct the maximum likelihood estimate $\hat{\theta}$ asymptotically normal with mean θ and covariance matrix $i^{-1}(\theta)/n$, then we can use the asymptotic pivot

$$\frac{\{Z - \tilde{m}(\hat{\theta})\}}{\{(\nabla_\theta \tilde{m})^{\mathrm{T}} i^{-1}(\nabla_\theta \tilde{m})/n + \tilde{v}(\hat{\theta})/\tilde{n}\}^{1/2}} \tag{9.9}$$

having asymptotic pivotal distribution N(0, 1), the first term in the denominator being evaluated at $\theta = \hat{\theta}$, and independence of $\hat{\theta}$ and Z being assumed. Note that (9.9) is invariant under reparameterization, but not under transformations of the target of predictions, so that careful choice of the latter is advisable.

The second possibility is that the value to be predicted is a single observation, or a function of a small number of observations to which asymptotic theory is not applicable. A discussion of this, probably the more common case, is given in section 9.4.

9.2 Estimating equations and quasi-likelihood

In the previous chapters we have deliberately put rather little emphasis on point estimation as such. That is, although the maximum likelihood estimate $\hat{\theta}$ plays a key role in the discussion, it is as the basis primarily for the calculation of confidence regions rather than for final data summarization.

Point estimation in its own right, i.e. the provision of single points $\tilde{\theta}$, without explicit indication of uncertainty, as indicative of the value of an unknown parameter θ, is in our view not usually the appropriate way to summarize the results of a statistical analysis in that some recognition of uncertainty via indication of precision achieved is a central feature of statistical inference. Point estimation without specification of precision is, however, of interest in certain circumstances:

1. when the precision is very high;
2. in decision problems in which the space of possible decisions is equivalent to the parameter space or to a subspace thereof. Here for a full specification a loss function and prior distribution need to be available;

3. in problems in which it is convenient to analyse data in two (or more) stages. This is particularly natural when the data are divided into independent sets. The data in each set may then be reduced to point estimates of one or more relevant parameters, and the variation between sets can be represented by a model for the way these parameters depend on explanatory variables characterizing those sets. If, as would often be the case, the second stage of analysis is via a linear model, it will be desirable that the first-stage estimates are approximately unbiased and have known variance, the latter in order to check on the need for weighted rather than ordinary least squares in the second stage of analysis.

Our brief discussion of point estimation and in particular of unbiased point estimation will be mainly aimed at the last of these issues.

When a point estimate is given by an explicit function of the observed random variables, the study of its properties is a matter of direct distributional calculation. Often, however, as for example with maximum likelihood estimates, the estimates under study are defined only implicitly, usually as the roots of nonlinear equations to be solved by iteration. From this point of view it is quite natural to consider the properties of estimating equations directly. Consider, then, an estimate $\tilde{\theta}$ of a parameter θ defined as the root, assumed unique, of the equation

$$v(y, \tilde{\theta}) = 0, \tag{9.10}$$

for some function v and where y is the observed vector of random variables Y with density $f_Y(y; \theta)$. Of course, we are often interested only in certain components of θ leading to equations of the form $v(y, \psi) = 0$. It is convenient to arrange that $v(y, \theta)$ behaves asymptotically like the sum of n independent random variables.

For instance, if $Y_1, \ldots, Y_n$ are independent and identically distributed symmetrically about μ with known scale, possible estimating equations for μ are

$$\sum(Y_j - \mu) = 0, \quad \sum \operatorname{sgn}(Y_j - \mu) = 0, \quad \sum g(Y_j - \mu) = 0, \tag{9.11}$$

where in the third case $g(\cdot)$ is an odd function, $g(z) = -g(-z)$, usually chosen to 'downweight' the more extreme observations; for example, the resulting estimates are the sample mean, the sample median and a 'robust' estimate of the type introduced by Huber (1964). If the 'scale' σ is unknown we can replace the third of them by

$$\sum g\{Y_j-\mu)/\sigma\}=0,\quad \sum[(Y_j-\mu)g\{(Y_j-\mu)/\sigma\}-\sigma]=0. \qquad (9.12)$$

If instead of identically distributed components we suppose that the location parameter μ_j for Y_j has the linear form $\mu_j=x_j^r\beta_r$ $(r=1,\ldots,d)$, then, for example, the last estimating equation in (9.11) may be replaced by the set of equations

$$\sum_{j=1}^{n} x_j^s g(Y_j-x_j^r\beta_r)=0 \quad (s=1,\ldots,d). \qquad (9.13)$$

We define the estimating equation to be exactly unbiased if, for all θ,

$$E\{v(Y,\theta);\theta\}=0. \qquad (9.14)$$

For some purposes it is enough that the expectation is $O_p(1)$ as $n\to\infty$. Note that if $v^{(1)},\ldots,v^{(k)}$ give unbiased estimating equations for θ then so, too, does $\sum a_j v^{(j)}$ for any constants $a_1,\ldots,a_k$.

To study the asymptotic properties of the estimates defined by $v(Y,\tilde{\theta})=0$ we follow essentially the route used for the maximum likelihood estimating equation, i.e. the special case where v is the score function.

Consider first the case of a scalar parameter and linearize by a first-order expansion around the true parameter value θ to give

$$\tilde{\theta}-\theta \doteq -v(Y,\theta)/\nabla_\theta v(Y,\theta). \qquad (9.15)$$

Now suppose that, asymptotically in n, $v(Y,\theta)/\sqrt{n}$ is asymptotically $N\{0,\nu_{1,1}(\theta)\}$ and $\nabla_\theta v(Y,\theta)/n$ converges in probability to $\nu_{11}(\theta)$ for some functions $\nu_{1,1},\nu_{11}$ (not in general connected with likelihood derivatives). Then $\tilde{\theta}$ is asymptotically normal with mean θ and variance

$$n^{-1}\nu_{1,1}(\theta)/\nu_{11}^2(\theta) \qquad (9.16)$$

or, in the vector parameter case, with covariance matrix

$$n^{-1}\nu_{11}^{-1}\nu_{1,1}\nu_{11}^{-1}. \qquad (9.17)$$

Note that in the particular case where $Y_1,\ldots,Y_n$ are independently and identically distributed with density $f_1(y;\theta)$ and v is the total score function

$$v(y,\theta)=\sum\partial\log f_1(y_j;\theta)/\partial\theta,$$

then

$$\nu_{1,1} = \text{var}\left\{\frac{\partial \log f_1(y_j, \theta)}{\partial \theta}\right\}, \quad \nu_{11} = E\left\{\frac{\partial^2 \log f_1(Y_j, \theta)}{\partial \theta^2}\right\},$$

$\nu_{1,1} = -\nu_{11}$ is the expected information per observation and the usual formulae for maximum likelihood are recovered.

Result (9.16) motivates

$$\rho_v(\theta) = [E\{\partial v(Y,\theta)/\partial\theta\}]^2/\text{var}\{v(Y,\theta)\} \tag{9.18}$$

as a measure of the sensitivity of any unbiased estimating equation $v(Y,\theta) = 0$, where Y is now the full vector of random variables. It can be shown that of all unbiased estimating equations $\rho_v(\theta)$ is maximized by the score function, this being an 'exact' rather than asymptotic result.

If $v^{(1)}, \ldots, v^{(k)}$ are unbiased estimating equations for the same parameter θ arising from independent sets of data, the optimal combined unbiased estimating equation $\sum a_j v^{(j)} = 0$, where the a_k are chosen to maximize the sensitivity measure $\rho_v(\theta)$ above, has $a_j = \nu_{11}^{(j)}/\nu_{1,1}^{(j)}$ and hence has sensitivity measure

$$\rho_v(\theta) = \sum \rho_{v^{(j)}}(\theta).$$

The above analysis can be extended in various ways, for example to the inclusion of higher-order terms and to the study of quasi-likelihood. For the former we carry the expansion (9.15) to one higher term thus obtaining instead of (9.15) the expression

$$\tilde{\theta} - \theta \doteq -v(Y,\theta)/\nabla_\theta v(Y,\theta) - \tfrac{1}{2}v^2(Y,\theta)\nabla_\theta^2 v(Y,\theta)/\{\nabla_\theta v(Y,\theta)\}^3. \tag{9.19}$$

To study this we write in the usual way

$$v(Y,\theta) = \sqrt{n}Z_0, \quad \nabla_\theta v(Y,\theta) = n\nu_{11}(\theta) + \sqrt{n}Z_1,$$
$$\nabla_\theta^2 v(Y,\theta) = n\nu_{11}(\theta) + \sqrt{n}Z_2,$$

where the Zs have zero mean and bounded standard deviations as $n \to \infty$. In addition to the previously defined moments, we write $\nu_{1,11}(\theta) = \text{cov}(Z_0, Z_1)$. Then (9.19) can be written as

$$\tilde{\theta} - \theta \doteq -\frac{Z_0}{\sqrt{n\nu_{11}}}\left(1 + \frac{Z_1}{\sqrt{n\nu_{11}}}\right) - \frac{1}{2}\frac{Z_0^2 \nu_{111}}{n\nu_{11}^3}$$

so that, to order n^{-1}, the bias in $\tilde{\theta}$ is

$$-\frac{\nu_{1,11}}{n\nu_{11}^2} - \frac{1}{2}\frac{\nu_{1,1}\nu_{111}}{n\nu_{11}^3}. \tag{9.20}$$

One fruitful use of the theory of estimating equations is in connection with quasi-likelihood estimation. For this, suppose that we have an ambient model M^* involving an unknown parameter θ leading to the maximum likelihood estimating equation

$$\nabla_\theta \ell^* = 0. \tag{9.21}$$

Suppose that under a broader family of models M defined by parameters (θ, τ), in which θ retains its physical interpretation under M^*,

$$E(\nabla_\theta \ell^*; \theta, \tau) = 0. \tag{9.22}$$

Then in a reasonable sense the formal maximum likelihood estimate $\hat{\theta}^*$ derived from (9.21) remains an estimate of θ.

Suppose, further, that τ can be defined in such a way that

$$E\{(\nabla_\theta \ell^*)\nabla_\theta^T \ell^*\} = (1+\tau)E(-\nabla_\theta \nabla_\theta^T \ell^*) \tag{9.23}$$

in which $\tau = 0$ corresponds to the original model M^*. Then it follows that $\hat{\theta}^*$ is, under the usual conditions, asymptotically normal with mean θ and covariance matrix $(1+\tau)$ times that obtaining under the ambient model M^*. Thus $(1+\tau)$ is a variance inflation factor.

Finally, note that in order for this conclusion to hold we need not the full model M but only the key first- and second-order requirements (9.22) and (9.23). For asymptotic confidence intervals for θ we need also to estimate τ. When the estimate $\hat{\theta}^*$ is used in the context of the incompletely specified model (9.22) and (9.23) it is called a **quasi likelihood estimate**.

In general there will be some loss of efficiency compared with full maximum likelihood estimation in the model M, although exploration of special cases shows that the loss is often rather small, and in any case is very small locally near the model M^*, $\tau = 0$.

The most familiar example of this argument arises when the model M^* is a normal theory linear model with, in the first place, known error variance σ_0^2. Then M is the same normal theory linear model with inflated error variance; if the error variance is unknown then M^* and M are identical. Now the equation (9.22) requires only that the linear part of the model is correctly specified and a sufficient condition for the requirement (9.23) is that the errors of different observations are uncorrelated and with constant variance. The quasi-likelihood estimates are the ordinary least squares estimates.

Under the incompletely specified model Gauss's theorem gives the fairly weak optimality property of minimum variance among all linear estimates that are unbiased (or equivalently of bounded mean squared error). For a general model M the maximum likelihood estimates will be nonlinear functions of the observations, although if the errors are an unknown multiple of independent and identically distributed random variables with a known continuous distribution the theory of transformation models shows that optimum inference is obtained via the conditional distribution of the least squares estimates.

One important application arises when M^* is based on a scalar parameter exponential family model, say with density

$$a(y)\exp\{y\phi - k(\phi)\}$$

and mean parameter $\eta = k'(\phi)$. We can write the variance of the canonical statistic $k''(\phi)$ as a function of η as $v(\eta)$, say, If now for $Y_1, \ldots, Y_n$ there is a model M^* specifying the mean parameter as a function of explanatory variables and we take as the corresponding partly specified model one in which the variance of the canonical statistic is $(1+\tau)v(\eta)$, then the previous discussion applies. A simple example is where the observed random variables have a Poisson distribution under M^* and a variance $(1+\tau)$ times their mean under the incompletely specified model.

The treatment of quasi-likelihood outlined here centres on the ambient model, whereas that described, for example, by McCullagh and Nelder (1989) and McCullagh (1991) starts directly with the variance–mean relation, $v(\mu)$, noting that this suggests a formal score function for Y_j, namely

$$U_j(\theta) = \nabla_\theta \mu_j(\theta) v^{-1}\{\mu_j(\theta)\}\{y_j - \mu_j(\theta)\}.$$

This has the key properties that

$$E\{U_j(\theta)\} = 0, \quad \text{var}\{U_j(\theta)\} = E\{-\nabla_\theta U_j^{\mathrm{T}}(\theta)\}$$

and leads to an estimating equation for θ obtained by equating the total score to zero.

9.3 Stochastic processes and time series

While the majority of the illustrative examples in the earlier chapters involve independent random variables, the primary results have been phrased directly in terms of the log-likelihood function and its derivatives. For our results to apply we need general regularity conditions and more specifically that cumulants of log-likelihood derivatives are asymptotically proportional to a large notional quantity, usually interpreted as a sample size or a time period of observation. These conditions allow local linearization and appeal to the central limit theorem and the associated asymptotic expansions.

When we turn to problems connected with stochastic processes and time series, that is to formal inferential problems where there is an essential dependence between the random variables observed, there are a number of distinct situations, namely those where

1. a log-likelihood can be found and the conditions for the previous theory shown to hold. These are broadly processes with short-range dependence and a relatively long run of data;
2. a log-likelihood can be found, but the previous theory does not apply, usually because the average observed information does not converge to its expected value, different realizations containing essentially different amounts of information;
3. short series of data are available on a large number of independent individuals, standard asymptotic considerations becoming available in the light of the independence of individuals provided that any nuisance parameters employed to describe differences between individuals are of finite dimensions, i.e. are specified not by a large number of independent nuisance parameters but either by a regression model or by a random effects model;
4. a log-likelihood is not available in useful form or, occasionally, is unsuitable as a base for inference in that it depends critically on unrealistic aspects of the model. In some cases simulation methods may be used to estimate a log-likelihood and in others more empirical methods not based on the likelihood may have to be used.

Problems of the third type usually raise no special issues in the present context and will be dealt with only briefly, and the last type are rather outside the present discussion. We therefore concentrate largely on the first two types of problems. Explicit availability of a log-likelihood function amounts in practice to having either a close connection with a multivariate normal distribution or a fairly explicit Markov structure, so that successive conditional densities in the full likelihood can be used (cf. section 2.2.3). While most stochastic models of interest in applications do have Markovian features, these are often latent rather than directly observed, and integration over the distribution of the unobserved features can cause considerable difficulties. In complex problems it may be unavoidable to use very empirical approaches in which a few properties of the process which can be calculated analytically are equated to their sample analogues for the purpose of estimation, and other properties used informally to assess model adequacy; see *Further results and exercise* 9.9.

Examples of the calculation of log-likelihood functions for simple stochastic models have already been given in Chapter 2. Here we give a few more illustrations, chosen to throw light upon some general issues. For processes with a stationary distribution there are two main forms of log-likelihood, depending on whether the initial state is to be regarded as a given constant or is taken from the stationary distribution of the process. The distinction is of most consequence for problems of the third type above. A further general point is that where stochastic processes can be defined in different but equivalent forms there can be appreciable gains in simplicity by judicious choice of specification; for instance, point-process models can be defined via counting, interval and intensity specifications and the second or third usually yield the most direct form of log-likelihood. In other cases suppression of irrelevant details is helpful; see Example 9.1.

Example 9.1 Birth process

In a pure birth process starting from a single individual at time $t = 0$, there is for each individual alive at time t a chance $\rho\Delta + o(\Delta)$ of splitting into two (or equivalently giving birth to a new individual) in $(t, t + \Delta)$, independently of past history and the evolution of other individuals. Let n_t be the number of individuals at time t. As shown in Example 2.7, the log-likelihood based on observing the process $\{M_t\}$ over $(0, t_0)$ is then

$$\ell(\rho) = n_{t_0} \log \rho - s_{t_0}\rho,$$

where n_{t_0} is the number of births in $(0, t_0)$ and s_{t_0} is the total lifetime of all individuals in $(0, t_0)$, i.e. for the corresponding random variables

$$N_{t_0} = M_{t_0} - 1, \quad S_{t_0} = \int_0^{t_0} M_t \, dt.$$

It follows that, unless the sampling is such as to fix either n_{t_0} or s_{t_0}, we have a (2, 1) exponential family with

$$\ell'(\rho) = n_{t_0}/\rho - s_{t_0}, \quad -\ell''(\rho) = n_{t_0}/\rho^2.$$

Thus the maximum likelihood estimate is $\hat{\rho} = n_{t_0}/s_{t_0}$, the observed information at the maximum likelihood point is $\hat{j} = s_{t_0}^2/n_{t_0} = s_{t_0}/\hat{\rho}$ and the expected information is $i(\rho) = (e^{\rho t_0} - 1)\rho^2$, since it is easily shown that $E(M_t) = e^{\rho t}$.

If the data from which the log-likelihood is calculated are, instead of observations of the process $\{M_t\}$ $(0 \leqslant t \leqslant t_0)$, the detailed tree showing the evolution of individuals, i.e. including information on which individual divided at each division point, there would be an additional term $\log n_{t_0}!$ added to the log-likelihood. Because this does not depend on ρ, inference about ρ is unaltered. Put differently, the process $\{M_t\}$ is sufficient, although not minimal sufficient, for ρ; the detailed information contained in the behaviour of individuals is relevant for testing the adequacy of the model but not for inference about ρ assuming the model to be correct.

A radical difference from the theory of earlier chapters now emerges essentially because, even for very large t, different realizations contain very different amounts of information; $\hat{j}/i(\rho)$ tends to a limiting non-degenerate random variable rather than to the constant 1. In terms of the general description of problems above, this one is of type 2. The limiting structure of the process is described to the first order by the results that as $t_0 \to \infty$

$$M_{t_0} \stackrel{\sim}{\to} e^{\rho t_0} V, \quad (S_{t_0}/\rho)^{1/2}(\hat{\rho} - \rho) \stackrel{\sim}{\to} W,$$

where (V, W) are independent random variables, V with a unit exponential distribution and W with a standard normal distribution (Keiding, 1974). Thus $-\ell''(\rho) \stackrel{\sim}{\to} (e^{\rho t_0}/\rho^2)V$. The limiting distribution of M_{t_0} follows immediately from the exact distribution obtained from the forward equation of the process; a more elaborate argument is needed to show that for large t the process $\{M_t\}$ has the form $\{e^{\rho t} V\}$, i.e. essentially grows deterministically along one of a randomly selected set of trajectories, this accounting for the substantial variation between realizations in the amount of information about ρ. It follows also that $S_t/E(S_t) \stackrel{\sim}{\to} V$.

A direct proof of these distributional results is not straightforward. The simplest approach (Waugh, 1970) is via the representation of the birth process as a transform of a Poisson process, namely that the process, defined at time t to be the number of individuals at time $\rho^{-1}\log(1 + t/V)$, is a Poisson process of unit rate. Here V is the unit exponential random variable as above, but now regarded as helping to define the whole process and not merely the behaviour as $t \to \infty$.

It follows from the above results that $\hat{\rho}$ is a consistent estimate of ρ. If the 'usual' results held we could normalize $\hat{\rho}-\rho$ by the square root of either the observed or the expected information, yielding respectively

$$(S_{t_0}/\hat{\rho})^{1/2}(\hat{\rho}-\rho) \overset{\sim}{\to} W,$$

$$\{E(S_{t_0})/\rho\}^{1/2}(\hat{\rho}-\rho) \overset{\sim}{\to} W/\sqrt{V},$$

so that observed information yields intervals based on the standard normal distribution and expected information intervals based on the Student t distribution with 2 degrees of freedom.

While higher-order developments have been studied by Jensen (1987b), it seems compelling even on general grounds that the random normalization via observed information is to be preferred. This is not because of the relatively great dispersion of the Student t distribution with 2 degrees of freedom, but rather because the taking of expectations over trajectories with very differing amounts of information is in conflict with notions of conditionality. While there is not an exactly ancillary statistic, the ratio of observed to expected information, evaluated at $\hat{\rho}$, is nearly ancillary and is indeed close to the random variable V. Conditional inference based on this approximately ancillary statistic can be shown to be equivalent, to the first order, to use of observed information to normalize $\hat{\rho}-\rho$ (Feigin and Reiser, 1979). To repeat, the key difference from the problems of earlier chapters is the substantial asymptotic dispersion of the ratio of observed to expected information. Confidence intervals based on the likelihood ratio statistic yield answers asymptotically equivalent to the normalization by observed information.

The disadvantages of the use of expected information can be seen in even more striking form if, instead of the birth process, we consider the birth–death process, i.e. we have for each individual a chance $\nu\Delta + o(\Delta)$ of death in $(t, t+\Delta)$. If $\nu \geqslant \rho$, ultimate extinction is certain and once that has occurred no further information is accumulated, so that if one took the limiting operations literally rather than as approximation-generating devices asymptotic considerations do not apply, at least to a single realization. Even if $\nu < \rho$, extinction is possible and indeed has appreciable probability if ν is close to ρ and the initial number of individuals is fairly small. Then the use of expected information, which grows exponentially, would be particularly inappropriate.

If instead of starting with a single individual the process starts with m_0 individuals, there results in effect the superposition of m_0 independent realizations and a standard asymptotic discussion is possible based on $m_0 \to \infty$. □

The features shown in the above example of the birth process are characteristic of the special class of non-ergodic processes for which $\hat{\imath}^{-\frac{1}{2}}\hat{\jmath}\hat{\imath}^{-\frac{1}{2}}$ tends in probability or in distribution to a non-degenerate random variable.

Example 9.2 Alternating Poisson process
Consider a two-state process observed in continuous time and let the times spent in the two states $\{0,1\}$ be independent random variables exponentially distributed with parameters ρ_0, ρ_1. Then, because of the strong connection with the Poisson process, the log-likelihood is

$$n_0 \log \rho_0 + n_1 \log \rho_1 - s_0\rho_0 - s_1\rho_1,$$

where (n_0, n_1) are the numbers of exits from states 0 and 1, respectively, and (s_0, s_1) are the total occupancy times; in general this is a $(4,2)$ curved exponential family. Now the stationary probability that the system is in state 1 is $\rho_0/(\rho_0+\rho_1)$, so that if the state at time zero, x_0, say, can be assumed to have the stationary distribution, we add to the log-likelihood

$$x_0 \log \rho_1 + (1-x_0)\log \rho_0 - \log(\rho_0+\rho_1).$$

For one long realization the addition of this term affects only the higher-order asymptotic theory, changing the maximum likelihood estimates by $O_p(1/t_0)$ for example, compared with their standard errors of $O_p(1/\sqrt{t_0})$, where t_0 is the period of observation. If, however, we observed m independent realizations of the same process, we are, as above, led to two forms of log-likelihood:

$$\ell_c = n_0 \log \rho_0 + n_1 \log \rho_1 - s_0\rho_0 - s_1\rho_1,$$
$$\ell_u = \ell_c + x_0 \log \rho_1 + (m-x_0)\log \rho_0 - m\log(\rho_0+\rho_1),$$

where now in the unconditional log likelihood, ℓ_u, the factor x_0 is the number of realizations starting in state 0.

Note that $-\partial^2\ell_c/\partial\rho_0^2 = n_0/\rho_0^2$ and it can be shown that, taking for simplicity the initial state to have the equilibrium distribution,

(a) $E(S_0) = t_0\rho_1/(\rho_0+\rho_1)$, $E(N_0) = E(N_1) = t_0\rho_0\rho_1/(\rho_0+\rho_1)$;

(b) (N_0, S_0) are asymptotically bivariate normal with variances of order t_0.

Property (b) is essentially a consequence of each entry into, say, state 0 forming a regeneration point and is enough to imply all the usual properties of maximum likelihood estimates.

The improvement in precision in using ℓ_u as compared to ℓ_c can be assessed by computing expected information matrices. These are

$$i_c = t_0(\rho_0 + \rho_1)^{-1}\,\mathrm{diag}(\rho_1/\rho_0, \rho_0/\rho_1) = E(N_0)\,\mathrm{diag}(1/\rho_0^2, 1/\rho_1^2),$$

$$i_u = i_c + m(\rho_0 + \rho_1)^{-2}\begin{pmatrix} \rho_1/\rho_0 & 1 \\ 1 & \rho_0/\rho_1 \end{pmatrix}.$$

It can be shown, as is clear on general grounds, that the contribution of information from the initial state is more marked for the parameter ρ_1/ρ_0 than for the individual rates ρ_0 and ρ_1, and that the contribution of the additional term is quite small as soon as t_0 is such that $E(N_0) > m$, i.e. as soon as the average number of transitions of each type per individual series is more than one.

There are a number of variants of this model. For example, the jth series may have its own pair of parameters (ρ_{j0}, ρ_{j1}), say, with $\rho_{j1} = \psi^2\rho_{j0}$. Write $\lambda_j^2 \sim \rho_{j1}\rho_{j0}$, so that $\rho_{j1} = \lambda_j\psi$, $\rho_{j0} = \lambda_j/\psi$ and, on taking the conditional likelihood for simplicity, we have that

$$\ell_c = \sum(n_{j1} + n_{j0})\log\lambda_j + \sum(n_{j1} - n_{j0})\log\psi - \sum(\lambda_j/\psi)s_{j0} - \sum\lambda_j\psi s_{j1}.$$

It can be shown by direct calculation that ψ and λ_j are orthogonal, so that an adjusted profile likelihood in the relatively simple form (8.31) is available. It is unclear whether the ordinary maximum likelihood estimate of ψ is consistent as $m \to \infty$ with the individual series lengths remaining short. □

Example 9.1, the birth process, illustrates the essential failure of 'standard' theory. In Example 9.2 'standard' theory does apply and the only special features arise from the dual form of likelihood and the possible presence of a large number of nuisance parameters. In Example 9.3 below, the value of the unknown parameter determines whether the system is of the first or of the second type.

Example 9.3 First-order autoregressive process

As in Example 2.3 we have that the log-likelihood for a Gaussian first-order autoregressive process is given by (2.19). This is the version corresponding to the unconditional log-likelihood of Example 9.2 in which the initial value has the stationary distribution of the process. Of course, the stationary distribution exists if and only if $|\rho| < 1$. For simplicity we concentrate here on the conditional likelihood in which y_1 is a constant and take also the special case of zero mean, $\mu = 0$, so that the log-likelihood is

$$-\tfrac{1}{2}(n-1)\log\tau - \frac{\sum(y_j - \rho y_{j-1})^2}{2\tau},$$

where the sum is over $j = 2, \ldots, n$. This forms a $(3, 2)$ exponential family with canonical statistic $(\sum Y_{j-1}Y_j, \sum Y_j^2, Y_n^2)$.

The score vector thus has components

$$U_\rho = \tau^{-1}(\sum Y_{j-1}Y_j - \rho \sum Y_{j-1}^2)$$
$$U_\tau = -\tfrac{1}{2}(n-1)/\tau + \tfrac{1}{2}\tau^{-2}\sum(y_j - \rho y_{j-1})^2$$

and the corresponding observed and expected information matrices are asymptotically

$$\hat{j} = \operatorname{diag}\{\sum y_{j-1}^2/\hat{\tau}, 2n/\hat{\tau}^2\},$$
$$i = \operatorname{diag}[\{n(1-\rho^2) + \rho^{2n})\}/(1-\rho^2)^2, 2n/\tau^2]$$

respectively, where for the expected information we have taken $Y_1 = 0$ for simplicity. Note the orthogonality of ρ and τ and that in fact $\hat{\rho}_\tau$ is independent of τ.

If $|\rho| < 1$, the expected information matrix is asymptotically

$$n \operatorname{diag}\{(1-\rho^2)^{-1}, 2\tau^{-2}\}.$$

It can be shown, for example by expressing $\hat{\rho}, \hat{\tau}$ in terms of the innovations of the process, that the 'standard' limiting results hold, although they are approached only slowly if $|\rho|$ is near one. On the other hand if $|\rho| > 1$, the observed information increases geometrically fast. In fact,

$$\sum Y_{j-1}^2 \sim \frac{\rho^{2n-2}}{(\rho^2-1)^2}\tau Z^2,$$

where Z has a standardized normal distribution. It follows that the ratio $\hat{j}/i(\hat{\rho})$ has a distribution close to chi-squared with 1 degree of freedom and is approximately ancillary. Here again conditioning via the ratio of observed to expected information or norming via the observed information is appropriate and

$$(\sum Y_{j-1}^2/\hat{\tau})^{1/2}(\hat{\rho}-\rho)$$

is asymptotically standard normal.

The special case $\rho = 1$, when the process is a random walk, is intermediate between the two regimes and has received extensive treatment in the econometric literature (Phillips, 1987; Banerjee, Dolador, Galbraith and Hendry, 1993).

Realistic time-series models have various complicating features, such as the inclusion of regression on explanatory variables and the presence of trends as well as serial correlation. The estimation of autocorrelation from a large number of individually short series of unknown means needs care because of the strong negative correlation induced into residuals by taking differences from a sample mean or more generally residuals from a fitted regression; see *Further results and exercises* 9.6. □

9.4 Prediction

Except for brief comments in sections 8.6 and 9.1, we have in the previous discussion taken as the focus of interest the unknown parameter involved in the probability distribution assumed to have generated the data. Such parameters, suitably chosen, are the basis for interpretation of the process underlying the data; in other contexts the parameter is best regarded as determining aspects of a large number of future observations drawn from the same or related random systems. Sometimes, however, interest lies in assessing the values of a small number of as yet unobserved random variables, in particular of a single such random variable, for instance the next observation in a time series. Then some new considerations arise.

It is convenient to use the term 'estimation' when the object of interest is an unknown constant and the term 'prediction' when the interest is in an as yet unobserved random variable. There is, however, an intermediate situation in which the model for the data contains random variables not directly observable and interest lies in the value of one or more of these random variables, as realized but not observed. Such problems may be called of 'empirical Bayes' form; we shall treat them alongside prediction problems. In a fully Bayesian treatment of such issues, all unknowns are regarded as random variables so that the mathematical distinction between estimation and prediction disappears; nevertheless, the qualitative distinction may still be helpful.

From the viewpoint of the present book, we formulate the prediction problems as follows. The data, y, are, as previously, observed values of a random variable, Y, with density $p_Y(y;\theta)$ and the value Z, to be predicted, has conditionally on $Y=y$, the density $q(z\mid y;\theta)$. Some examples to illustrate this formulation are now given.

Example 9.4 Prediction for a random sample
Suppose that Y is formed from independent and identically distributed random variables $Y_1,\ldots,Y_n$ and that Z is an independent further observation from the same distribution. That is, Z is independent of Y. In the normal theory version of this problem, a simple solution via a pivotal argument has been given in section 9.1. □

Example 9.5 Time-series prediction
The simplest form of time-series prediction problems refers to the Gaussian first-order autoregression, namely

$$Y_{r+1}=\mu+\rho(Y_r-\mu)+\varepsilon_{r+1},$$

where the innovations $\varepsilon_1,\ldots,\varepsilon_{n+1}$ are independently normally distributed with zero mean and variance σ^2. Here $Y=(Y_1,\ldots,Y_n)$ and in the simplest one-step ahead prediction problem Z given $Y=y$ is normal with mean

$$\mu+\rho(y_n-\mu)$$

and variance σ^2. Fairly direct modifications deal with k-step ahead prediction and with the prediction of the average of several values. Prediction with more complex models is usually best via an autoregressive representation or approximation. □

Example 9.6 Empirical Bayes estimation
Simple problems of parametric empirical Bayes estimation can be quite directly associated with a variety of familiar statistical problems. We mention two. Let $Y_1,\ldots,Y_n$ have a Poisson distribution compounded with a gamma distribution. That is, Y_j has conditionally on μ_j a Poisson distribution of mean μ_j, where μ_j is an unobserved random variable with density $\alpha(\alpha x)^{\beta-1}e^{-\alpha x}/\Gamma(\beta)$, denoted by $\Gamma(\alpha,\beta)$. Suppose that $Z=\mu_1$. A direct calculation with Bayes's theorem shows that the conditional distribution of Z given $Y=y$ is $\Gamma(\alpha+1,$

$\beta + y_1$). If (α, β) were known, this would be a complete solution for inference about μ_1; in applications (α, β) would virtually always have to be estimated from the available data.

The simplest normal theory analogue is of the one-way random effects model, as specified in, for example, *Further results and exercises* 1.2. That is, the observed random variables Y_{is} $(i = 1, \ldots, m;$ $s = 1, \ldots, r)$ have the structure

$$Y_{is} = \mu + \xi_i + \eta_{is}$$

with the ξs and ηs independently normally distributed with zero mean and variances σ_ξ^2 and σ_η^2. The system is described by unknown parameters $(\mu, \sigma_\xi^2, \sigma_\eta^2)$. Suppose that, instead of interest being focused on them, the quantity of direct interest is $\mu + \xi_1$, the mean of 'group' 1, or a contrast of means such as $\xi_2 - \xi_1$. To complete the correspondence with the general formulation we have to specify the conditional distributions of the values to be predicted given $Y = y$, where $Y = \{Y_{is}: i = 1, \ldots, m; s = 1, \ldots, r\}$. These are easily seen to be, respectively, normal distributions of means

$$\mu + \frac{\sigma_\xi^2}{\sigma_\xi^2 + \sigma_\eta^2/r}(\bar{y}_{1+} - \mu), \quad \frac{\sigma_\xi^2}{\sigma_\xi^2 + \sigma_\eta^2/r}(\bar{y}_{2+} - \bar{y}_{1+})$$

and variances

$$\frac{\sigma_\xi^2\sigma_\eta^2/r}{\sigma_\xi^2 + \sigma_\eta^2/r}, \quad \frac{2\sigma_\xi^2\sigma_\eta^2/r}{\sigma_\xi^2 + \sigma_\eta^2/r}. \qquad \square$$

There are broadly four theoretical approaches to the prediction problem in addition to the fully Bayesian solution, i.e. one in which the full vector of the parameters has a well-defined prior distribution. These approaches are as follows:

1. the construction of a 'likelihood' for the variate to be predicted, providing an extension of the notion of likelihood for an unknown parameter to that for a random variable;
2. the use of a parallel with the Neyman–Pearson theory of testing hypotheses in which Z is regarded having a distribution with a parameter θ^* and the hypothesis $\theta^* = \theta$ is to be tested from data (Y, Z). All those Z which together with Y are consistent with the null hypothesis at the significance level in question constitutes a prediction region for Z;

3. a solution via a pivotal quantity, a function of (Y, Z) with a fixed and known distribution thus yielding prediction intervals or limits for Z with the relevant frequency interpretation;
4. a somewhat empirical approach in which unknown parameters are replaced by estimates and the resulting prediction limits modified to achieve the required interpretation asymptotically.

We concentrate here on the last approach. A difficulty with the predictive likelihood approach, 1, is that, as a device for generating prediction limits or intervals the purpose and properties of the latter are not quite clear; see, however, Butler (1986). Method 2, the exploitation of the link with testing hypothesis, is a valuable device for solving special cases but is somewhat outside the scope of this book. The use of pivotal quantities has already been illustrated in section 9.1; it is restricted largely to problems in which a simple transformational structure allows the construction of a pivot.

We concentrate on the calculation of an upper α prediction limit for z, i.e. a value $v_\alpha(y)$ such that, for all θ,

$$P\{Z < v_\alpha(Y); \theta\} = 1 - \alpha, \tag{9.24}$$

where the distribution is over (Y, Z). If there is an ancillary statistic for the estimation of θ from Y, the above probability can be taken conditionally.

Suppose that $z_\alpha(\theta; y)$ is the upper α point of the distribution of Z given $Y = y$, i.e.

$$G\{z_\alpha(\theta; y) | y\} = 1 - \alpha,$$

where $G(z|y)$ is the distribution function of Z given $Y = y$. In the important case where Z and Y are independent the dependence of y can be suppressed and the notation $z_\alpha(\theta)$ used. A direct approach is to estimate θ by maximum likelihood and to replace $z_\alpha(\theta; y)$ by $z_\alpha(\hat{\theta}; y)$ and to take this as a first-order approximation to the required prediction limit $v_\alpha(y)$. We consider asymptotic theory in which the dimension of Y increases while Z remains one-dimensional, i.e. we suppose that errors of estimation of θ have a small effect in predicting Z as compared with the variations in Z for fixed θ. Of course, there are other possibilities which may occasionally be relevant.

As an example, in predicting a single observation from a normal distribution of unknown mean μ and known variance σ_0^2 we take as the first-order prediction limit $\bar{y} + k_\alpha^* \sigma_0$, where $\Phi(-k_\alpha^*) = \alpha$, the value ignoring the error of estimation of μ.

To examine more closely the properties of this limit, we take first the case where Y and Z are independent. Then

$$P\{Z < z_\alpha(\hat{\theta}); \theta\} = E[G\{z_\alpha(\hat{\theta}); \theta\}; \theta] = E\{H_\alpha(\hat{\theta}; \theta)\}, \quad (9.25)$$

say, where the expectation is over the distribution of Y. Thus if, to order n^{-1}, $E(\hat{\theta}) = \theta + b(\theta)/n$, and $\text{var}(\hat{\theta}) = \bar{\imath}^{-1}(\theta)/n$, where $b(\theta)$ is a $d_\theta \times 1$ vector and $\bar{\imath}(\theta)$ is the expected information matrix per observation, then, again to order n^{-1},

$$\begin{aligned} P\{Z < z_\alpha(\hat{\theta}); \theta\} &= H_\alpha(\theta; \theta) + n^{-1}[b^{\mathrm{T}}(\theta)\nabla_{\hat{\theta}} H_\alpha(\hat{\theta}; \theta) \\ &\quad + \tfrac{1}{2}\text{tr}\{\bar{\imath}^{-1}(\theta)\nabla_{\hat{\theta}}\nabla_{\hat{\theta}}^{\mathrm{T}} H_\alpha(\hat{\theta}; \theta)\}]_{\hat{\theta}=\theta} \\ &= 1 - \alpha + c_\alpha(\theta)/n, \end{aligned} \quad (9.26)$$

say.

Note that although $z_\alpha(\hat{\theta})$ differs from $Z_\alpha(\theta)$ by $O_{\mathrm{p}}(n^{-1/2})$, the coverage probability attached to $z_\alpha(\hat{\theta})$ differs from α only by $O(n^{-1})$.

There are now various ways of modifying $z_\alpha(\hat{\theta})$ to achieve a coverage probability $\alpha + O(n^{-3/2})$. The most direct is to write

$$\alpha_1 = \alpha + c_\alpha(\hat{\theta})/n \quad (9.27)$$

and then to take $z_{\alpha_1}(\hat{\theta})$ as the prediction limit. Another possibility is to amend $z_\alpha(\hat{\theta})$ to $z_\alpha(\hat{\theta}) + d_\alpha(\hat{\theta})/n$. Then, to order n^{-1},

$$P\{Z < z_\alpha(\hat{\theta}) + d_\alpha(\hat{\theta})/n\} = 1 - \alpha + n^{-1}[c_\alpha(\theta) + d_\alpha(\theta)g\{z_\alpha(\theta); \theta\}],$$

where $g(z; \theta)$ is the density of Z. This suggests replacing $z_\alpha(\hat{\theta})$ by

$$v_{\alpha_2}(y) = z_\alpha(\hat{\theta}) - [c_\alpha(\hat{\theta})/g\{z_\alpha(\hat{\theta}); \hat{\theta}\}]^{n-1}. \quad (9.28)$$

The application of (9.27) and (9.28) lacks exact invariance in two senses. One is under the replacement of Z by a monotonic nonlinear function of Z, and the other is under nonlinear transformation of θ. Near-linearity of H as a function of $\hat{\theta}$ is desirable but may be difficult to assess. Study of particular examples suggests that often the correction terms in (9.27) and (9.28) are of negligible practical importance.

Example 9.7 Prediction for a random normal sample

As the simplest illustration suppose that $Y_1, \ldots, Y_n$ are independently normally distributed with unknown mean μ and known variance σ_0^2 and that Z is an independent further observation from the same distribution. Then, as noted above $\hat{\mu} = \bar{Y}, z_\alpha(\mu) = \mu + k_\alpha^* \sigma_0$, and we have

$$G(z;\mu)=\Phi\{(z-\mu)/\sigma_0\},$$
$$H_\alpha(\hat{\mu};\mu)=G\{z_\alpha(\hat{\mu});\mu\}=\Phi\{(\bar{y}-\mu)/\sigma_0+k_\alpha^*\},$$
$$b(\hat{\mu})=0, \bar{\imath}^{-1}(\hat{\mu})=\sigma_0^2.$$

Thus

$$c_\alpha(\hat{\mu})=-\tfrac{1}{2}k_\alpha^*\phi(k_\alpha^*)$$

so that (9.28) and (9.27) give, respectively,

$$\alpha_1=\alpha-\tfrac{1}{2}k_\alpha^*\phi(k_\alpha^*)/n,$$
$$v_{\alpha_2}=\bar{y}+k_\alpha^*\sigma_0+\tfrac{1}{2}k_\alpha^*\sigma_0/n,$$

and hence the latter differs from the exact solution only in replacing $(1+n^{-1})^{1/2}$ by $1+\frac{1}{2}n^{-1}$. □

In principle, in a complicated example, the expected information can be replaced by the observed information and the partial derivatives of H_α can be computed numerically, provided that H_α can be computed for a suitable grid of values of $\hat{\theta}$, fixing the second argument θ at the observed $\hat{\theta}$.

To extend the discussion to time-series prediction and to empirical Bayes problems, we have to allow Y and Z to be dependent. Thus we have to evaluate

$$E[G\{z_\alpha(\hat{\theta};Y)\};\theta]$$

see (9.25). If the dependence on $\hat{\theta}$ and Y is sufficiently simple it may be possible to evaluate the expectations directly. For many applications, however, the dependence of $z_\alpha(\theta;Y)$ on Y is via only a low-dimensional random variable W; indeed, W is often one-dimensional. Then conditionally on $W=w$, the information matrix $\bar{\imath}(\theta)$ is to the first-order unchanged, the bias $b(\theta)$ is a function of w, $b(\theta,w)$, say, and (9.26) generalizes to

$$P\{Z<z_\alpha(\hat{\theta},W);\theta\,|\,W=w\}=1-\alpha+c_\alpha(\theta,w)/n,$$

where

$$c_\alpha(\theta,w)=[b^{\mathrm{T}}(\theta,w)\nabla_{\hat{\theta}}H_\alpha(\hat{\theta},w;\theta)+\tfrac{1}{2}N\{\bar{\imath}^{-1}(\theta)\nabla_{\hat{\theta}}\nabla_{\hat{\theta}}^{\mathrm{T}}H_\alpha(\hat{\theta},w;\theta)\}]_{\hat{\theta}=\theta}. \tag{9.29}$$

This enables a conditional correction to be made; in some cases it may be feasible to take expectations of (9.29) over the distribution of W.

Example 9.8 Normal theory empirical Bayes problem
In the components of variance problem of Example 9.6, we consider the calculation of prediction limits for the mean of the first 'group', $\mu + \xi_1$. If all parameters are known the prediction limit is

$$\mu + \frac{\sigma_\xi^2}{\sigma_\xi^2 + \sigma_\eta^2/r}(\bar{Y}_{1+} - \mu) + k_\alpha^* \frac{\sigma_\xi \sigma_\eta/\sqrt{r}}{(\sigma_\xi^2 + \sigma_\eta^2/r)^{1/2}}. \tag{9.30}$$

Suppose, for simplicity of illustration, that only μ is unknown. We replace it in (9.30) by $\bar{Y}_{++}$ for which

$$\bar{\imath}^{-1} = (\sigma_\xi^2 + \sigma_\eta^2/r)^{-1},$$

where we take the average information per 'group' and the asymptotic argument is as $m \to \infty$. Here $W = \bar{Y}_{1+}$ and the bias in $\bar{Y}_{++}$ given $W = w$ is $m^{-1}(\bar{y}_{1+} - \mu)$. Direct calculation shows that

$$c_\alpha(\hat{\mu}, w) = m^{-1}\left\{(\bar{y}_{1+} - \hat{\mu})\frac{\sigma_\eta^2/r}{(\sigma_\xi^2 + \sigma_\eta^2/r)^{3/2}} - \frac{1}{2}\frac{\sigma_\eta^4/r^2}{(\sigma_\xi^2 + \sigma_\eta^2/r)^2} k_\alpha^* \phi(k_\alpha^*)\right\} \tag{9.31}$$

and in this case we can take expectations with respect to $W = \bar{Y}_{1+}$, noting this to be a non-asymptotic calculation, to obtain $c_\alpha(\hat{\mu})$ given by the second term of (9.31). □

9.5 Asymptotic randomization theory

The main emphasis of this book has been on formal methods of parametric inference under an assumed parametric model. A different approach, applicable to problems of sampling and experimental design, rests on the randomization used at the design stage of the study, that is to say, on representations in which the only stochastic element in the data arises from that randomization. There are at least two stages to such a theoretical development.

1. the establishment of exact results for the randomization means, variances (and sometimes higher moments) of estimates, usually linear functions of the data, and for the expectations of quadratic functions;
2. the development of approximate results for the distribution of certain statistics, often based on limiting normality.

Asymptotic considerations enter almost inevitably into stage 2.

While 'exact' randomization distributions can in principle be found by computer enumeration, their precise form will nearly always be very complicated. Asymptotic considerations do not commonly enter into stage 1, although there are situations, exemplified below, where asymptotic arguments enter even there.

In the very simplest version of the problem there is a finite population of N individuals, and the value for the jth individual of the variable of interest is the unknown constant c_j. Suppose that a sample of n individuals is drawn randomly without replacement from the population giving, in order of selection, the values $Y_1, \ldots, Y_n$. Then, for example, if $I_1, I_2, \ldots$ denote the labels of the first, second, ... individuals chosen, then in the light of the randomization

$$\begin{aligned} &P_{\mathrm{R}}(I_1 = j, Y_1 = c_j) = N^{-1} \quad (j = 1, \ldots, N), \\ &P_{\mathrm{R}}(I_2 = k, Y_2 = c_k | I_1 = j) = (N-1)^{-1} \quad (k \neq j = 1, \ldots, N), \end{aligned} \tag{9.32}$$

etc., where P_{R} denotes the probability induced by the randomization. Note that the randomness arises, in this formulation, solely from the randomized sampling procedure, not from any stochastic mechanism that may have generated $\{c_j\}$.

To estimate the finite population mean $\bar{Y} = \sum Y_r / N$, we may consider

$$k_1 = \bar{y} = \sum_{j=1}^{n} y_j / n.$$

It can be shown that

$$E_{\mathrm{R}}(k_1) = \bar{Y} = K_1, \quad \mathrm{var}_{\mathrm{R}}(k_1) = K_2(1-f)n^{-1}, \tag{9.33}$$

where $f = n/N$, the sampling fraction, and

$$K_2 = \sum (Y_r - \bar{Y})^2 / (N-1). \tag{9.34}$$

The quantity K_2 plays in many ways the role of the variance of the finite population. Further, if $k_2 = \sum (Y_j - \bar{Y})^2 / (n-1)$, then $E_{\mathrm{R}}(k_2) = K_2$. For general discussion it is convenient to define population characteristics via the unbiased estimate of infinite population cumulants that would be appropriate were the populations formed by a random sample from that infinite population; this is the reason for the notation $K_1, K_2, \ldots$.

The above results are of type (a) above, i.e. are mathematically exact under the assumptions stated. An example of the second type of result arises in the study of the distributions of

$$\frac{k_1 - K_1}{\sqrt{\{k_2(1-f)/n\}}} \tag{9.35}$$

as the basis for the derivation of confidence limits for K_1. Asymptotic study of (9.35) involves a notional sequence of problems with $N, n \to \infty$ and with $f = n/N$ fixed, $0 < f < 1$. For first-order theory it is not difficult to show that under weak assumptions about K_4, the finite population fourth cumulant, we have $\text{var}(k_2) \to 0$ so that the denominator of (9.35) converges in probability to the standard deviation of the numerator. For the numerator a special form of central limit theorem for dependent random variables (Renyi, 1970, p. 460) proves asymptotic normality so that (9.35) is asymptotically N(0, 1). Rigorous development of an asymptotic expansion appears not to be possible without further assumptions because of the discreteness of the randomization distribution and the complicated structure of the support of (9.35) and indeed of k_1 itself. Nevertheless an Edgeworth-type adjustment, based on appropriate third and fourth moments, to the standard normal distribution for (9.35) is likely to result in improved approximation.

The arguments sketched above for random sampling without replacement from a finite population can be extended to more complex forms of sampling involving stratified and multi-stage sampling. A conceptually different although mathematically quite closely related extension is to the randomization theory of experimental design, some aspects of which we now sketch.

Suppose that we have n experimental units $U_1, \ldots, U_n$ and t treatments $T_1, \ldots, T_t$, that one treatment chosen by the investigator can be applied to each unit and that if T_r is applied to U_j the resulting response has the form

$$\tilde{Y}_{j,r} = \tau_r + \xi_j,$$

regardless of the allocation of treatments to the other units. This is called the assumption of unit–treatment additivity; contrasts among the τ_r are the focus of interest, the ξ_j being constants characterizing the experimental units. In the randomization formulation the ξ_j are unknown constants, not random variables; probabilistic considerations enter from the randomization used in treatment allocation. Because only one treatment can be applied to each experimental unit, for each value of j only one value of $\tilde{Y}_{j,r}$ can be observed, namely

$$Y_j = \tilde{Y}_{j,T_{(j)}},$$

where $T_{(j)}$ is the (random) treatment applied to U_j.

For many 'standard' designs, i.e. systems of random allocation of treatment to units, there are defined via a 'standard' analysis of variance, mean squares for treatments and residuals, MS_{tr} and MS_{res}, estimated treatment constrasts $\sum \ell_r \hat{\tau}_r$, where $\sum \ell_r = 0$, together with an estimated variance for such a contrast, namely that

$$\text{evar}(\sum \ell_r \hat{\tau}_r)$$

can be calculated as

$$MS_{\text{res}} \sum \ell_r^2 c_{tn},$$

where c_{tn} is a constant, often t/n.

The 'exact' results corresponding to those for the mean of a random sample from a finite population are that

$$E_{\mathbf{R}}(\sum \ell_r \hat{\tau}_r) = \sum \ell_r \tau_r,$$
$$\text{var}_{\mathbf{R}}(\sum \ell_r \hat{\tau}_r) = E_{\mathbf{R}}\{\text{evar}(\sum \ell_r \hat{\tau}_r)\},$$
$$E_{\mathbf{R}}(MS_{\text{tr}}) \geqslant E_{\mathbf{R}}(MS_{\text{res}})$$

with equality if and only if the treatments are equivalent, i.e. if and only if all the τ_r are the same.

The randomized choice of a design can be regarded in one of two essentially equivalent ways. We may take the units as fixed and the treatment allocation to be chosen at random from a specified set of possible designs. Alternatively, there may be a specified treatment allocation for units numbered $1, \ldots, n$. A permutation is then chosen at random from a given set (finite group) of permutations to attach the units $U_1, \ldots, U_n$ to the numbers $1, \ldots, n$, i.e. the units are permuted rather than the treatments. This second formulation has the advantage that it stresses the key role played by the group of permutations of the units that leaves the design procedure, and hence its randomization properties, invariant.

For example, in a randomized block design the group is that of all permutations of individuals within the same block, together with permutations of blocks as a whole. These considerations lead to a simple proof of the 'exact' properties listed above. The statistics considered are linear or quadratic functions of the observations, the randomization properties considered are easily seen to be linear or quadratic functions of the ξs, and their invariance under the group of permutations establishes that left- and right-hand sides are equal except for a constant of proportionality; that the constant is unity is then not difficult to show.

Asymptotic considerations enter either, as in the sampling case, if we want approximations to the 'exact' randomization distribution of pivotal quantities such as

$$(\sum \ell_r \hat{\tau}_r - \sum \ell_r \tau_r)/\{\text{evar}(\sum \ell_r \hat{\tau}_r)\}^{1/2},$$

or because we consider estimates more complicated than simple polynomial functions of the data. The simplest, most familiar instances of the latter are the estimates arising in analysis of covariance. We conclude by illustrating the ideas involved in the simplest case of the comparison of two treatments.

Consider an experiment with $n = 2m$ experimental units comparing two treatments, which it is now convenient to denote by T_-, T_+, and write the corresponding responses in the unit–treatment additivity model for unit U_j in the form $\pm\frac{1}{2}\theta + \xi_j$, that is, take the values of τ for the two treatments to be $\pm\frac{1}{2}\theta$.

The simplest form of randomization is to choose m of the $2m$ units at random to receive T_+, the remainder receiving T_-. Then, in a natural notation,

$$\begin{aligned}\hat{\theta} &= \bar{Y}_+ - \bar{Y}_-,\\ MS_{\text{res}} &= (n-2)^{-1}\{\textstyle\sum_+(Y_j - \bar{Y}_+)^2 + \sum_-(Y_j - \bar{Y}_-)^2\},\\ \text{evar}_{\mathbf{R}}(\hat{\theta}) &= 2MS_{\text{res}}/m.\end{aligned}$$

A key result of the 'exact' randomization theory is that

$$\text{var}_{\mathbf{R}}(\hat{\theta}) = E_{\mathbf{R}}\{\text{evar}_{\mathbf{R}}(\hat{\theta})\}.$$

This is proved, as indicated above, by noting that the design procedure is unchanged by an arbitrary permutation of the units. Both left- and right-hand side of the above equation are thus symmetric functions of degree 2 in the ξs invariant under addition of a constant and are thus multiples of $\sum(\xi_j - \bar{\xi})^2$ and the identity can then be proved, for example by considering special cases. The argument hinges on the fact that averages of quadratic functions are themselves quadratic functions and on the invariance.

A similar argument works if the units are divided into a number of blocks each containing an even number of units and the allocation is randomized subject to each treatment appearing equally often in each block. The grouping into blocks enables some aspects of the assumption of unit–treatment additivity to be tested.

Suppose now that there is a concomitant variable z_j $(j = 1, \ldots, 2m)$ known before the treatment allocation. In general the concomitant variable is a vector although for simplicity of exposition we shall mostly suppose it to be scalar.

We start as before from the assumption of unit–treatment additivity; the availability of z_j enables some aspects of this assumption to be tested. We make no specific assumption about the relation between ξ_j and z_j, although the motivation for the following analysis hinges on there being a roughly linear relation between them. This suggests examining the randomization properties of the analysis of covariance estimates

$$\tilde{\theta} = \bar{Y}_+ - \bar{Y}_- - \hat{\beta}(\bar{z}_+ - \bar{z}_-),$$
$$\text{evar}(\tilde{\theta}) = MS_{\text{res}}\{2/m + (\bar{z}_+ - \bar{z}_-)^2/SSD_{\text{res}}(z)\},$$

where

$$SSD_{\text{res}}(z) = \textstyle\sum_+(z_j - \bar{z}_+)^2 + \sum_-(z_j - \bar{z}_-)^2,$$
$$\hat{\beta} = \{\textstyle\sum_+(Y_j - \bar{Y}_+)z_j + \sum_-(Y_j - \bar{Y}_-)z_j\}/SSD_{\text{res}}(z),$$
$$MS_{\text{res}.z} = (2m-3)^{-1}[\textstyle\sum_+(Y_j - \bar{Y}_+)^2 + \sum_-(Y_j - \bar{Y}_-)^2 - \hat{\beta}^2 SSD_{\text{res}}(z)].$$

Three kinds of randomization analysis can be considered. First, we may wish to explore the general consequences of choosing a design at random, ignoring the values of z_j and then applying the above formulae or some modification thereof. For this we need the unconditional randomization moments of the above expressions. Because these are rational functions of the z_j the previous arguments break down; the randomization average of such functions has no simple structure in general. An asymptotic discussion is the only route to a general result.

The second possibility is that a design has been chosen ignoring the z_j, data obtained, and it then being decided to apply the above adjustments.

The third possibility is that the z_j are used in selecting a design, in which case it is natural to choose a design at random subject to $\bar{z}_+ - \bar{z}_-$ being zero (or small).

In the third case it is clear that we need not the unconditional randomization analysis but a conditional one, given the constraint on $\bar{z}_+ - \bar{z}_-$. This applies also in the second case by an extension of the notion of an ancillary statistic to the observed value of $\bar{z}_+ - \bar{z}_-$. The value of this gives no direct information about the value of θ, but large and small values of $\bar{z}_+ - \bar{z}_-$ indicate respectively relatively low and high precisions of estimation.

In some special cases such as when the z_j are indicator variables for a subdivision into blocks the exact conditional randomization distribution can be reconstructed by the previous arguments; the necessary invariance arguments apply to permutations within blocks. In general, however, when the z_j are continuous, the number of arrangements with a given value of $\bar{z}_+ - \bar{z}_-$ or its multi-dimensional equivalent is likely to be very small.

To analyse this we suppose, without loss of generality, that $\sum z_j = 0$ and write

$$\xi_j = \mu_\xi + \beta z_j + \eta_j,$$

where

$$\beta = \sum \xi_j z_j / \sum z_j^2,$$

$$\sum z_j \eta_j = \sum \eta_j = 0.$$

Then, on noting that $\bar{z}_+ + \bar{z}_- = 0$, we have that

$$\tilde{\theta} - \theta = 2\bar{\eta}_+ / \{1 - \bar{z}_+^2 / s_{02}\},$$

where $s_{02} = \sum z_j^2/(2m)$. Thus, conditionally on $\bar{z}_+$, we have that

$$\operatorname{var}(\tilde{\theta}) = 4\operatorname{var}(\bar{\eta}_+ | \bar{z}_+)/(1 - \bar{z}_+^2/s_{02})^2.$$

We now see the role for an asymptotic argument. We need the conditional variance of $\bar{\eta}_+$ given $\bar{z}_+$. In general, calculation of this is not possible; indeed the conditional distribution may be virtually degenerate. To make progress we assume that the discrete randomization distribution is smoothed into a form to which an Edgeworth expansion may be applied. For simplicity we concentrate on the most important case, namely where by design $\bar{z}_+ = 0$.

Now the joint marginal distribution of $(\bar{\eta}_+, \bar{z}_+)$ is asymptotically bivariate normal with exactly zero correlation and variances

$$s_{20}(2m)^{-1}\{1 + (2m)^{-1}\}, \quad s_{02}(2m)^{-1}\{1 + (2m)^{-1}\},$$

with an error $O(m^{-2})$. Here $s_{20} = \sum \eta_j^2/(2m)$. Because the sample size is half the population size the third-order cumulants are zero and so we need the standardized fourth-order cumulants, the leading terms for which are

$$-R_{40}/m, -R_{31}/m, -R_{22}/m, -R_{13}/m, -R_{04}/m,$$

where R_{st} is the standardized version of the joint (s,t) moment of (η_j, z_j), i.e.

$$\sum \eta_j^s z_j^t/(2m).$$

Thus, in particular,

$$R_{22} = \{\sum \eta_j^2 z_j^2/(2m)\}/\{\sum \eta_j^2/(2m)\}\{\sum z_j^2/(2m)\}.$$

It then follows from the bivariate conditional Edgeworth expansion that given $\bar{z}_+ = 0$ the variance of $\bar{\eta}_+$ can be found and it follows that, to this order, the required conditional randomization variance of $\tilde{\theta}$ is

$$2s_{20}m^{-1}\{1 + R_{22}/(2m) + 1/(2m)\} + O(1/m^2).$$

A similar calculation gives that the conditional randomization expectation of the estimated variance $2MS_{\text{res}.z}/m$ is

$$2s_{20}m^{-1}(1 + m^{-1}).$$

As a check, note that if the conditions of normal theory apply, then $R_{22} \doteq 1$ and the last variance and estimated variance agree. The worst case, where the randomization variance might systematically exceed its estimate, is where R_{22} is large, as would happen when the plot of ξ_j against z_j has substantial outliers at the two extremes of z_j. Otherwise the asymptotic randomization theory lends support to the standard analyses.

Further results and exercises

9.1 Let $Y_{j1}, \ldots, Y_{jn_j}$ be independent and identically distributed exponentially with mean μ_j $(j = 1, 2)$. It the two sample means are $\bar{Y}_1, \bar{Y}_2$ show that $\psi\bar{Y}_1/\bar{Y}_2$ is a pivot for inference about $\psi = \mu_2/\mu_1$ and that the pivotal distribution is the F distribution with $(2n_1, 2n_2)$ degrees of freedom.

[Section 9.1]

9.2 Let $p_1(Y_1, \psi), \ldots, p_k(Y_k, \psi)$ be independent pivots for the same parameter ψ. Show that, provided monotonicity in ψ is preserved, any function of $p_1, \ldots, p_k$ will also be a pivot. Discuss how an optimal combined single pivot might be defined.

[Section 9.1]

9.3 Let (Y_{j1}, Y_{j2}) be independently normally distributed with variance σ^2 and $E(Y_{j1}) = \chi_j$, $E(Y_{j2}) = \psi\chi_j\,(j = 1, \ldots, m)$. Show that

$$(Y_{j2} - \psi Y_{j1})/\{\sigma\sqrt{(1 + \psi^2)}\}$$

has the distribution N(0, 1) for all ψ. Examine the condition for monotonicity in ψ. Derive a new 'combined' pivot by taking an arbitrary linear combination of the separate pivots, examine the sense in which an equally weighted combination is optimal and derive confidence limits for ψ.

[Section 9.1]

9.4 On the basis of observations on $Y_1, \ldots, Y_n$ independent and distributed as $N(\mu, \sigma^2)$, obtain prediction limits for (a) the median (b) the largest value of an independent sample of size $\tilde{n}$ from the same distribution.

[Section 9.4]

9.5 Obtain the log-likelihood for the Gaussian first-order autoregressive process via the multivariate normal form of the joint distribution. Examine the corresponding problem for a first-order moving average process.

[Section 9.3]

9.6 For each of m independent Gaussian first-order autoregressions, a small number r of observations are obtained, the initial state having the equilibrium distribution of the process. Write down the log-likelihood, assuming that the series have different means $\mu_1, \ldots, \mu_m$ but that the innovation variance σ^2 and correlation parameter ρ are the same for all series. Show that ρ is orthogonal to the parameters

$$(\mu_1, \ldots, \mu_m, \log\sigma - (2r)^{-1}\log(1 - \rho^2))$$

and obtain an adjusted profile likelihood for ρ.

In the special case $r = 3$, suggest a simple test for the null hypothesis $\rho = 0$.

[Cruddas, Reid and Cox, 1989; Cox and Solomon, 1988]

9.7 Consider a normal theory linear model with correlated errors, the correlation matrix of the errors being a smooth function of unknown parameters. Show that the marginal likelihood for the correlational parameters is the same as the adjusted profile likelihood.

[Bellhouse, 1990]

9.8 Suppose that a large number m of independent realizations of length t are observed for the immigration–death process of immigration rate α and death rate ν. (Equivalently this is the so-called M/M/1 queueing process of arrival rate α and service rate ν). Show that the stationary distribution of the process is geometric with parameter $\rho = \alpha/\nu$. Write down the log-likelihood functions (a) ℓ_c taking the initial states as given and arbitrary and (b) ℓ_u taking the initial states as having the stationary distribution of the process. Show from the corresponding information matrices that

$$\frac{\operatorname{var}(\hat{\alpha}_u)}{\operatorname{var}(\hat{\alpha}_c)} = \frac{\operatorname{var}(\hat{\nu}_u)}{\operatorname{var}(\hat{\nu}_c)} = \frac{1 + \dfrac{\rho}{(1-\rho)^2\eta}}{1 + \dfrac{2\rho}{(1-\rho)^2\eta}},$$

where $\eta = \alpha t$ is the expected number of arrivals per period of observation, and that

$$\frac{\operatorname{var}(\hat{\rho}_u)}{\operatorname{var}(\hat{\rho}_c)} = \frac{1}{1 + \dfrac{2\rho}{(1-\rho)^2\eta}}.$$

[Cox, 1968; Section 9.3]

9.9 A complex stochastic model is to be fitted to extensive data, it being recognized that not all aspects of the data are satisfactorily represented by the model. The model contains d adjustable parameters and analytical solutions are available for $k > d$ features of the process, unbiased estimates of which can be constructed from the data. Suggest how simulation could be used to obtain an efficient procedure for fitting the model and for assessing adequacy of fit as regards the features under examination.

As a simple example consider the discrete-time shot noise process

$$Y_{j+1} = \rho Y_j + Z_{j+1},$$

where $P(Z_{j+1} = 0) = \theta$ while, conditional on it being positive, the innovation Z_{j+1} is exponentially distributed with parameter γ. Suppose that the exactly geometric form of the decays is unrealistic but that it is not judged sensible to modify the model. Suggest how to fit the model using the mean, median, variance, lag-one autocorrelation and the proportion of observations below a given threshold.

[Section 9.3]

9.10 Suppose that the pure birth process starts with a single individual and is observed over the interval $[0, t]$. Show that the directed likelihood R for the birth intensity ρ has an expansion of the form

$$P(R \leqslant r) = \Phi(r) + (\tfrac{1}{2}\gamma - \tfrac{1}{3})\sqrt{\pi}\, e^{-\frac{1}{2}\gamma}\phi(r) + O(\gamma^3 e^{-\gamma})$$

for $\gamma = \rho t \to \infty$.

[Jensen, 1987b; Section 9.3]

9.11 Extend the result of the previous exercise to a general class of non-ergodic models, including the Ornstein–Uhlenbeck process.

[Jensen, 1987b; Section 9.3]

9.12 Explain in general terms why in (9.26) the error in the coverage probability is only $O(n^{-1})$.

[Cox and Reid, 1987b]

9.13 Develop the theory for predicting a single observation from a normal distribution when both mean and variance are unknown. Compare the answer with that from an exact pivotal Student t distribution.

[Section 9.4]

Bibliographic notes

Pivotal distributions were in effect introduced by R.A. Fisher. See also Barnard (1977a; 1977b), who in later unpublished work gave them a central role in the theory of inference.

A general theory of estimating equations was introduced by Godambe (1960) and by Durbin (1960); see the book edited by Godambe (1991) for a thorough account of the topic. The idea of quasi-likelihood is due to Wedderburn (1974); see McCullagh (1991) for a review. For the connections with martingale theory, see Godambe and Heyde (1987).

For more detailed discussion of non-ergodic processes and methods for inference about such processes, see Basawa and Scott (1983), Sweeting (1989; 1992a), Jensen (1987b) and for branching processes the book of Guttorp (1991). In particular, Sweeting (1986; 1992a) has proposed an interesting concept of asymptotic ancillarity rather different from that used in the present book, namely that of asymptotic constancy of the likelihood based on a in shrinking neighbourhoods of the true parameter point.

For the link between estimating equations and inference in stochastic processes, see Godambe and Heyde (1987).

Predictive likelihood is reviewed and extended by Butler (1986). See also Lauritzen (1974), Hinkley (1979), Barndorff-Nielsen (1981) and Davison (1986). For the treatment of prediction via tests of hypotheses, see Guttman (1970) and Cox and Hinkley (1974). The treatment here follows closely that of Cox (1973). There is an extensive literature on the effect of errors of parameter estimation on time-series forecasting, mostly in terms of the effect on mean squared error. For other approaches, see Aitchison and Dunsmore (1975) and Rao (1977).

Randomization was introduced by R.A. Fisher in his early work on experimental design (Fisher, 1926). A famous passage in Fisher (1935b), on the lady tasting tea, explained randomization as the basis for an 'exact' test of significance. Yates (1951) summarized and extended earlier work on the randomization justification of the analysis of variance for basic experimental designs. Grundy and Healy (1950) explained the group-theoretic basis of restricted randomization (Yates, 1951), a procedure in which certain undesirable patterns are excluded yet the key properties of the randomization distribution are retained. For important recent work, see Bailey (1981), Bailey and Rowley (1987) and Speed (1987). Cox (1982b; 1982c) has considered asymptotic aspects, the second paper being concerned with adaptive sequential randomization schemes designed to avoid imbalance over an unspecified number of trials.

References

Aitchison, J. and Dunsmore, I.R. (1975) *Statistical Prediction Analysis.* Cambridge University Press, Cambridge.

Akahira, M. (1991) Asymptotic theory of statistical estimation (collection of 10 papers). University of Tsukuba: Ibaraki, Japan.

Akahira, M. (1992) Contributions to statistical inference (collection of 19 papers). University of Tsukuba: Ibaraki, Japan.

Amari, S.-I. (1985) *Differential-Geometric Methods in Statistics.* Lecture Notes in Statistics 28. Springer-Verlag, Heidelberg.

Amari, S.-I. (1987a) Differential geometrical theory of statistics – towards new developments. *Differential Geometry in Statistical Inference.* Lecture Notes – Monograph Series 10. Institute of Mathematical Statistics, Hayward, Calif.: pp. 10–94.

Amari, S.-I. (1987b) Dual connections on the Hilbert bundles of statistical models. In *Geometrization of Statistical Theory* (ed. C.T.J. Dodson). UDLM Publications, Department of Mathematics, University of Lancaster, pp. 123–52.

Amari, S.-I. and Kumon, M. (1983) Differential geometry of Edgeworth expansion in curved exponential family. *Ann. Inst. Statist. Math.*, **35**, 1–24.

Andersen, E.B. (1970). Asymptotic properties of conditional maximum likelihood estimators. *J.R. Statist. Soc. B*, **32**, 283–301.

Andersen, E.B. (1971) Asymptotic properties of conditional likelihood ratio tests. *J. Am. Statist. Assoc.*, **66**, 630–633.

Andersen, E.B. (1973) *Conditional Inference and Models for Measuring.* Mentalhygiejnisk Forlag, Copenhagen.

Andersen, P.K. and Borgan, Ø. (1985) Counting process models for life history data: a review. *Scand. J. Statist.*, **12**, 97–158.

Andersen, P.K., Borgan, Ø., Gill, R.D. and Keiding, N. (1993) *Survival Data and Martingales.* Springer-Verlag, Heidelberg.

Anderson, T.W. (1973) Asymptotically efficient estimation of covariance matrices with linear structure. *Ann. Statist.*, **1**, 135–41.

Andrews, D.F. and Stafford, J. (1993) Tools for the symbolic computation of asymptotic expansions. *J. R. Statist. Soc. B*, **55**, 613–27.

Anscombe, F.J. (1961) Examination of residuals. *Proc. 4th Berkeley Symp.* **1**, 1–36.

Asmussen, S. and Edwards, D. (1983) Collapsibility and response variables in contingency tables. *Biometrika*, **70**, 567–78.

Atkinson, A.C. (1970). A method of discriminating between models (with discussion). *J. R. Statist. Soc. B*, **32**, 323–53.

Atkinson, A.C. and Lawrance, A.J. (1989) A comparison of asymptotically equivalent test statistics for regression transformation. *Biometrika*, **76**, 223–9.

Azzalini, A. (1983) Maximum likelihood estimation of order *m* for stationary stochastic processes. *Biometrika*, **70**, 381–7.

Bailey, R.A. (1981) A unified approach to design of experiments. *J. R. Statist. Soc. A*, **144**, 214–23.

Bailey, R.A. and Rowley, C.A. (1987) Valid randomization. *Proc. Roy. Soc. (London) A*, **410**, 105–24.

Banerjee, A., Dolador, J.J., Galbraith, J.K. and Hendry, D.F. (1993) *Co-integration, Error-Correction and the Econometric Analysis of Nonstationary Data.* Oxford University Press, Oxford.

Barnard, G.A. (1963) Some logical aspects of the fiducial argument. *J. R. Statist. Soc. B*, **25**, 111–14.

Barnard, G.A. (1974) Conditionality, pivotals and robust estimation. In *Proc. Conf. Foundational Questions in Statistical Inference. Memoir 1* (eds O.E. Barndorff-Nielsen, P. Blæsild and G. Schou). Dept. Theor. Statistics, Aarhus University, pp. 61–80.

Barnard, G.A. (1976) Conditional inference is not inefficient. *Scand. J. Statist.*, **3**, 132–4.

Barnard, G.A. (1977a) Problems of statistical inference in the life sciences. In *Mathematics and the Life Sciences* (ed. D.E. Matthews). Lecture Notes in Biomathematics 18. Springer-Verlag, Heidelberg, pp. 1–65.

Barnard, G.A. (1977b) Pivotal inference and the Bayesian controversy. *Bull. Int. Statist. Inst.*, **47**, 543–51.

Barnard, G.A., Jenkins, G.M. and Winsten, C.B. (1962) Likelihood inference and time series (with discussion). *J. R. Statist. Soc. A*, **125**, 321–75.

Barnard, G.A. and Sprott, D.A. (1971) A note on Basu's examples of anomalous ancillary statistics (with discussion). In *Foundations of Statistical Inference* (eds V.P. Godambe and D.A. Sprott). Holt, Rinehart and Winston, Toronto, pp. 163–76.

Barndorff-Nielsen, O.E. (1978) *Information and Exponential Families in Statistical Theory*. Wiley, Chichester.

Barndorff-Nielsen, O.E. (1980) Conditionality resolutions. *Biometrika*, **67**, 293–310.

Barndorff-Nielsen, O.E. (1981) Likelihood prediction. *Symposia Mathematica*, **XXV**, 11–24.

Barndorff-Nielsen, O.E. (1983) On a formula for the distribution of the maximum likelihood estimator. *Biometrika*, **70**, 343–65.

Barndorff-Nielsen, O.E. (1984) On conditionality resolution and the likelihood ratio for curved exponential models. *Scand. J. Statist.*, **11**, 157–70. Corrigendum: *Scand. J. Statist.*, **12** (1985), 191.

Barndorff-Nielsen, O.E. (1985) Properties of modified profile likelihood. In *Contributions to Probability and Statistics in Honour of Gunnar Blom* (eds J. Lanke and G. Lindgren). Dept. Math. Statist., Lund University, pp. 25–38.

Barndorff-Nielsen, O.E. (1986a) Inference on full and partial parameters based on the standardized signed log likelihood ratio. *Biometrika*, **73**, 307–22.

Barndorff-Nielsen, O.E. (1986b) Likelihood and observed geometries. *Ann. Statist.*, **14**, 856–73.

Barndorff-Nielsen, O.E. (1986c) Strings, tensorial combinants, and Bartlett adjustments. *Proc. Roy. Soc. (London) A*, **406**, 127–37.

Barndorff-Nielsen, O.E. (1987a) Differential and integral geometry in statistical inference. *Differential Geometry in Statistical Inference.* Lecture Notes – Monograph Series 10. Institute of Mathematical Statistics, Hayward, Calif. pp. 95–162.

Barndorff-Nielsen, O.E. (1987b) Differential geometry and statistics: some mathematical aspects. *Indian J. Math.*, **29**, 335–350. Ramanujan Centenary Volume.

Barndorff-Nielsen, O.E. (1988a) *Parametric Statistical Models and Likelihood.* Lecture Notes in Statistics 50. Springer-Verlag, Heidelberg.

Barndorff-Nielsen, O.E. (1988b) Contribution to discussion of paper by N. Reid (1988). *Statistical Science*, **3**, 228–9.

Barndorff-Nielsen, O.E. (1989) Contribution to the discussion of paper by R.E. Kass (1989). *Statistical Science*, **4**, 222–7.

Barndorff-Nielsen, O.E. (1990a) A note on the standardized signed log likelihood ratio. *Scand. J. Statist.*, **17**, 157–60.

Barndorff-Nielsen, O.E. (1990b) p^* and Laplace's method. *REBRAPE*, **4**, 89–103.

Barndorff-Nielsen, O.E. (1990c) Approximate interval probabilities. *J. R. Statist. Soc. B*, **52**, 485–96.

Barndorff-Nielsen, O.E. (1990d) Likelihood theory. In *Statistical Theory and Modelling* (eds D.V. Hinkley, N. Reid and E.J. Snell). Chapman & Hall, London, pp. 232–64.

Barndorff-Nielsen, O.E. (1991) Modified signed log likelihood ratio. *Biometrika*, **78**, 557–63.

Barndorff-Nielsen, O.E. (1992) Adjusted likelihood inference about interest parameters. *Theory Probab. Its Appl.*, (to appear).

Barndorff-Nielsen, O.E. (1993) Adjusted versions of profile likelihood and directed likelihood, and extended likelihood. *J. R. Statist. Soc. B*, (to appear).

Barndorff-Nielsen, O.E. and Blæsild, P. (1988) Combination of reproductive models. *Ann. Statist.*, **16**, 323–41.

Barndorff-Nielsen, O.E. and Blæsild, P. (1992) A type of second order asymptotic independence. *J. R. Statist. Soc. B*, **54**, 897–901.

Barndorff-Nielsen, O.E. and Blæsild, P. (1993a) Likelihood yokes and stable combinations of cumulants. *Scand. J. Statist.* **20**, 83–8.

Barndorff-Nielsen, O.E. and Blæsild, P. (1993b) Orthogeodesic models. *Ann. Statist.*, **21**, 1018–39.

Barndorff-Nielsen, O.E. Blæsild, P. and Eriksen, P.S. (1989) *Decomposition and Invariance of Measures, with a View to Statistical Transformation Models*. Lecture Notes in Statistics 58. Springer-Verlag, Heidelberg.

Barndorff-Nielsen, O.E., Blæsild, P., Pace, L. and Salvan, A. (1991) Formulas for geometric expansions. Research Report 207. Dept. Theor. Statist., Aarhus University.

Barndorff-Nielsen, O.E. and Chamberlin, S.R. (1991) An ancillary invariant modification of the signed log likelihood ratio. *Scand. J. Statist.*, **18**, 341–52.

Barndorff-Nielsen, O.E. and Chamberlin, S.R. (1994) Stable and invariant adjusted directed likelihood. *Biometrika* (to appear).

Barndorff-Nielsen, O.E. and Cox, D.R. (1979) Edgeworth and saddle-point approximations with statistical applications (with discussion). *J. R. Statist. Soc. B*, **41**, 279–312.

Barndorff-Nielsen, O.E. and Cox, D.R. (1984a) Bartlett adjustments to the likelihood ratio statistic and the distribution of the maximum likelihood estimator. *J. R. Statist. Soc. B*, **46**, 483–95.

Barndorff-Nielsen, O.E. and Cox, D.R. (1984b) The effect of sampling rules on likelihood statistics. *Int. Statist. Rev.*, **52**, 309–326.

Barndorff-Nielsen, O.E. and Cox, D.R. (1989) *Asymptotic Techniques for Use in Statistics*. Chapman & Hall, London.

Barndorff-Nielsen, O.E., Cox, D.R. and Reid, N. (1986) The role of differential geometry in statistical theory. *Int. Statist. Rev.*, **54**, 83–96.

Barndorff-Nielsen, O.E. and Hall, P. (1988) On the level-error after Bartlett adjustment of the likelihood ratio statistic. *Biometrika*, **75**, 374–8.

Barndorff-Nielsen, O.E. and Jørgensen, B. (1991) Some parametric models on the simplex. *J. Mult. Anal.*, **39**, 106–16.

Barndorff-Nielsen, O.E. and Jupp, P.E. (1987) Differential geometry, profile likelihood, *L*-sufficiency and composite transformation models. *Ann. Statist.*, **16**, 1009–43.

Barndorff-Nielsen, O.E., Jupp, P.E. and Kendall, W.S. (1991) Stochastic calculus, statistical asymptotics, Taylor strings and phyla. *Ann. Toulouse* (to appear).

Barndorff-Nielsen, O.E. and McCullagh, P. (1993) A note on the relation between modified profile likelihood and the Cox–Reid adjusted profile likelihood. *Biometrika*, **80**, 321–8.

Barndorff-Nielsen, O.E., Peters, D. and Pierce, D.A. (1993). Modified directed likelihood: decomposition of the correction term (to appear).

Barndorff-Nielsen, O.E. and Sørensen, M. (1993) A review of some aspects of asymptotic likelihood theory for stochastic processes. *Int. Statist. Rev.*, **62** (to appear).

Bartlett, M.S. (1936a) The information available in small samples. *Proc. Camb. Phil. Soc.*, **32**, 560–6.

Bartlett, M.S. (1936b) Statistical information and properties of sufficiency. *Proc. Roy. Soc. (London) A*, **154**, 124–37.

Bartlett, M.S. (1937) Properties of sufficiency and statistical tests. *Proc. Roy. Soc. (London) A*, **160**, 268–82.

Bartlett, M.S. (1953a) Approximate confidence intervals. *Biometrika*, **40**, 12–19.

Bartlett, M.S. (1953b) Approximate confidence intervals II: More than one unknown parameter. *Biometrika*, **40**, 306–17.

Bartlett, M.S. (1955) Approximate confidence intervals III. A bias correction. *Biometrika*, **42**, 201–4.

Basawa, I.V. and Scott, D.J. (1983) *Asymptotic Optimal Inference for Non-ergodic Models*. Lecture Notes in Statistics 17. Springer-Verlag, New York.

Basu, D. (1964) Recovery of ancillary information. *Sankhyā A*, **26**, 3–16.

Basu, D. (1975) Statistical information and likelihood (with discussion). *Sankhyā A*, **37**, 1–71.

Basu, D. (1977) On the elimination of nuisance parameters. *J. Am. Statist. Assoc.*, **72**, 355–66.

Becker, N. and Gordon, I. (1983) On Cox's criterion for discriminating between alternative ancillary statistics. *Int. Statist. Rev.*, **51**, 89–92.

Bellhouse, D.R. (1990) On the equivalence of marginal and approximate conditional likelihoods for correlation parameters under a normal model. *Biometrika*, **77**, 743–6.

Berger, J.O. and Wolpert, R.L. (1984) *The Likelihood Principle*. IMS Lecture Notes – Monograph Series, Institute of Mathematical Statistics, Hayward, California.

Besag, J. (1974) Spatial interaction and the statistical analysis of lattice systems. *J. R. Statist. Soc. B*, **36**, 192–236.

Besag, J. (1977) Efficiency of pseudolikelihood estimation for simple Gaussian fields. *Biometrika*, **64**, 616–18.

Besag, J. (1978) Some methods of statistical analysis for spatial data. *Bull. Int. Statist. Inst.*, **47**, 77–92.

Besag, J. (1986) On the statistical analysis of dirty pictures (with discussion). *J. R. Statist. Soc. B*, **48**, 259–302.

Birnbaum, A. (1962) On the foundations of statistical inference (with discussion). *J. Am. Statist. Assoc.*, **57**, 269–306.

Birnbaum, A. (1969) Concepts of statistical evidence. In *Philosophy, Science and Method* (eds S. Morgenbesser, P. Suppes and M. White). St Martin's Press, New York, pp. 112–43.

Birnbaum, A. (1970) On Durbin's modified principle of conditionality. *J. Am. Statist. Assoc.*, **65**, 402–3.

Boothby, W.M. (1975) *An Introduction to Differentiable Manifolds and Riemannian Geometry*. Academic Press, New York.

Box, G.E.P. (1949) A general distribution theory for a class of likelihood criteria. *Biometrika*, **36**, 317–46.

Box, G.E.P. and Cox, D.R. (1964) An analysis of transformations (with discussion). *J. R. Statist. Soc. B*, **26**, 211–52.

Brown, L.D. (1988) *Fundamentals of Statistical Exponential Families*. IMS Lecture Notes – Monograph Series 9. Institute of Mathematical Statistics, Hayward, Calif.

Brown, P.J. and Sundberg, R. (1987) Confidence and conflict in multivariate calibration. *J. R. Statist. Soc. B*, **49**, 46–57.

Buehler, R.J. (1982) Some ancillary statistics and their properties (with discussion). *J. Am. Statist. Assoc.*, **77**, 581–94.

Butler, R.W. (1986) Predictive likelihood inference with applications (with discussion). *J. R. Statist. Soc. B*, **48**, 1–38.

Butler, R.W. (1989) Approximate predictive pivots and densities. *Biometrika*, **76**, 489–501.

Chandra, T.K. and Mukherjee, R. (1985) Comparison of the likelihood ratio, Wald's and Rao's tests. *Sankhyā A*, **47**, 271–84.

Cheng, R.C.H. and Iles, T.C. (1987) Corrected maximum likelihood in non-regular problems. *J. R. Statist. Soc. B*, **49**, 95–101.

Cobb, C.W. (1978) The problem of the Nile: conditional solution to a change-point problem. *Biometrika*, **78**, 113–21.

Cordeiro, G.M. (1983) Improved likelihood-ratio tests for generalized linear models. *J. R. Statist. Soc. B*, **45**, 404–13.

Cordeiro, G.M. (1987) On the corrections to the likelihood ratio statistics. *Biometrika*, **74**, 265–74.

Cormack, R.M. (1992) Interval estimation for mark-recapture studies of closed populations. *Biometrika*, **48**, 567–76.

Cox, D.R. (1958) Some problems connected with statistical inference. *Ann. Math. Statist.*, **29**, 357–72.

Cox, D.R. (1961) Tests of separate families of hypotheses. *Proc. 4th Berkeley Symp.*, **1**, 105–23.

Cox, D.R. (1968) Some problems of statistical analysis connected with congestion. In *The Analysis of Congestion.* W.L. Smith and W.E. Wilkinson (Eds.). University of North Carolina Press, Chapel Hill, pp. 289–316.

Cox, D.R. (1971) The choice between alternative ancillary statistics. *J. R. Statist. Soc. B*, **33**, 251–5.

Cox, D.R. (1972) Regression models and life tables (with discussion). *J. R. Statist. Soc. B*, **34**, 187–220.

Cox, D.R. (1973) Prediction intervals and empirical Bayes confidence intervals. In *Perspectives in Probability and Statistics* (ed. J. Gani). Academic Press, London, pp. 47–55.

Cox, D.R. (1975) Partial likelihood. *Biometrika*, **62**, 269–76.

Cox, D.R. (1980) Local ancillarity. *Biometrika*, **67**, 273–8.

Cox, D.R. (1982a) Combination of data. In *Encyclopedia of Statistical Science 2* (eds S. Kotz and N.L. Johnson). New York, Wiley.

Cox, D.R. (1982b) Randomization and concomitant variables in the design of experiments. In *Statistics and Probability: Essays in Honor of C.R. Rao* (eds G. Kallianpur, P.R. Krishnaiah and J.K. Ghosh). North-Holland, Amsterdam, pp. 197–202.

Cox, D.R. (1982c) A remark on randomization in clinical trials. *Utilitas Mathematica*, **21A**, 245–52.

Cox, D.R. (1988) Some aspects of conditional and asymptotic inference: a review. *Sankhyā A*, **50**, 314–37.

Cox, D.R. (1990) Role of models in statistical analysis. *Statistical Science*, **5**, 169–74.

Cox, D.R. and Hinkley, D.V. (1974) *Theoretical Statistics*. Chapman & Hall, London.

Cox, D.R. and Isham, V. (1980) *Point Processes*. Chapman & Hall, London.

Cox, D.R. and Oakes, D. (1984) *Analysis of Survival Data*. Chapman & Hall, London.

Cox, D.R. and Reid, N. (1987a) Parameter orthogonality and approximate conditional inference (with discussion). *J. R. Statist. Soc. B*, **49**, 1–39.

Cox, D.R. and Reid, N. (1987b) Approximation to noncentral distributions. *Can. J. Statist.*, **15**, 105–14.

Cox, D.R. and Reid, N. (1992) A note on the difference between profile and modified profile likelihood. *Biometrika*, **79**, 408–11.

Cox, D.R. and Solomon, P.J. (1988) On testing for serial correlation in large numbers of small samples. *Biometrika*, **75**, 145–8.

Cox, D.R. and Wermuth, N. (1990) An approximation to maximum likelihood estimates in reduced models. *Biometrika*, **77**, 747–61.

Cramér, H. and Leadbetter, M.R. (1967) *Stationary and Related Stochastic Processes*. Wiley, New York.

Cressie, N. and Lahiri, S.N. (1993) The asymptotic distribution of REML estimators. *J. Mult. Anal.*, **45**, 217–33.

Cruddas, A.M., Reid, N. and Cox, D.R. (1989) A time series illustration of approximate conditional likelihood. *Biometrika*, **76**, 231–7.

Daniels, H.E. (1954) Saddlepoint approximations in statistics. *Ann. Math. Statist.*, **25**, 631–50.

Daniels, H.E. (1956) The approximate distribution of serial correlation coefficients. *Biometrika*, **43**, 169–85.

Daniels, H.E. (1958) Contribution to the discussion of paper by D.R. Cox. *J. R. Statist. Soc. B.*, **20**, 236–238.

Daniels, H.E. (1980) Exact saddlepoint approximations. *Biometrika*, **67**, 53–8.

Daniels, H.E. (1983) Saddlepoint approximations for estimating equations. *Biometrika*, **70**, 89–96.

Daniels, H.E. (1987) Tail probability approximations. *Internat. Statist. Rev.*, **55**, 37–48.

Davies, R.B. (1977) Hypothesis testing when a nuisance parameter is present only under the alternative. *Biometrika*, **64**, 247–54.

Davies, R.B. (1987) Hypothesis testing when a nuisance parameter is present only under the alternative. *Biometrika*, **74**, 33–43.

Davison, A.C. (1986) Approximate predictive likelihood. *Biometrika*, **73**, 323–32.

Davison, A.C. and Smith, R.L. (1990) Models for exceedances over high thresholds (with discussion). *J. R. Statist. Soc. B*, **52**, 393–442.

Dempster, A. (1972) Covariance selection. *Biometrics*, **28**, 157–75.

DiCiccio, T.J., Field, C.A. and Fraser, D.A.S. (1990) Approximations of marginal tail probabilities and inference for scalar parameters. *Biometrika*, **77**, 77–96.

DiCiccio, T.J. and Martin, M.A. (1993) Simple modifications for signed roots of likelihood ratio statistics. *J. R. Statist. Soc. B*, **55**, 305–16.

Durbin, J. (1960) Estimation of parameters in time series regression models. *J. R. Statist. Soc. B*, **22**, 139–53.

Durbin, J. (1970) On Birnbaum's theorem on the relation between sufficiency, conditionality and likelihood. *J. Am. Statist. Assoc.*, **65**, 395–8.

Durbin, J. (1980) Approximations for densities of sufficient estimators. *Biometrika*, **67**, 311–33.

Edwards, A.W.F. (1972) *Likelihood*. Cambridge University Press, Cambridge.

Efron, B. (1975) Defining the curvature of a statistical problem (with application to second order efficiency) (with discussion). *Ann. Statist.*, **3**, 1189–1242.

Efron, B. (1978) The geometry of exponential families. *Ann. Statist.*, **6**, 362–76.

Efron, B. and Hinkley, D.V. (1978) Assessing the accuracy of the maximum likelihood estimator: observed versus expected Fisher information (with discussion). *Biometrika*, **65**, 457–487.

Eriksen, P.S. (1987) Proportionality of covariance matrices. *Ann. Statist.*, **15**, 732–48.

Feigin, P.D. and Reiser, B. (1979) On asymptotic ancillarity and inference for Yule and regular nonergodic processes. *Biometrika*, **66**, 279–83.

Ferguson, H., Reid, N. and Cox, D.R. (1991) Estimating equations from modified profile likelihood. In *Estimating Functions* (ed. V.P. Godambe). Clarendon Press, Oxford, pp. 279–93.

Field, C. (1982) Small sample asymptotic expansions for multivariate M-estimates. *Ann. Statist.*, **10**, 672–89.

Field, C. and Ronchetti, E. (1990) *Small Sample Asymptotics*. Lecture Notes – Monograph Series, 13. Institute of Mathematical Statistics, Hayward, California.

Fisher, R.A. (1922) On the mathematical foundations of theoretical statistics. *Phil. Trans. Roy. Soc. A*, **222**, 309–68.

Fisher, R.A. (1925) Theory of statistical estimation. *Proc. Camb. Phil. Soc.*, **22**, 700–25.

Fisher, R.A. (1926) The arrangement of field experiments. *J. Ministry of Agric.*, **33**, 503–13.

Fisher, R.A. (1934) Two new properties of mathematical likelihood. *Proc. Roy. Soc. (London) A*, **144**, 285–307.

Fisher, R.A. (1935a) The logic of inductive inference. *J. R. Statist. Soc.*, **98**, 39–54.

Fisher, R.A. (1935b) *Design of Experiments*. Oliver and Boyd, Edinburgh.

Fisher, R.A. (1950) The significance of deviations from expectations in a Poisson series. *Biometrics*, **6**, 17–24.

Fisher, R.A. (1956) *Statistical Methods and Scientific Inference*. Oliver and Boyd, Edinburgh. (Third edn 1973, Collier Macmillan, London.)

Fisher, R.A. and Yates, F. (1963) *Statistical Tables for Biological, Agricultural and Medical Research*, 6th edn. Oliver and Boyd, Edinburgh.

Fraser, D.A.S. (1956) Sufficient statistics with nuisance parameters. *Ann. Math. Statist.*, **27**, 838–42.

Fraser, D.A.S. (1968) *The Structure of Inference*. Wiley, New York.

Fraser, D.A.S. (1979) *Inference and Linear Models*. McGraw-Hill, New York.

Fraser, D.A.S. (1988) Normed likelihood as saddlepoint approximation. *J. Mult. Anal.*, **27**, 181–93.

Fraser, D.A.S. (1990) Tail probabilities from observed likelihoods. *Biometrika*, **77**, 65–76.

Fraser, D.A.S. (1991) Statistical inference: likelihood to significance. *J. Am. Statist. Assoc.*, **86**, 258–65.

Fraser, D.A.S. and Reid, N. (1988) On conditional inference for a real parameter: A differential approach on the sample space. *Biometrika*, **75**, 251–64.

Frydenberg, M. (1990) Marginalization and collapsibility in graphical interaction models. *Ann. Statist.*, **18**, 790–805.

Frydenberg, M. and Lauritzen, S.L. (1989) Decomposition of maximum likelihood in mixed graphical interaction models. *Biometrika*, **76**, 539–55.

Gelfand, A.E. and Dalal, S.R. (1990) A note on overdispersed exponential families. *Biometrika*, **77**, 55–64.

Ghosh, J.K. (ed.) (1988) *Statistical Information and Likelihood. A Collection of Critical Essays by Dr. D. Basu.* Lecture Notes in Statistics 45. Springer-Verlag, Heidelberg.

Gill, R.D. and Johansen, S. (1990) Product integrals and counting processes. *Ann. Statist.*, **18**, 1501–55.

Godambe, V.P. (1960) An optimum property of regular maximum likelihood estimation. *Ann. Math. Statist.*, **31**, 1208–12.

Godambe, V.P. (ed.) (1991) *Estimating Functions.* Oxford University Press, Oxford.

Godambe, V.P. and Heyde, C.C. (1987) Quasi-likelihood and optimal estimating equations. *Int. Statist. Rev.*, **55**, 231–44.

Gong, G. and Samaniego, F.J. (1981) Pseudo maximum likelihood estimation: theory and applications. *Ann. Statist.*, **9**, 861–9.

Gordon, L. and Hudson, M. (1977) A characterization of the von Mises distribution. *Ann. Statist.*, **5**, 813–4.

Grambsch, P. (1983) Sequential sampling based on the observed Fisher information to guarantee the accuracy of the maximum likelihood estimator. *Ann. Statist.*, **11**, 68–77.

Green, P.J. (1987) Penalized likelihood for general semiparametric regression models. *Int. Statist. Rev.*, **55**, 245–60.

Grundy, P.M. and Healy, M.J.R. (1950) Restricted randomization and quasi-Latin squares. *J.R. Statist. Soc.*, **12**, 286–91.

Guttman, I. (1970) *Statistical Tolerance Regions.* Griffin, High Wycombe.

Guttorp, P. (1991) *Statistical Inference for Branching Processes.* Wiley, New York.

Haberman, S.J. (1977) Maximum likelihood estimation in exponential response models. *Ann. Statist.*, **5**, 815–41.

Halmos, P.R. and Savage, L.J. (1949) Application of the Radon–Nikodym theorem to the theory of sufficient statistics. *Ann. Math. Statist.*, **20**, 225–41.

Hayakawa, T. (1977) The likelihood ratio criterion and the asymptotic expansion of its distribution. *Ann. Inst. Statist. Math.*, **29**, 359–78.

Hinkley, D.V. (1979) Predictive likelihood. *Ann. Statist.*, **7**, 718–28. (Corrigendum, **8**, 694).

Hinkley, D.V. (1980) Likelihood as approximate pivotal distribution. *Biometrika*, **67**, 287–92.

Hinkley, D.V. and Hinkley, E.A. (1970) Inference about the change-point in a sequence of binomial variables. *Biometrika*, **57**, 477–88.

Hinkley, D.V., Reid, N. and Snell, E.J. (eds) (1991) *Statistical Theory and Modelling*. Chapman & Hall, London.

Holm, S. (1985) Implications and equivalence among statistical inference rules. In *Contributions to Probability and Statistics in Honour of Gunnar Blom* (eds J. Lanke and G. Lindgren). Dept. Math. Statist., Lund University.

Hougaard, P. (1982) Parametrizations of non-linear models. *J. R. Statist. Soc. B*, **44**, 244–52.

Hougaard, P. (1985) Saddlepoint approximations for curved exponential families. *Statist. Prob. Letters*, **3**, 161–6.

Huber, P.J. (1964) Robust estimation of a location parameter. *Ann. Math. Statist.*, **35**, 73–101.

Huzurbazar, V.S. (1950) Probability distributions and orthogonal parameters. *Proc. Camb. Phil. Soc.*, **46**, 281–4.

Huzurbazar, V.S. (1956) Sufficient statistics and orthogonal parameters. *Sankhyā*, **17**, 217–20.

Jeffreys, H. (1939) *Theory of Probability*. Oxford University Press, Oxford. (Third edn, 1961).

Jensen, J.L. (1986a) Similar tests and the standardized log-likelihood ratio statistic. *Biometrika*, **73**, 567–72.

Jensen, J.L. (1986b) Inference for the mean of a gamma distribution with unknown shape parameter. *Scand. J. Statist.*, **13**, 135–51.

Jensen, J.L. (1987a) Standardized log-likelihood ratio statistics for mixtures of discrete and continuous observations. *Ann. Statist.*, **15**, 314–24.

Jensen, J.L. (1987b) On asymptotic expansions in non-ergodic models. *Scand. J. Statist.*, **14**, 305–18.

Jensen, J.L. (1988) Uniform saddlepoint approximations. *Adv. Appl. Prob.*, **20**, 622–34.

Jensen, J.L. (1991) Uniform saddlepoint approximations and log-concave densities. *J. R. Statist. Soc. B*, **53**, 157–72.

Jensen, J.L. (1992) The modified signed likelihood statistic and saddlepoint approximations. *Biometrika*, **79**, 693–703.

Joe, H. and Reid, N. (1985) Estimating the number of faults in a system. *J. Am. Statist. Assoc.*, **80**, 222–6.

Jørgensen, B. (1982) *Statistical Properties of the Generalized Inverse Gaussian Distribution*. Springer-Verlag, Berlin.

Jørgensen, B. (1987) Exponential dispersion models (with discussion). *J.R. Statist. Soc. B*, **49**, 127–62.

Kalbfleisch, J.D. (1975) Sufficiency and conditionality (with discussion). *Biometrika*, **62**, 251–68.

Kalbfleisch, J.D. and Sprott, D.A. (1970) Applications of likelihood methods to models involving large numbers of parameters (with discussion). *J.R. Statist. Soc. B*, **32**, 175–208.

Keiding, N. (1974) Estimation in the birth process. *Biometrika*, **61**, 71–80.

Kass, R.E. (1989) The geometry of asymptotic inference (with discussion). *Statistical Science*, **4**, 187–263.

Kendall, W.S. (1993) Computer algebra in probability and statistics. *Statist. Neerlandica*, **47**, 2–25.

Lai, T.L. and Siegmund, D. (1983) Fixed accuracy estimation of an autoregressive parameter. *Ann. Statist.*, **11**, 478–85.

Lauritzen, S.L. (1974) Sufficiency, prediction and extreme models. *Scand. J. Statist.*, **1**, 128–34.

Lawley, D.N. (1956) A general method for approximating to the distribution of the likelihood ratio criteria. *Biometrika*, **43**, 295–303.

Le Cam, L. (1986) *Asymptotic Methods in Statistical Decision Theory*. Springer-Verlag, Berlin.

Le Cam, L. (1990a) On the standard asymptotic confidence ellipsoids of Wald. *Int. Statist. Rev.*, **58**, 129–52.

Le Cam, L. (1990b) Maximum likelihood: an introduction. *Int. Statist. Rev.*, **58**, 153–72.

Le Cam, L. and Yang, G.L. (1990) *Asymptotics in Statistics. Some Basic Concepts*. Springer-Verlag, New York.

Lehmann, E.L. (1983) *Theory of Point Estimation*. Wiley, New York.

Lehmann, E.L. (1986) *Testing Statistical Hypotheses*. Wiley, New York.

Lehmann, E.L. (1990) Model specification: the views of Fisher and Neyman, and later developments. *Statistical Science*, **5**, 160–8.

Liang, K.L. (1987) Estimating functions and approximate conditional likelihood. *Biometrika*, **74**, 695–702.

Lindsay, B.G. (1988) Composite likelihood methods. *Contemporary Mathematics*, **80**, 221–39.

Liptser, R.S. and Shiryaev, A.N. (1978) *Statistics of Random Processes. 2. Applications.* Springer-Verlag, New York. (Translation of 1974 Russian edn.)

Lloyd, C.J. (1992) Effective conditioning. *Austral. J. Statist.*, **34**, 241–60.

Lugannani, R. and Rice, S. (1980) Saddlepoint approximation for the distribution of the sum of independent random variables. *Adv. Appl. Probab.*, **12**, 475–90.

Mantel, N. (1987) Understanding Wald's test for exponential families. *Amer. Statist.*, **41**, 147–8.

McCullagh, P. (1982) Discussion of paper by A.C. Atkinson, *J.R. Statist. Soc. B*, **44**, 29–30.

McCullagh, P. (1984a) Local sufficiency. *Biometrika*, **71**, 233–44.

McCullagh, P. (1984b) Tensor notation and cumulants of polynomials. *Biometrika*, **71**, 461–76.

McCullagh, P. (1987) *Tensor Methods in Statistics.* Chapman & Hall, London.

McCullagh, P. (1991) Quasi-likelihood and estimating functions. In *Statistical Theory and Modelling* (eds D.V. Hinkley, N. Reid and E.J. Snell). Chapman & Hall, London, pp. 265–85.

McCullagh, P. (1992) Conditional inference and Cauchy models. *Biometrika*, **79**, 247–59.

McCullagh, P. and Cox, D.R. (1986) Invariants and likelihood ratio statistics. *Ann. Statist.*, **14**, 1419–30.

McCullagh, P. and Nelder, J.A. (1989) *Generalized Linear Models.* (2nd edn). Chapman & Hall, London.

McCullagh, P. and Tibshirani, R. (1990) A simple method for the adjustment of profile likelihoods. *J.R. Statist. Soc. B*, **52**, 325–44.

Michel, R. (1978) Higher order asymptotic sufficiency. *Sankhyā A*, **40**, 76–84.

Mora, M. (1992) Geometric expansions for the distributions of the score vector and the maximum likelihood estimator. *Ann. Inst. Statist. Math.*, **44**, 63–83.

Murray, M.K. and Rice, J.W. (1993) *Differential Geometry and Statistics.* Chapman & Hall, London.

Neyman, J. and Pearson, E.S. (1928) On the use and interpretation of certain test criteria for purposes of statistical inference. *Biometrika*, **20A**, 175–240; 263–94.

Neyman, J. and Pearson, E.S. (1936) Sufficient statistics and uniformly most powerful tests of statistical hypotheses. *Statistical Research Memoirs*, **1**, 113–37.

Neyman, J. and Scott, E.L. (1948) Consistent estimates based on partially consistent observations. *Econometrica*, **16**, 1–32.

Oakes, D. (1988) Partial likelihood: Applications, ramifications, generalizations. *Contemporary Mathematics*, **80**, 67–77.

Ogawara, M. (1951) A note on the test of serial correlation coefficients. *Ann. Math. Statist.*, **22**, 115–18.

Owen, A.B. (1988) Empirical likelihood ratio confidence intervals for a single functional. *Biometrika*, **75**, 237–49.

Owen, A.B. (1991) Empirical likelihood for linear models. *Ann. Statist.*, **19**, 1725–47.

Pace, L. and Salvan, A. (1992) A note on conditional cumulants in canonical exponential families. *Scand. J. Statist.*, **19**, 185–191.

Patefield, W.M. (1977) On the maximized likelihood function. *Sankhyā B*, **39**, 92–6.

Patterson, H.D. and Thomson, R. (1971) Recovery of inter–block information when block sizes are unequal. *Biometrika*, **58**, 545–54.

Pázman, A. (1984) Probability distribution of the multivariate nonlinear least squares estimates. *Kybernetika*, **20**, 209–30.

Pázman, A. (1987) On formulas for the distribution of nonlinear l.s. estimates. *Statistics*, **18**, 3–15.

Pázman, A. (1990) Small-sample distributional properties of nonlinear regression estimators (a geometric approach). With discussion. *Statistics*, **21**, 323–68.

Pedersen, B.V. (1981) A comparison of the Efron–Hinkley ancillary and the likelihood ratio ancillary in a particular example. *Ann. Statist.*, **9**, 1328–33.

Peers, H.W. (1971) Likelihood ratio and associated test criteria. *Biometrika*, **58**, 577–87.

Pfanzagl, J. (1985) *Asymptotic Expansions for General Statistical Models*. Lecture Notes in Statistics 31. Springer-Verlag, Berlin.

Pfanzagl, J. (1990) *Estimation in Semiparametric Models*. Lecture Notes in Statistics 63. Springer-Verlag, Heidelberg.

Phillips, P.C.B. (1987) Time series regression with a unit root. *Econometrica*, **55**, 277–301.

Pierce, D.A. (1982) The asymptotic effect of substituting estimators for parameters in certain types of statistics. *Ann. Statist.*, **10**, 475–8.

Pierce, D.A. and Peters, D. (1992) Practical use of higher order asymptotics for multiparameter exponential families (with discussion). *J.R. Statist. Soc. B*, **54**, 701–37.

Pierce, D.A. and Schafer, D.W. (1986) Residuals in generalized linear models. *J. Am. Statist. Assoc.*, **81**, 977–86.

Rao, C.R. (1947) Large sample tests of statistical hypotheses concerning several parameters with applications to problems of estimation. *Proc. Camb. Phil. Soc.*, **44**, 40–57.

Rao, C.R. (1962) Efficient estimates and optimum inference procedures in large samples (with discussion). *J.R. Statist. Soc. B*, **24**, 46–72.

Rao, C.R. (1973) *Linear Statistical Inference and Its Applications*, 2nd edn. Wiley, New York.

Rao, C.R. (1977) Prediction of future observations with special reference to linear models. In *Multivariate Analysis, IV* (ed. P.R. Krishnaiah). North-Holland, Amsterdam, pp. 193–208.

Rao, C.R. and Chakravarti, I.M. (1956) Some small sample tests of significance for a Poisson distribution. *Biometrics*, **12**, 264–82.

Reeds, J.A. (1985) Asymptotic number of roots of Cauchy location likelihood equations. *Ann. Statist.*, **13**, 775–84.

Reid, N. (1988) Saddlepoint methods and statistical inference (with discussion). *Statistical Science*, **3**, 213–38.

Rémon, M. (1984) On a concept of partial sufficiency: L-sufficiency. *Int. Statist. Rev.*, **52**, 127–36.

Renyi, A. (1970) *Probability Theory*. North-Holland, Amsterdam.

Ripley, B.D. (1988) *Statistical Inference for Spatial Processes*. Cambridge University Press, Cambridge.

Royall, R.M. (1986) Model robust confidence intervals using maximum likelihood estimators. *Int. Statist. Rev.*, **54**, 221–6.

Salamin, P.-A. (1993). Personal communication.

Sandved, E. (1972) Ancillary statistics in models without and with nuisance parameters. *Skand. Aktuar.*, **55**, 81–91.

Searle, S.R., Casella, G. and McCulloch, C. (1992) *Variance components*. Wiley, New York.

Sen, P.K. and Singer, J.M. (1993) *Large Sample Methods in Statistics*. Chapman & Hall, London.

Severini, T.A. (1990) Conditional properties of likelihood-based significance tests. *Biometrika*, **77**, 343–52.

Silvey, S.D. (1961) A note on maximum-likelihood in the case of dependent random variables. *J.R. Statist. Soc. B*, **23**, 444–52.

Skovgaard, I.M. (1985a) A second order investigation of asymptotic ancillarity. *Ann. Statist.*, **13**, 534–51.

Skovgaard, I.M. (1985b) Larger deviation approximations for maximum likelihood estimators. *Probab. Math. Statist.*, **6**, 89–107.

Skovgaard, I.M. (1986) A note on the differentiation of cumulants of log likelihood derivatives. *Int. Statist. Rev.*, **54**, 29–32.

Skovgaard, I.M. (1987) Saddlepoint expansions for conditional distributions. *J. Appl. Prob.*, **24**, 875–87.

Skovgaard, I.M. (1990a) On the density of minimum contrast estimators. *Ann. Statist.*, **18**, 779–89.

Skovgaard, I.M. (1990b) *Analytic Statistical Models*. Lecture Notes – Monograph Series 15, Institute of Mathematical Statistics, Hayward, Calif.

Smith, R.L. (1989) A survey of nonregular problems. *Bull. Int. Statist. Inst.*, **53**, 353–72.

Solomon, P.J. and Cox, D.R. (1992) Nonlinear components of variance models. *Biometrika*, **79**, 1–16.

Sørensen, M. (1983) On maximum likelihood estimation in randomly stopped diffusion-type processes. *Int. Statist. Rev.*, **51**, 93–110.

Speed, T. (1987) What is an analysis of variance? (with discussion). *Ann. Statist.*, **15**, 885–941.

Sprott, D.A. (1973) Normal likelihoods and their relation to large sample theory of estimation. *Biometrika*, **60**, 457–65.

Sprott, D.A. (1975) Applications of maximum likelihood methods to finite samples. *Sankhyā B*, **37**, 259–270.

Stern, F. (1977) An independence in Brownian motion with drift. *Ann. Prob.*, **5**, 571–2.

Sverdrup, E. (1966) The present state of the decision theory and the Neyman–Pearson theory. *Rev. Int. Statist. Inst.*, **34**, 309–33.

Sweeting, T.J. (1980) Uniform asymptotic normality of the maximum likelihood estimator. *Ann. Statist.*, **8**, 1375–81.

Sweeting, T.J. (1986) Asymptotic conditional inference for the offspring mean of a supercritical Galton–Watson process. *Ann. Statist.*, **14**, 925–933.

Sweeting, T.J. (1989) On conditional weak convergence. *J. Theor. Prob.*, **2**, 461–74.

Sweeting, T.J. (1992a) Asymptotic ancillarity and conditional inference for stochastic processes. *Ann. Statist.*, **20**, 580–9.

Sweeting, T.J. (1992b) Parameter-based asymptotics. *Biometrika*, **79**, 219–30.

Temme, N.M. (1982) The uniform asymptotic expansion of a class of integrals related to cumulative distribution functions. *SIAM J. Math. Anal.*, **13**, 239–53.

Tweedie, M.C.K. (1957) Statistical properties of Inverse Gaussian distributions I, II. *Ann. Math. Statist.*, **28**, 362–77; 696–705.

Venzon, D.J. and Moolgavkar, S.H. (1988) A method for computing profile-likelihood-based confidence intervals. *Appl. Statist.*, **37**, 87–94.

Væth, M. (1985) On the use of Wald's test in exponential families. *Int. Statist. Rev.*, **53**, 199–214.

Waugh, W.A.O'N. (1970) Transformation of a birth process into a Poisson process. *J.R. Statist. Soc. B*, **32**, 418–31.

Wedderburn, R.W.M. (1974) Quasi-likelihood functions, generalised linear models, and the Gauss–Newton method. *Biometrika*, **61**, 439–47.

Welch, B.L. (1939) On confidence limits and sufficiency with particular reference to parameters of location. *Ann. Math. Statist.*, **10**, 58–64.

Welch, B.L. (1949) Appendix to Mrs Aspin's tables for use in comparisons whose accuracy involves two variances, separately estimated, and on certain approximations to the tabled function. *Biometrika*, **36**, 293–6.

Welch, B.L. and Peers, H.W. (1963) On formulae for confidence points based on integrals of weighted likelihoods. *J.R. Statist. Soc. B*, **25**, 318–29.

Wendel, J.G. (1980) Hitting spheres with Brownian motion. *Ann. Prob.*, **8**, 164–9.

Wilks, S.S. (1938) The large-sample distribution of the likelihood ratio for testing composite hypotheses. *Ann. Math. Statist.*, **9**, 60–2.

Woodroofe, M. (1982) *Nonlinear Renewal Theory in Sequential Analysis*. SIAM, Philadelphia.

Yates, F. (1951) Bases logiques de la planification des expériences. *Ann. Inst. H. Poincaré*, **12**, 97–112.

Zellner, A. (1962) An efficient method of estimating seemingly unrelated regressions and tests for aggregation bias. *J. Am. Statist. Assoc.*, **57**, 348–68.

Author index

Subject index